INTERNATIONAL

REVIEW OF CYTOLOGY

VOLUME 68

ADVISORY EDITORS

INTERNATIONAL

Review of Cytology

EDITED BY

G. H. BOURNE
St. George's University School of Medicine
St. George's, Grenada
West Indies

J. F. DANIELLI
Worcester Polytechnic Institute
Worcester, Massachusetts

ASSISTANT EDITOR
K. W. JEON
Department of Zoology
University of Tennessee
Knoxville, Tennessee

VOLUME 68

1980

ACADEMIC PRESS   *A Subsidiary of Harcourt Brace Jovanovich, Publishers*
New York   London   Toronto   Sydney   San Francisco

ACADEMIC PRESS, INC.
111 Fifth Avenue, New York, New York 10003

United Kingdom Edition published by
ACADEMIC PRESS, INC. (LONDON) LTD.
24/28 Oval Road, London NW1 7DX

LIBRARY OF CONGRESS CATALOG CARD NUMBER: 52–5203

ISBN 0–12–364468–2

PRINTED IN THE UNITED STATES OF AMERICA

80 81 82 83   9 8 7 6 5 4 3 2 1

Contents

Moisture Content as a Controlling Factor in Seed Development and Germination

C. A. ADAMS AND R. W. RINNE

Applications of Protoplasts to the Study of Plant Cells

LARRY C. FOWKE AND OLUF L. GAMBORG

Control of Membrane Morphogenesis in Bacteriophage

GREGORY J. BREWER

Scanning Electron Microscopy of Intracellular Structures

KEIICHI TANAKA

The Relevance of the State of Growth and Transformation of Cells to Their Patterns of Metabolite Uptake

RUTH KOREN

Intracellular Source of Bioluminescence

BEATRICE M. SWEENEY

Differentiation of MSH-, ACTH-, Endorphin-, and LPH-Containing Cells in the Hypophysis during Embryonic and Fetal Development

JEAN-PAUL DUPOUY

Cell Death: The Significance of Apoptosis

A. H. WYLLIE, J. F. R. KERR, AND A. R. CURRIE

List of Contributors

Numbers in parentheses indicate the pages on which the authors' contributions begin.

C. A. ADAMS (1), *USDA-SEA-AR, Department of Agronomy, University of Illinois, Urbana, Illinois 61801*

GREGORY J. BREWER (53), *Department of Medical Microbiology and Immunology, Southern Illinois University School of Medicine, Springfield, Illinois 62708*

A. R. CURRIE (251), *Department of Pathology, University of Edinburgh, Edinburgh, Scotland*

JEAN-PAUL DUPOUY (197), *Laboratory of Animal Physiology, Faculty of Sciences, University of Picardie, 80 Amiens, France*

LARRY C. FOWKE (9), *Department of Biology, University of Saskatchewan, Saskatoon, Saskatchewan, Canada S7N 0W0*

OLUF L. GAMBORG* (9), *Prairie Regional Laboratory, National Research Council, Saskatoon, Saskatchewan, Canada S7N 0W9*

J. F. R. KERR (251), *Department of Pathology, University of Queensland, Brisbane, Australia*

RUTH KOREN (127), *Department of Pharmacology, The Hebrew University, Hadassah Medical School, Jerusalem, Israel*

R. W. RINNE (1), *USDA-SEA-AR, Department of Agronomy, University of Illinois, Urbana, Illinois 61801*

BEATRICE M. SWEENEY (173), *Department of Biological Sciences, University of California, Santa Barbara, California 93106*

KEIICHI TANAKA (97), *Department of Anatomy, Tottori University School of Medicine, 683 Yonago, Tottori-Ken, Japan*

A. H. WYLLIE (251), *Department of Pathology, University of Edinburgh, Edinburgh, Scotland*

---

*Present address: International Plant Research Institute, San Carlos, California.

INTERNATIONAL REVIEW OF CYTOLOGY, VOL. 68

Moisture Content as a Controlling Factor in Seed Development and Germination

C. A. ADAMS AND R. W. RINNE

USDA-SEA-AR, Department of Agronomy, University of Illinois, Urbana, Illinois

I. Introduction

The metabolic activities of seeds can be divided conveniently into four phases. The first is initiated by pollination leading to formation of the embryonic axis and of cellular structures that accumulate reserve materials. Rapid cell division occurs at this time. The second phase is characterized most obviously by a rapid increase in seed mass when both fresh weight and dry weight increase. During this phase, cell division does not occur, but there is active biosynthesis of reserve materials. These reserves are laid down in discrete subcellular organelles such as protein bodies (Pernollet, 1978), lipid bodies, and starch grains. The third stage is seed maturation when dry weight accumulation slows down and ultimately ceases and fresh weight declines markedly. During this phase, there is a net loss of water from the seed. The fourth and final stage is germination of the mature seed, accompanied by rapid hydration, leading to destruction of the seed reserve tissue and establishment of a new plant.

Seed development and maturation are associated with an overall loss of moisture. Usually the very young seed has the highest moisture content. In soybeans and lima beans, moisture decreased from about 85 to 15% (Klein and Pollock, 1968; Rubel et al., 1972) (Table I); in wheat, it declined from 68 to 10% (McWha, 1975); in castor beans, the range was 89 to 7% (Canvin, 1963); and corn showed a decline of 84 to 23% (Sprague, 1936). In any other part of the plant such a decrease in moisture content would be disastrous, yet the seed structures are able to survive this level of dehydration.

Germination has an inverted moisture regime compared to that of develop-

1

TABLE I

THE MOISTURE CONTENT IN DEVELOPING AND GERMINATING
SOYBEAN SEEDS

Developing seeds		Germinating seeds	
Days after flowering	Percentage of moisture[a] (whole seed)	Days after planting	Percentage of moisture[a] (2 cotyledons)
17	81.4	1	52.4
24	75.8	2	59.8
31	68.3	3	62.6
38	63.0	4	64.9
45	58.3	5	71.9
52	21.7	8	85.9
		9	87.8

[a] Percentage of moisture expressed on a fresh weight basis.

ment. In germinating seeds, the moisture content is initially low and then increases rapidly during the first few days after planting. The moisture content in the seed tissues can reach a level higher than that attained during development. Table I illustrates these changes in moisture content of soybean seeds during development and germination.

Seeds not only survive severe desiccation, but apparently, in many cases, require it as a necessary step in their developmental process. As early as 1890, Brown and Morris reported that immature barley embryos were incapable of secreting amylase activity if they were removed from grains that had not been dehydrated. Embryos taken from dried grains readily produced amylolytic activity. Sprague (1936) showed that there was a strong relationship between the loss of water from corn kernels and their ability to germinate normally. Immature seeds harvested as early as 10 days after flowering achieved a 50% germination rate if dried to 17% moisture before planting. Grains with more than 25% moisture, if planted immediately, had a reduced percentage germination and an increased variability in germination. Immature barley aleurone layers will not secrete amylase activity in response to gibberellic acid unless they are first dried (Bilderbeck, 1971). The hormonal control system in barley aleurone layers appears to operate fully only after the grain has lost moisture (Evans *et al.*, 1975). This correlates with germination, as undried, intact barley grains did not germinate (Evans *et al.*, 1975). However, when dried for 2 days and then rewetted, the grains readily germinated. Clearly, the final stage in production of viable seeds by several species is a dehydration of the tissue well below that normally encountered by other plant tissues.

II. Concept of a Water Gradient

The continuous decline in moisture content during development through dehydration, followed by rehydration and increasing moisture content could constitute a chemical gradient. Such gradients may play a major role in developmental physiology, but the chemical entity involved is not known (Crick, 1970).

A water gradient could be further enhanced by an increase in the proportion of structured or vicinal water over bulk or free water during seed development (Etzler and Drost-Hansen, 1979). Cells in a developing seed would probably generate increasing surface and interphase regions by virtue of the insoluble reserve materials accumulated as protein bodies, oil droplets, and starch grains. The water in a developing seed should become increasingly structured, which may control or influence the biological activity of the cells (Etzler and Drost-Hansen, 1979).

As the fertilized ovule develops into a seed undergoing rapid cell division, a declining moisture regime becomes apparent. This could convey information to the cells signaling the necessity of synthesizing reserve materials and of stopping cell division. HeLa cells grown in hypertonic media do show a slower division rate (Rao, 1968).

Hydration of a mature seed during germination changes the polarity of the water gradient. As the moisture content rapidly increases, the proportion of vicinal to free water in the tissue is reduced. This condition may cause growth and utilization of cellular reserves. Rapidly dividing cancer cells have a very high state of hydration (Beall, 1979). Damadian (1971) concluded that in malignant tissue there was a significant decrease in structured or vicinal water.

Developing seeds produce reserve materials and germinating seeds utilize these materials. Clearly, at each developmental stage, the tissue's cells are informed as to their appropriate biochemical response. A reversal of the polarity of a water gradient would be an unambiguous source of this information. This is somewhat analogous to the concept of positional information proposed by Holder (1979).

III. Physiological Response to a Water Gradient

Moisture status would not only dictate the overall metabolic pattern but would also influence the expression of the cell's perceived status. Observations on soybeans and lima beans illustrate this influence of water content on metabolism (Klein and Pollock, 1968; Rubel et al., 1972). In these seeds, mass increased linearly while moisture content dropped from 85 to 60%. At about 60% moisture, the seeds enter their maturation phase when dry weight accumulation ceases and rapid dehydration commences. In wheat, water continued to enter the grain until maximum dry weight was reached (Sofield *et al.*, 1977), and starch formation stopped soon after the beginning of water loss (Radley, 1976). Once the water content has declined to somewhere around 60%, biochemical expression of the general metabolic plan is affected.

In any living cell, there is a balance of synthesis and degradation of cellular products. Siekevitz (1972) suggested that nucleic acid and protein turnover are necessary procedures to protect the individualistic integrity of the cell and to maintain the efficiency of usage of informational energy. Developing seeds are characterized as having very active synthesis of specialized reserve materials that do not turn over. Presumably the enzyme and nucleic acids involved in that synthesis and in maintaining general metabolic activity do turn over. Desiccation seems more likely to interrupt the synthetic rather than the degradative processes of metabolic turnover.

The amino acid incorporating activity of maize kernels declines as the grain dehydrates (Rabson *et al.*, 1961). Water stress in a drought-tolerant moss causes ribosomes to detach from the mRNA and fail to reinitiate (Dhindsa and Bewley, 1976a). Early growth and differentiation of seeds is associated with a proliferation of endoplasmic reticulum, and the proportion of bound ribosomes increases. At dehydration, there is a breakup of the endoplasmic reticulum and of the polysomes (Öpik, 1968). Further studies on this aspect by Payne and Boulter (1969) showed that during seed dehydration there is a loss of membrane-bound ribosomes and a concomitant increase in free ribosomes. Since storage protein likely is synthesized by these membrane-bound ribosomes, then loss of the functional machinery during desiccation would obviously terminate the seed's synthetic activities. The activity of glucose-6-phosphate dehydrogenase in the range of water contents of 5 to 20%, which occurs in dry seeds, is reduced 100–1000-fold compared to the rates that occur when water is not limiting (Stevens and Stevens, 1977). The pentose phosphate pathway is of major importance in producing NADPH for biosynthetic reactions, and dehydration severely curtails this process. High energy phosphate production in both mitochondria and chloroplasts is reduced during dehydration (Todd, 1972). Degradative processes could continue, however, because levels of enzymes involving hydrolysis usually remain constant or increase under dehydration (Todd, 1972). Hydrolytic enzymes do not decrease until fairly severe desiccation has taken place. RNase activity was found to increase during dehydration of plant tissue (Dhindsa and Bewley, 1976b). Hydrolases tend to be rather stable enzymes and several persist into the mature seed (Mayer and Poljakoff-Mayber, 1963). Carbohydrate metabolizing enzymes, by contrast, show a more varied response. Duffus and Rosie (1973) found that in developing barley seeds activity of the enzymes of the glycolytic pathway increased through development and then declined to a low level at maturity. Enzymes of the pentose phosphate pathway, however, remained at a high level into maturity, but presumably this activity would not be expressed under the prevailing moisture conditions (Stevens and Stevens, 1977).

IV. Moisture in Seed Maturation and Storage

The final phase of seed development (dehydration to air dryness) results in cessation of synthesis of reserve materials originally dictated by a declining water gradient. The seed now becomes committed to a degradative or controlled senescence program that is temporarily interrupted by insufficient moisture.

This concept of seed dehydration promoting senescence and inhibiting synthesis is also consistent with observations on seed longevity and aging. Many seeds remain viable for as long as 30 years when stored dry (Haferkamp *et al.*, 1953). Nevertheless, seed damage that does lead to lowered germination rates is apparent during dry storage (Villiers, 1974). Probably degradative activities can still continue, albeit at a slow rate, even in air dry seeds, although counterbalancing synthetic activities are inhibited. When seeds are maintained under conditions of high humidity and temperature, these deteriorative changes are greatly accelerated and viability falls rapidly (Berjak and Villiers, 1972). It would appear that low moisture contents are compatible with degradative activities but not with synthetic processes. Consequently, storage under humid conditions severely accelerates degradative activities that soon render the seed nonviable. These effects are reduced by lowering the moisture content and, as a general rule, a decrease of 1% in moisture content may approximately double the life span of a seed sample (Harrington, 1973). Further support for this concept of seed maturation has been provided in a series of observations by Villiers (1972, 1974) who showed that the seeds of *Fraxinus excelsior, Fraxinus americana,* and lettuce had greater viability when stored fully imbibed under nongerminating conditions than when stored under low moisture conditions or at ambient temperatures. In air dry or partially hydrated tissues, enzyme turnover and repair is probably suspended in favor of degradative activities. In seeds stored fully imbibed but not germinating, macromolecular repair and membrane synthesis may occur at a normal rate to balance the deteriorative changes.

As pointed out by Villiers (1974), in a natural environment, the majority of seeds are air-dry for only a short time. After maturation on the plant, the seeds are dispersed onto the soil surface where they soon become rehydrated. If they do not germinate immediately but remain dormant, they are able to survive for long periods of time with a high water content. The dehydration process still plays a necessary role in switching the polarity of the cell water gradient. Consequently, the seeds maintain viability but do not start accumulating food reserves as in a developing seed.

V. Conclusions

The theory developed here proposes that moisture content plays a role in seed physiology in at least two ways. First, it may convey information about the expected metabolic activity of the seed, i.e., development, resting, or germination. The proportion of structured or vicinal water in the tissue compared to the bulk water, as well as the change in absolute water content, may play a role here. A dehydration to air dryness would be necessary for the unequivocal switching of the water gradient polarity. This may explain why Sprague (1936) found it necessary to dry immature seeds below 60% moisture for good viability, even though germination occured at 60% moisture. Clearly, the absolute moisture content does not dictate a specific physiological activity. Second, the expression of the perceived information is also probably controlled by moisture status. Severe dehydration irreversibly switches off production of reserve materials and switches on a controlled senescence program that continues at variable rates until the seed reserve structure is destroyed.

VI. Applications

These two functions of water in the control of seed metabolism have a long-standing practical application in the malting industry. We have already described the lack of response to gibberellic acid of undried barley aleurone tissue (Bilderbeck, 1971; Evans et al., 1975). The change in polarity of the water gradient in the hydration of dried barley seeds may predispose the aleurone cells to respond to gibberellic acid. Such observations agree well with the theory of hormone action discussed by Holder (1979). The polarity of the water gradient would specify the status of the aleurone cell. The hormone then influences the interpretation of this perceived status by stimulating the production of hydrolytic enzymes that speed up the malting process. Moisture content still exerts a control on the expression of hormonal action. Water stress also inhibits the induction of hydrolytic enzymes in barley aleurone cells (Armstrong and Jones, 1973). Not only must the aleurone tissue be dried first, but it must be maintained in a sufficiently hydrated state to be able to respond to a hormonal stimulus. The basic principle of malting is that germination of grain is allowed to occur under a restricted moisture regime. Insufficient moisture inhibits vigorous seedling growth, but the moisture gradient still informs the seed of its germinative status. Consequently, the controlled senescence program continues, manifested by the increased production of hydrolytic enzymes and a high degree of breakdown or modification of the reserve materials. The grain is then dried by heating, which kills the seedling tissues but does not destroy the hydrolytic enzymes in the malt.

Several predictions and questions may be formulated based on this theory of moisture content exerting control over seed physiological behavior:

1. The seed embryonic axis matures considerably in advance of the seed, and the major part of seed growth is taken up by accumulation of reserve materials.

2. Reducing seed moisture content below about 50% irrevocably switches the seed into a controlled senescence program that is only suspended by dehydration. It continues on rewetting, eventually leading to destruction of the major portion of the original seed.

3. Seed dehydration reduces the activity of biosynthetic enzymes more severely than that of degradative enzymes.

4. Since ultimate seed size is a function of the rate and duration of seed growth, then maintaining seed moisture levels above 60 to 70% in developing seeds may increase seed size.

5. Normal seed size is not a necessity for producing viable seeds. Immature seeds, if dried, are frequently viable. Therefore, the growing season for some species could be decreased by harvesting immature seeds and drying them under controlled conditions.

6. Enzymes concerned with accumulation of reserves are produced in limited quantities, so that enzyme amount controls the reaction. Dehydration rapidly inactivates these enzymes and thus terminates the accumulation of reserves.

7. Enzymes involved in general metabolism are produced in large quantities, so their activity is regulated by substrates and products. These enzymes are partially inhibited by dehydration.

8. What are the physiological differences between a system losing moisture as a developing seed and one gaining moisture as a germinating seed?

9. What changes go on in dehydrating barley aleurone cells to predispose them to respond to gibberellic acid?

10. How does dehydration irrevocably switch off the genetic information for biosynthesis of reserve materials?

11. Is dehydration triggered by restricted supply of water to the seed or more rapid water loss from the seed?

12. Is there some level of desiccation that seeds must reach before the switch to germination will occur on rehydration?

These predictions and questions are amenable to experimental investigation. Consequently, they should provide a working framework for further study of the phenomenon of seed development and seed germination. This whole process of seed metabolism can be considered a particular case of morphogenesis that is to some extent controlled by plant hormones.

REFERENCES

Armstrong, J. E., and Jones, R. L. (1973). *J. Cell Biol.* **59,** 444.
Beall, P. T. (1979). *In* "Cell-Associated Water" (W. Drost-Hansen and J. S. Clegg, eds.), p. 440. Academic Press, New York.
Berjak, P., and Villiers, T. A. (1972). *New Phytol.* **71,** 1075.
Bilderbeck, D. E. (1971). *Plant Physiol.* **48,** 331.
Brown, H. T., and Morris, G. H. (1890). *J. Chem. Soc.* **57,** 458.
Canvin, D. T. (1963). *Can. J. Biochem. Physiol.* **41,** 1879.
Crick, F. (1970). *Nature (London)* **225,** 420.
Damadian, R. (1971). *Science* **171,** 1151.
Dhindsa, R. S., and Bewley, J. D. (1976a). *J. Exp. Bot.* **27,** 513.
Dhindsa, R. S., and Bewley, J. D. (1976b). *Science* **191,** 181.
Duffus, C. M., and Rosie, R. (1977). *New Phytol.* **78,** 391.
Etzler, F. M., and Drost-Hansen, W. (1979). *In* "Cell-Associated Water" (W. Drost-Hansen and J. S. Clegg, eds.), p. 440. Academic Press, New York.
Evans, M., Black, M., and Chapman, J. (1975). *Nature (London)* **258,** 144.
Haferkamp, M. E., Smith, L., and Nilan, R. A. (1953). *Agron. J.* **45,** 434.
Harrington, J. F. (1973). *In* "Seed Ecology" (W. Heyerdecker, ed.), p. 251. Butterworths, London.
Holder, N. (1979). *J. Theor. Biol.* **77,** 195.
Klein, S., and Pollock, B. M. (1968). *Am. J. Bot.* **55,** 658.
Mayer, A. M., and Poljakoff-Mayber, A. (1963). *In* "The Germination of Seeds," p. 236. Macmillan, New York.
McWha, J. A. (1975). *J. Exp. Bot.* **26,** 823.
Öpik, H. (1968). *J. Exp. Bot.* **19,** 64.
Payne, P. I., and Boulter, D. (1969). *Planta* **84,** 263.
Pernollet, J. C. (1978). *Phytochemistry,* **17,** 1473.
Rabson, R., Mans, R. J., and Novelli, G. D. (1961). *Arch. Biochem. Biophys.* **93,** 555.
Radley, M. (1976). *J. Exp. Bot.* **27,** 1009.
Rao, P. N. (1968). *Science* **160,** 774.
Rubel, A., Rinne, R. W., and Canvin, D. T. (1972). *Crop Sci.* **12,** 739.
Siekevitz, P. (1972). *J. Theor. Biol.* **37,** 321.
Sofield, I., Wardlaw, I. F., Evans, L. T., and Zee, S. Y. (1977). *Aust. J. Plant Physiol.* **4,** 799
Sprague, G. F. (1936). *J. Am. Soc. Agron.* **28,** 472.
Stevens, E., and Stevens, L. (1977). *J. Exp. Bot.* **28,** 292.
Todd, G. W. (1972). *In* "Water Deficits and Plant Growth" (T. T. Kozlowski, ed.), Vol. III, p. 368. Academic Press, New York.
Villiers, T. A. (1972). *New Phytol.* **71,** 145.
Villiers, T. A. (1974). *Plant Physiol.* **53,** 875.

INTERNATIONAL REVIEW OF CYTOLOGY, VOL. 68

Applications of Protoplasts to the Study of Plant Cells

Larry C. Fowke

Department of Biology, University of Saskatchewan, Saskatoon, Saskatchewan, Canada

Oluf L. Gamborg[1]

Prairie Regional Laboratory, National Research Council, Saskatoon, Saskatchewan, Canada

I. Introduction

Plant cells are surrounded by a complex cell wall composed primarily of polysaccharides. Techniques have been developed for the routine removal of this

[1]Current address: International Plant Research Institute, San Carlos, California.

wall from a wide range of plant cells (Cocking, 1972; Gamborg and Miller, 1973; Constabel, 1975; Gamborg, 1976; Vasil, 1976; Bajaj, 1977). The resulting naked cells or protoplasts provide a unique experimental system that is proving very useful for studying the structure and function of plant cells. Many manipulations that are not possible with intact plant cells are feasible with protoplasts. These include surface labeling of the plasma membrane, complete cell wall regeneration, and cell fusion. Protoplasts are particularly important for examining the plant plasma membrane, a component of the plant cell that is normally inaccessible due to the presence of a cell wall. This chapter examines the major applications of plant protoplasts to the study of plant cells. Emphasis will be placed on structural studies of higher plant protoplasts especially by electron microscopic methods. Brief reference will also be made to research on protoplasts from other groups of plants and relevant work with animal cells.

II. Structure of Isolated Protoplasts

A. GENERAL STRUCTURE

Protoplasts have been isolated from a wide variety of plant species and when cultured have produced callus and in some cases complete plants. Table I lists examples of plants that have been regenerated from protoplasts. Although a number of different plant tissues have been used for isolating protoplasts, most research to date has been carried out with protoplasts derived from cells of either suspension culture or from leaf mesophyll (Fig. 1). Perhaps the most extensively studied protoplasts are those from tobacco (*Nicotiana tabacum*) leaves. Protoplasts are isolated by digestion of the cell wall with a mixture of fungal hydrolytic enzymes (e.g., cellulases, pectinases, hemicellulases) in solutions containing osmotic stabilizers. Detailed information concerning plant sources, enzyme solutions, and isolation procedures has been previously provided (Cocking, 1972; Gamborg, 1975, 1976; Vasil, 1976; Bajaj, 1977; Eriksson *et al.*, 1978).

Enzyme digestion results in the production of high yields of spherical protoplasts (Figs. 2–4). Freshly isolated protoplasts from many different species have been examined by light and electron microscopy. The appearance of protoplasts in published micrographs varies from those apparently well preserved to others exhibiting a loss of spherical shape, a marked increase in cytoplasmic density, and a fragmentation of the central vacuole. It is difficult to determine the cause of these problems. Some may be due to the general viability of plant material, whereas others may be due to inadequate protoplast isolation techniques or problems with preparation for microscopy. If care is taken in the processing of healthy protoplasts, most of these difficulties can be eliminated (Fowke, 1975; Torrey and Landgren, 1977). The best results have been achieved with double

TABLE I
SPECIES FROM WHICH PROTOPLASTS HAVE REGENERATED PLANTS

Plant species	Origin of protoplasts	References
Asparagus officinalus	Callus, cladodes	Bui-dang-ha and MacKenzie (1973)
Atropa belladonna	Cultured cells	Gosch et al. (1975)
	Mesophyll cells	Lörz and Potrykus (1979)
Brassica napus	Mesophyll cells	Kartha et al. (1974); Thomas et al. (1976)
Bromus inermis	Cultured cells	Kao et al. (1973)
Datura innoxia	Mesophyll cells	Schieder (1975)
	Cultured cells	Furner et al. (1978)
Daucus carota	Cultured cells	Grambow et al. (1972); Dudits et al. (1976a)
Hyoscyamus muticus	Mesophyll and cultured cells	Lörz et al. (1979)
	Mesophyll cells	Wernicke et al. (1979)
Medicago sativa	Mesophyll cells	Kao and Michayluk (1979)
Nicotiana species	Mesophyll cells	Takebe et al. (1971); Nitsch and Ohyama (1971); Gleba et al. (1974); Raveh and Galun (1975); Facciotti and Pilet (1979)
	Epidermal cells	Davey et al. (1974)
	Callus	Koblitz (1978)
	Cultured cells (albino)	Gamborg et al. (1979)
Petunia species	Mesophyll cells	Durand et al. (1973); Frearson et al. (1973); Binding (1974); Hayward and Power (1975); Power et al. (1976a); Sink and Power (1977)
Ranunculus sceleratus	Mesophyll cells	Dorion et al. (1975)
Solanum species	Mesophyll cells	Binding and Nehls (1977); Butenko et al. (1977); Shepard and Totten (1977); Nehls (1978); Binding et al. (1978); Grun and Chu (1978)

fixation in glutaraldehyde and osmium tetroxide, gradual dehydration and infilt-ration, and embedding in epoxy resins (Fig. 4). The most critical step in the procedure appears to be the initial fixation using a low concentration of glutaral-dehyde in either culture medium or a solution of comparable osmoticum (Fowke, 1975).

Most protoplasts are released from single cells and thus contain one nucleus. However, spontaneous fusion can occur to produce homokaryons (also referred to as multinucleate protoplasts or spontaneous fusion bodies). Multinucleate protoplasts have been described in a variety of plants, e.g., soybean (Miller *et al.*, 1971; Fowke *et al.*, 1975a), *Ammi visnaga* (Fowke *et al.*, 1973, 1974a), carrot (Grambow *et al.*, 1972), tobacco (Withers and Cocking, 1972; Morel *et al.*, 1973; Takebe *et al.*, 1973), and rice (Maeda and Hagiwara, 1974). Ultra-

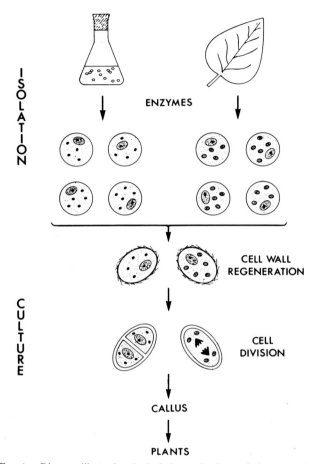

Fig. 1.   Diagram illustrating the isolation and culture of plant protoplasts.

Fig. 2.   Light micrograph of a pea leaf protoplast showing the large central vacuole and chloroplasts (arrows) in the thin peripheral layer of cytoplasm. Bar, 5 μm.

Fig. 3.   Light micrograph of tobacco cell culture protoplasts. Nomarski optics. Bar, 10 μm.

Fig. 4.   Electron micrograph showing pinto bean cell culture protoplast. Note the small vacuoles and leucoplasts (arrows). Bar, 3 μm.

Fig. 5.   Electron micrograph of platinum–palladium replica showing surface of freshly isolated *Vicia* cell culture protoplast. Microfibrils are not present on the protoplast surface. Note the protrusions of the membrane. Bar, 1 μm. (From Williamson *et al.*, 1977.)

Fig. 6.   Electron micrograph of platinum–palladium replica showing edge of *Vicia* cell culture protoplast after 20 hours in culture. Note the extensive network of microfibrils. Bar, 1 μm. (Kindly provided by Dr. F. A. Williamson.)

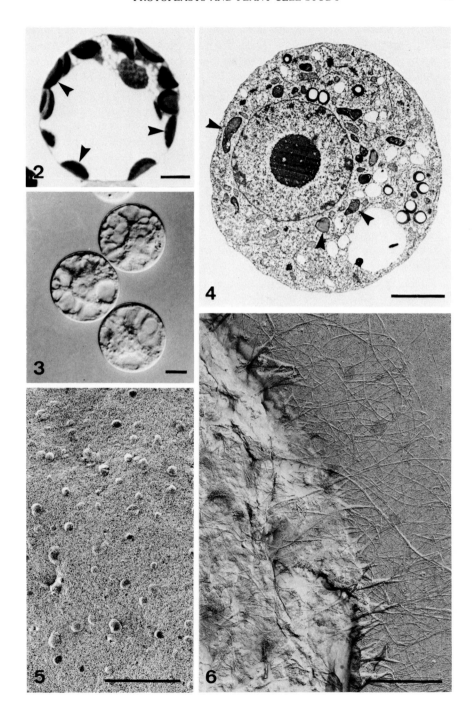

structural studies of protoplast isolation from tobacco leaf and oat root tissues indicate that spontaneous fusion results from an expansion of plasmodesmata connecting adjacent cells. Multinucleate protoplasts containing from 2 to approximately 30 nuclei can result (Miller *et al.*, 1971; Motoyoshi, 1971; Morel *et al.*, 1973; Cocking, 1974). In rare cases, isolated protoplasts in suspension will aggregate and undergo spontaneous fusion (Robenek and Peveling, 1978).

Leaf protoplasts usually contain a large central vacuole with disc-shaped chloroplasts packed tightly against the plasma membrane in a thin peripheral layer of cytoplasm (e.g., Fig. 2 and Roland and Prat, 1973; Takebe *et al.*, 1973; Fowke *et al.*, 1975b, 1977; Hughes *et al.*, 1976). Only limited ultrastructural information is available concerning the changes occurring during isolation of leaf protoplasts. Tobacco protoplasts have been compared to source tissue and a number of changes were observed. The most striking change involved the appearance of semicrystalline bodies in the chloroplasts (Milne, 1972; Takebe *et al.*, 1973; Gigot *et al.*, 1975; Taylor and Hall, 1978). These structures are believed to be fraction I protein (Takebe *et al.*, 1973; Taylor and Hall, 1978). Since similar structures were observed in both tobacco protoplasts and leaf tissue treated with osmoticum and in other stressed leaves, Taylor and Hall (1978) suggest that their appearance is a general response to water stress. Dehydration of leaves by plasmolysis or wilting tends to favor their formation (Gunning *et al.*, 1968). The appearance of numerous lipid droplets in the cytoplasm and chloroplasts of freshly isolated protoplasts is also thought to result from physiological stress (Taylor and Hall, 1978). Fragmentation of chloroplasts during protoplast isolation has been reported (Gigot *et al.*, 1975) but does not seem to be a general phenomenon.

Protoplasts derived from suspension cultures are usually characterized by many scattered vacuoles rather than the single large vacuole seen in leaf mesophyll protoplasts. They also normally contain leucoplasts with starch grains and few internal lamellae rather than chloroplasts (Fig. 4). Only one comparative study dealing with changes during isolation of cell culture protoplasts is available (Fowke *et al.*, 1973). Freshly isolated protoplasts showed a general increase in organelles when compared to initial cells, with a marked increase in endoplasmic reticulum and associated polysomes. These changes possibly reflect a general activation of the protoplasts in preparation for cell wall regeneration.

The results to date, while rather meager, do suggest some basic differences between leaf and cell culture protoplasts not only in structure but also in their response to enzyme treatment. Protoplasts derived from cell cultures do not seem to show deleterious effects to the same extent as those from leaves. This may reflect the basic difference in source material since leaves are highly differentiated with little or no cell division, whereas cells in culture are usually growing and dividing rapidly.

B. The Protoplast Surface

With appropriate choice of conditions of isolation, viable protoplasts with no detectable cell wall can be produced (Fig. 5). The surface of protoplasts is often characterized by projections, the most common being spherical structures ranging in size from 100 to 1000 nm with most being approximately 200 nm in diameter (Burgess and Linstead, 1977b; Williamson et al., 1977; Willison and Grout, 1978). Based on both transmission electron microscopy (TEM) and scanning electron microscopy (SEM), Burgess and Linstead (1977b) conclude that these projections result from small vacuoles evaginating the plasma membrane. The presence of these structures may indicate weak regions in the plasma membrane. In contrast, Willison and Grout (1978) suggest that the projections may be plasmalemmasomes. However, this seems unlikely since plasmalemmasomes are usually larger in size and are not normally present in such large numbers (Fowke and Setterfield, 1969).

A variety of techniques have been used to monitor the removal of cell wall material during protoplast isolation. Since many studies with protoplasts (e.g., examination of the plasma membrane, cell wall regeneration, protoplast fusion) require the removal of most if not all of the cell wall, the techniques for monitoring wall removal become very important. The following discussion examines these techniques and indicates their major advantages and limitations.

At the light microscopic level, Calcofluor white, which binds to cellulose and other β-linked glucans (Hughes and McCully, 1975), has been used with fluorescence microscopy to stain plant cell walls. The absence of Calcofluor staining on freshly isolated protoplasts is considered a good indication of wall removal. The stain, however, cannot be used to demonstrate the complete removal of the cell wall because it is not sufficiently sensitive to detect small numbers of microfibrils (Williamson et al., 1977).

A variety of electron microscopic techniques have also been used to demonstrate the absence of cell wall on protoplasts. Standard TEM has been widely applied but it also suffers from a lack of sensitivity since it usually cannot distinguish small numbers of microfibrils on the surface of the plasma membrane. SEM has recently been used to examine freshly isolated protoplasts (Burgess and Linstead 1976b; Burgess et al., 1977a,b; Fowke, 1978). It is now apparent that SEM is subject to serious limitations for such studies. Fine detail on the surface of protoplasts may be completely obscured by the coating unless it is raised above the surface (Burgess et al., 1977b; Burgess and Linstead, 1979). Protoplasts that appear naked may in fact be covered by fine structures such as microfibrils. Furthermore, the coating procedures have been shown to substantially increase the sizes of structures on the surface of protoplasts. Even the initial fixation procedures are important for SEM. The presence or absence of holes in the plasma membrane seems to be directly related to the fixation procedure (Burgess *et al.*, 1977b). It is clear from these studies that results of SEM must be interpreted with caution. Future SEM of plant protoplasts should be accompanied by comparable observations using TEM techniques.

Two other techniques that have proved more reliable for examining the surfaces of protoplasts are freeze-etching and surface replica formation (Figs. 5 and 6). Both techniques offer the distinct advantage of being able to examine the surface of the plasma membrane with high resolution TEM (Abermann *et al.*, 1972; Henderson and Griffiths, 1972). Replicas are perhaps more useful since they permit the examination of very large surface areas. Actual size measurements from shadowed material must be interpreted carefully since resolution is affected by the metals used and the thickness of the film. Shadowing with platinum–carbon is particularly useful for high resolution work (Abermann *et al.*, 1972). A number of studies using these techniques have demonstrated the absence of microfibrils on freshly isolated protoplasts [freeze-etch (Willison and Cocking, 1975; Grout, 1975; Willison and Grout, 1978; Robenek, 1979) and shadowed replicas (Fig. 5 and Williamson *et al.*, 1976, 1977)].

Another electron microscopic technique that has proved both rapid and very sensitive is negative staining. This approach has been particularly useful for examining the size and structure of microfibrils during wall regeneration (see Section IV,B,1). The presence or absence of cell wall polysaccharides can also be studied by using modifications of the periodic acid–Schiff method at the electron microscopic level (Roland, 1978). Freshly prepared protoplasts stained in this manner show little or no stained material on their surfaces (Prat and Roland, 1970; Fowke *et al.*, 1973; Roland and Prat, 1973).

III. Studies of the Plasma Membrane

Information concerning the structure, chemistry, and function of the plasma membrane in plants is very limited. The paucity of knowledge about this membrane is primarily due to the difficulty in isolating and characterizing the plasma membrane. Isolation of the plasma membrane may be facilitated by the use of plant protoplasts. With the cell wall removed, the plasma membrane can be surface-labeled and thus more easily identified when fractionated (Boss and Ruesink, 1979). The progress in this area and the particular problems associated with this type of research are critically reviewed by Quail (1979). The removal of the cell wall during protoplast isolation and exposure of the plasma membrane also provides an excellent opportunity to study the plasma membrane *in situ*. A number of different approaches have been used recently to examine the plasma membrane of freshly isolated protoplasts.

A. Carbohydrate Stains

Results of studies using specific stains confirm the presence of carbohydrates in the plant plasma membrane. For example, both the phosphotungstic acid–chromic acid (PTA–CrO_3) and the silicotungstic acid–chromic acid (STA–CrO_3) procedures are used to detect carbohydrates in plants (Roland, 1978). These stains bind strongly to the plasma membrane of plants and under certain conditions are believed to be specific for components of this membrane (Quail, 1979). The use of these stains with freshly isolated protoplasts has been very limited. Taylor and Hall (1978, 1979) applied STA–CrO_3 to the surface of tobacco protoplasts and despite somewhat variable results they did provide evidence for the presence of carbohydrates in the protoplast plasma membrane. Plant plasma membranes also react with the modified periodic acid–Schiff (PAS) procedure of Thiery (see Roland, 1978), which is used to stain carbohydrates. When this latter stain is applied to freshly isolated protoplasts, the plasma membrane gives a positive reaction (Prat and Roland, 1971). The results described indicate the presence of a carbohydrate component in the plasma membrane. These studies, however, fail to establish whether the carbohydrates are an integral part of the membrane or either remnants or precursors of the cell wall.

B. Lectin Studies

Lectins are proteins or glycoproteins that bind specifically to sugar residues. The best known lectin is concanavalin A (Con A) from the seeds of Jack bean. Con A is a metalloprotein that binds specifically to α-D-glucose and α-D-mannose sugars (Nicolson, 1974; Knox and Clarke, 1978). Near neutral pH, Con A exists as a tetramer with four sugar binding sites. Con A will bind to the surface of plant protoplasts causing agglutination (Glimelius *et al.*, 1974; Burgess and Linstead, 1976a; Williamson *et al.*, 1976; Chin and Scott, 1979).

The ability of Con A to bind to protoplasts makes it a suitable probe for examining the surface of plant protoplasts. Four different labels visible in the electron microscope have recently been attached to freshly isolated protoplasts by Con A (Table II, Fig. 7). The results of these studies indicate the presence of binding sites with a high affinity for Con A on the plasma membrane. When labels are attached to protoplasts prefixed with glutaraldehyde or to live protoplasts in the cold, an even binding pattern is observed (Williamson *et al.*, 1976; Glimelius *et al.*, 1978a; Burgess and Linstead, 1977b). Incubation of protoplasts at room temperature following labeling either in the cold or at room temperature results in a clustering of Con A binding sites. These observations have been interpreted to indicate that the plant plasma membrane, like its counterpart in a variety of animal cells (Nicolson, 1974), complies with the fluid mosaic model

TABLE II

ELECTRON MICROSCOPIC STUDIES OF CONCANAVALIN A BINDING SITES ON THE PLASMA MEMBRANE
OF HIGHER PLANT PROTOPLASTS

Electron microscopic label	Origin of protoplasts	Technique	References
Hemocyanin	Cultured soybean cells	TEM: thin sections and replicas	Williamson *et al.* (1976)
	Leek stem	TEM: thin sections	Williamson (1979)
Colloidal gold	Tobacco mesophyll cells, cultured Grapevine cells	TEM: thin sections	Burgess and Linstead (1976a)
	Tobacco mesophyll cells	TEM: thin sections, SEM	Burgess and Linstead (1977b)
Ferritin	Tobacco mesophyll cells	TEM: thin sections	Burgess and Linstead (1976a)
	Leek stem	TEM: thin sections	Williamson (1979)
Peroxidase–DAB	Cultured carrot cells	TEM: thin sections	Glimelius *et al.* (1978a)

proposed by Singer and Nicolson (1972). The nature of the lectin binding sites has not been elucidated. It is encouraging that similar results have been obtained with protoplasts of different origins and a variety of markers since variations in binding of Con A with different cytochemical markers have been reported with animal cells (Temmink *et al.*, 1975).

It appears that at least five other lectins with sugar specificities differing from that of Con A will bind to plant protoplasts (Larkin, 1978; Chin and Scott, 1979). These lectins might be employed in experiments similar to those described in

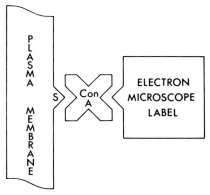

FIG. 7. Diagram illustrating the use of concanavalin A (Con A) to attach an electron microscopic label to a sugar (S) in the protoplast plasma membrane.

order to expand our understanding of the structure and physiology of the plasma membrane. In future experiments designed to study sugar groups by electron microscopic labeling, it is perhaps advisable to utilize hemocyanin labeling and replica techniques (Williamson *et al.*, 1976). Replicas permit the study of large areas of the plasma membrane at high resolution. SEM also encompasses large areas of the protoplast surface, but resolution is limited and problems do arise with interpretation of the image (Burgess and Linstead, 1977b). Standard TEM of thin sections is of course limited by the small sample size.

C. THE INNER SURFACE OF THE PLASMA MEMBRANE

Protoplasts are providing an interesting system for studying the distribution of cortical microtubules associated with the inner surface of the plasma membrane. The techniques for examining these microtubules were first reported by Marchant (1978) for protoplasts of the green alga *Mougeotia* and have recently been applied to studies of higher plant protoplasts (Figs. 8 and 9). The method involves attaching protoplasts to glass coverslips with polylysine and gently bursting them osmotically. The plasma membrane is washed, fixed, and then treated sequentially with antibodies to porcine brain tubulin and a second antibody labeled with fluorescein isothiocyanate (FITC). When examined in a fluorescence microscope, microtubules are visualized on the inner surface of the plasma membrane (Fig. 8). The fixed membrane preparations can also be negatively stained and examined by TEM (Fig. 9 and Marchant, 1978, 1979; Marchant and Hines, 1979) to provide further information regarding the size, structure, and distribution of microtubules. An alternative immunofluorescence method for examining cortical microtubules in plants involving the use of Triton X-100 has recently been described (Lloyd *et al.*, 1979).

IV. Cell Wall Formation

Protoplasts can be cultured in defined media and are capable of regenerating new cell walls. Cultured protoplasts therefore provide a unique system to study the complete assembly of the cell wall at the surface of the cell. This system should be very useful for examining the role of the plasma membrane in the process of wall formation, particularly with respect to the deposition of cellulose microfibrils. Protoplasts may also improve our understanding of the role of cell organelles in primary cell wall synthesis by plant cells. Isolation and characterization of organelles such as Golgi bodies during active wall formation should be facilitated with naked plant cells. Studies of the inner surface of the protoplast plasma membrane during wall regeneration may also clarify the role of microtubules in the process of wall formation. Finally, as a word of caution, it is

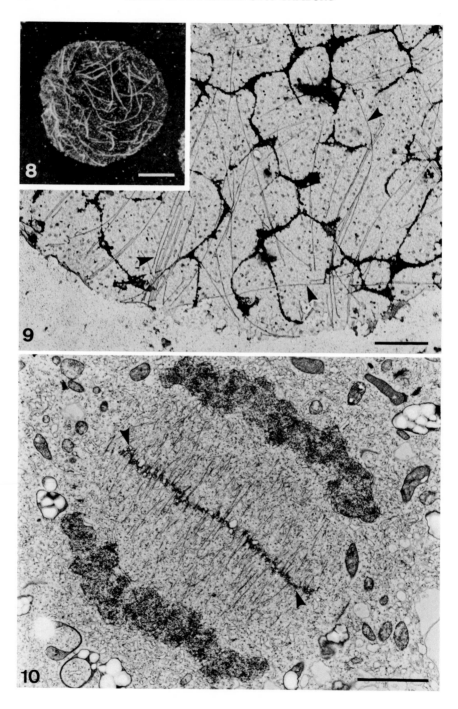

important to remember that wall regeneration by isolated protoplasts is an *in vitro* system and that both the polymer chemistry and assembly of the new walls could differ somewhat from the original walls.

A. TIME COURSE OF WALL FORMATION

Protoplasts seem to differ in their capacity to regenerate cell walls in culture. Whereas some protoplasts are unable to form a substantial wall even with prolonged culture [e.g., carrot (Hellmann and Reinert, 1971) and "Pauls Scarlet" rose (Pearce and Cocking, 1973)], most are able to regenerate complete walls resembling those of normal cells within a week (see Willison, 1976). The failure of some protoplasts to form cell walls likely is a direct result of the viability of protoplasts and/or the lack of adequate culture conditions rather than any inherent quality of the protoplasts.

Protoplasts vary markedly in the speed of initiation of wall formation. In general, it appears that protoplasts from meristematic cells are capable of initiating wall synthesis much more rapidly than protoplasts from differentiated cells. Protoplasts from rapidly growing suspension cultures, for example, initiate wall regeneration almost immediately following isolation and produce a network of microfibrils within 24 hours [Figs. 5 and 6; soybean (Williamson *et al.*, 1976) and *Vicia* (Williamson *et al.*, 1977)]. Protoplasts from leaf mesophyll on the other hand take considerably longer. With tobacco leaf protoplasts, which have been studied the most thoroughly, lag periods ranging from 3 to 24 hours have been reported (Willison and Cocking, 1975; Grout, 1975; Burgess *et al.*, 1978; Burgess and Linstead, 1979). It is perhaps not surprising that differentiated leaf protoplasts require time to adjust their metabolism before initiating wall synthesis and deposition. Once adjusted, leaf mesophyll protoplasts are capable of almost immediate microfibril deposition (Burgess *et al.*, 1978). Prat and Williamson (1976) also reported a difference between wall regeneration on protoplasts from meristematic and mature cells of *Allium porum* (leek). Protoplasts from young leaves initiated wall regeneration within 1.5 hours as compared to 4 hours for protoplasts from mature leaves.

FIG. 8. Light micrograph showing fluorescent staining of microtubules on the inner surface of a tobacco protoplast plasma membrane. The membrane preparation was treated with antibodies to porcine brain tubulin followed by a second antibody labeled with FITC. Bar, 10 μm.

FIG. 9. Electron micrograph of a negatively stained membrane preparation similar to the one in Fig. 8. Many microtubules (arrows) are present on the inner surface of the plasma membrane. Bar, 1 μm. (Kindly provided by Dr. P. Van der Valk.)

FIG. 10. Electron micrograph showing early telophase stage of mitosis in a soybean protoplast. The newly formed nuclei are separated by a phragmoplast (arrows) typical of dividing cells of higher plants. Bar, 3 μm. (From Fowke *et al.*, 1974b.)

B. Nature of the Regenerated Cell Wall

The nature of the newly formed wall has been examined by a variety of electron microscope techniques, e.g. polysaccharide staining (Roland and Prat, 1973; Fowke *et al.*, 1974a; Prat and Williamson, 1976), freeze-etching (Willison and Cocking, 1975; Grout, 1975; Robenek and Peveling, 1977; Willison and Grout, 1978), metal replicas (Prat and Williamson, 1976; Williamson *et al.*, 1977), negative staining (Prat and Williamson, 1976; Herth and Meyer, 1977; Burgess and Linstead, 1979), and SEM (Burgess *et al.*, 1978; Burgess and Linstead, 1976b, 1977a, 1979). Only a limited number of biochemical studies have been reported (Hanke and Northcote, 1974; Herth and Meyer, 1977; Asamizu *et al.*, 1977; Takeuchi and Komamine, 1978; Klein and Delmer, 1979).

The new cell wall is composed of a loosely organized net of microfibrils (e.g., Fig. 6 and Prat and Roland, 1971; Burgess and Fleming, 1974a; Willison and Cocking, 1975; Grout, 1975; Williamson *et al.*, 1977; Burgess *et al.*, 1978). Initially, most of the matrix polysaccharides are apparently lost to the culture medium (Hanke and Northcote, 1974; Takeuchi and Komamine, 1978). The loss of matrix polysaccharides to the medium may be due to an insufficient number of microfibrils. With further deposition of microfibrils, more matrix would be retained with the eventual production of a compact wall similar to the original cell wall.

1. *Microfibrils*

It has generally been assumed that the microfibrils formed on the surface of cultured plant protoplasts are composed of cellulose. This assumption was based primarily on evidence of a circumstantial nature (e.g., fibrillar appearance, positive reaction with carbohydrate stains, cellulase digestability). Direct evidence for the cellulosic nature of newly formed microfibrils has recently been provided. In combined ultrastructural and biochemical studies of wall regeneration by protoplasts, microfibrils consisting of crystalline cellulose have been reported on cultured protoplasts derived from both tobacco mesophyll (Herth and Meyer, 1977) and suspension-cultured soybean cells (Klein and Delmer, 1979, and personal communication).

The size of microfibrils reported on cultured protoplasts varies with the technique used to examine them. Microfibrils examined with conventional TEM techniques or after negative staining range in diameter from 12 to 60 Å (Prat and Williamson, 1976; Herth and Meyer, 1977; Burgess and Linstead, 1979), whereas those observed in shadowed replicas or freeze-etch preparations range in diameter from 80 to 150 Å (Willison and Cocking, 1972; Prat and Williamson, 1976).

Fibers observed on the surface of tobacco protoplasts by SEM consist of bundles of microfibrils and approach values of 400 Å in diameter (Willison and Grout, 1978; Burgess and Linstead, 1979).

Cellulose microfibril deposition is generally believed to occur at the plasma membrane of plant cells (Chrispeels, 1976; Robinson, 1977) and results from studies with protoplasts regenerating cell walls are consistent with this idea. Microfibrils are first detected on the surface of the plasma membrane of cultured protoplasts. Results of freeze-etching studies suggest that preformed microfibrils of indeterminate length rise out of the plasma membrane or crystallize at the plasma membrane surface (Willison and Cocking, 1975; Grout, 1975; Willison, 1976). In contrast, Burgess and Linstead (1976b) have suggested on the basis of SEM observations that microfibrils form on the surface by end synthesis. Their interpretation must, however, be reexamined since it is now clear that the SEM techniques used tend to obliterate the very fine microfibrils on the protoplast surface (Willison and Grout, 1978; Burgess and Linstead, 1979).

One current model for synthesis and deposition of cellulose microfibrils by plant cells involves the presence in the plasma membrane of mobile enzyme complexes. Such complexes have been observed at the ends of microfibrils in the plasma membrane in both algal and higher plant cells (Montezinos and Brown, 1976; Mueller *et al.*, 1976; Willison and Grout, 1978; Giddings *et al.*, 1980; Mueller and Brown, 1980). Similar structures have not been observed in the plasma membrane of protoplasts. Smaller membrane particles have been observed in many protoplasts (Willison and Cocking, 1972, 1975; Robenek and Peveling, 1977; Davey and Mathias, 1979). These particles have not been clearly implicated in the process of cell wall formation. Robenek and Peveling (1977) reported some interesting correlations between the distribution and number of particles and the onset of wall formation on protoplasts of *Skimmia japonica*. However, Davey and Mathias (1979) argue that the particle arrangements described by Robenek and Peveling likely are artifacts induced by glycerol treatment.

Microtubules may play an important role in orienting cellulose microfibrils during cell wall formation in plants (Hepler and Palevitz, 1974; Heath, 1974; Robinson, 1977). With plant protoplasts, fluorescent antibody techniques and negative staining can be used to detect microtubules on the inner surface of the plasma membrane (see Section III,C and Figs. 8 and 9). Studies of cultured protoplasts of the green alga *Mougeotia* using these techniques (Marchant, 1979; Marchant and Hines, 1979) provide evidence to support the hypothesis that microtubules control cellulose microfibril orientation. These methods will be most useful for providing information regarding microtubule number, size, and orientation in freshly isolated protoplasts, as well as during the early stages of cell wall regeneration.

2. *Matrix*

As mentioned previously, most of the matrix polysaccharides appear to be lost to the culture medium during the initial stages of cell wall regeneration. Very little is known about the nature of these polysaccharides or their modes of synthesis and deposition. It is well established that Golgi bodies are responsible for the synthesis and deposition of matrix components during primary cell wall growth in higher plants (Chrispeels, 1976; Robinson, 1977). The evidence to support a similar role for Golgi bodies in protoplasts regenerating cell walls is at best circumstantial and somewhat contradictory. Both the number of Golgi bodies and their activity as judged by associated vesicles seem to vary considerably in cultured protoplasts (e.g., Fowke *et al.,* 1974a; Gigot *et al.,* 1975; Robenek and Peveling, 1977). Furthermore, attempts to stain presumed cell wall precursors with carbohydrate stains have yielded conflicting results (Prat and Roland, 1971; Roland and Prat, 1973; Fowke *et al.,* 1974a).

The suggestion that endoplasmic reticulum may also be important in the process of cell wall formation is similarly supported by circumstantial evidence. Arguments have been based on the marked increase in endoplasmic reticulum during wall regeneration and the presence of sheets of endoplasmic reticulum parallel to and closely associated with the plasma membrane (Burgess and Flemming, 1974a; Robenek and Peveling, 1977; Robenek, 1979).

It is clear from the foregoing that little is known about the functions of Golgi bodies or endoplasmic reticulum in the regeneration of cell walls by cultured protoplasts. There is need for a more rigorous examination of these organelles, perhaps by combining autoradiographic and biochemical studies.

3. *Abnormal Cell Walls and Budding*

A variety of abnormal cell walls have been reported on cultured plant protoplasts. Some seem to be a direct result of the culture conditions. For example, when tobacco protoplasts are cultured in a saline medium, they produce a nonrigid "pseudo-wall" that cannot be separated from the cytoplasm by plasmolysis (Meyer and Abel, 1975). Other strange cell walls may relate to the origin of the plant tissue used to isolate protoplasts (e.g., tomato locule). Protoplasts derived from this tissue do not regenerate a normal wall in culture but produce a multilamellar cell wall as a prelude to the synthesis of cellulose microfibrils (Willison and Cocking, 1972; Cocking, 1974). Tomato locule protoplasts also contain very strange membrane-bounded accumulations of presumed wall polysaccharides that have been termed "wall-bodies" (Pearce *et al.,* 1974). These structures may develop as a wounding response (Willison, 1976).

The formation of cytoplasmic protrusions or budding at the surface of cultured protoplasts is a common phenomenon (see review by Willison, 1976). Budding seems to occur in weakened areas of the newly formed cell wall, possibly as a result of excess microfibril slippage due to the loss of matrix components (Hanke and Northcote, 1974) and tends to be prevalent under conditions that are suboptimal for protoplast growth and cell wall development.

V. Cell Division

Protoplasts that lack a normal complex cell wall may be very useful for studying the process of cell division in plants. A great deal of basic information regarding the structure and function of the mitotic spindle and the formation of the phragmoplast and cell plate in plants has been derived from studies of the natural protoplasts found in the endosperm of the African blood lily, *Haemanthus katherinae* (Hepler and Jackson, 1968; Bajer and Molè-Bajer, 1972; Newcomb, 1978). Protoplasts isolated from a variety of plants will exhibit sustained division in culture. Cell divisions are usually initiated within the first 2–3 days of culture [e.g., carrot (Grambow *et al.*, 1972; Wallin and Eriksson, 1973), soybean (Kao *et al.*, 1971; Fowke *et al.*, 1974), and tobacco (Nagata and Takebe, 1970)] when protoplasts are enclosed within a thin, loosely organized cell wall. Protoplasts derived from actively dividing cells usually initiate cell division earlier than protoplasts from nondividing cells (Torrey and Landgren, 1977).

Cultured protoplasts offer a number of distinct advantages for studies of cell division. The presence of very little wall material at the cell surface permits high resolution studies of cell division with the light microscope. The amount of cell wall present may be minimized by using inhibitors of cell wall formation such as coumarin (Burgess and Linstead, 1977a; Meyer and Herth, 1978). Young regenerating cells may also permit studies of the spindle using indirect immunofluorescent techniques. By using such techniques, it may be possible, for example, to achieve uptake of fluorescent labeled antitubulin to study microtubule distribution during cell division. Finally, the absence of a fully developed cell wall should facilitate the isolation and characterization of intact spindles and chromosomes from plant cells.

A. Division in Uninucleate Protoplasts

Cell division has been reported in many protoplasts, but few detailed studies of the process have been reported. The fact that a number of protoplasts can regenerate to plants (Table I) suggests that protoplasts are capable of normal cell division. A comparative light and electron microscopic study of cell division in protoplasts and cultured cells of soybean confirms this suggestion (Fowke *et al.*, 1974b). Apart from the absence of preprophase bands of microtubules, the formation, structure, and functioning of the spindle is characteristic of higher plants in general (e.g., Bajer and Molè-Bajer, 1972; Pickett-Heaps, 1974). The absence of preprophase bands may reflect the general lack of organized growth in cultured cells and protoplasts. Preprophase bands are typical of organized plant tissues (Pickett-Heaps, 1974; Gunning *et al.*, 1978). Cytokinesis in soybean protoplasts is accomplished by a phragmoplast (Fig. 10) and cell plate and the newly formed crosswall contains plasmodesmata (Fowke *et al.*, 1974b).

Cultured protoplasts tend to regenerate some cell wall prior to mitosis, and it has been suggested that wall regeneration is a necessary prerequisite for division (Schilde-Rentschler, 1977). Tobacco protoplasts cultured in cellulase or treated with coumarin to prevent wall regeneration are unable to divide (Schilde-Rentschler, 1977; Burgess and Linstead, 1977a). In contrast, Meyer and Herth (1978) reported that mitosis proceeded in tobacco protoplasts even when cell wall formation was blocked with inhibitors. Further studies are necessary to resolve this apparent controversy. The results to date clearly indicate that protoplasts with rather poorly developed cell walls are capable of normal cell division (Fowke *et al.*, 1974b).

Protoplasts in culture are prone to a number of abnormalities in cell division. While abnormalities have been reported during mitosis (Landgren, 1976), the stage of cell division most susceptible to irregularities seems to be cytokinesis. In some protoplasts, cytokinesis is apparently totally lacking and multinucleate protoplasts are formed as a result of mitosis (see Section V,B). In others, cytokinesis proceeds abnormally to produce crosswalls with many perforations (Prat and Poirier-Hamon, 1975). Pearce and Cocking (1973) reported the formation and subsequent disappearance of a phragmoplast in cultured protoplasts of "Pauls Scarlet" rose. However, they failed to demonstrate the phragmoplast with TEM. Herth and Meyer (1978) reported similar findings in an electron microscopic study of tobacco protoplasts cultured in a saline medium. Budding of protoplasts frequently occurred with one daughter nucleus moving into the bud. They propose that cytokinesis is accomplished by a cleavage process similar to cleavage in yeast, but they do not provide sufficient evidence in support of a dynamic cleavage process.

It is difficult to determine the cause of the above observations. The abnormalities in tobacco protoplasts cultured in saline medium are perhaps not too surprising since under these culture conditions other abnormalities have been reported (Meyer and Abel, 1975). In fact, it has been suggested that a major cause of irregularities in division is an unfavorable nutrient or hormone balance in the culture medium (Prat and Poirier-Hamon, 1975; Torrey and Landgren, 1977).

B. DIVISION IN MULTINUCLEATE PROTOPLASTS

As mentioned previously, multinucleate protoplasts generally are produced by spontaneous fusion during protoplast isolation. In some cases, they may also arise by continued mitosis without subsequent cytokinesis (e.g., Eriksson and Jonasson, 1969; Hellman and Reinert, 1971; Torrey and Landgren, 1977). The details of this process have not been clearly documented. Torrey and Landgren (1977) propose that such an uncoupling of mitosis and cytokinesis is due to unfavorable culture conditions.

It has been demonstrated by both light and electron microscopy that multinucleate protoplasts arising by spontaneous fusion are apparently capable of undergoing normal cell division (Miller *et al.*, 1971; Motoyoshi, 1971; Fowke *et al.*, 1975a). Nuclei may fuse prior to division, although in most multinucleate protoplasts nuclei remain separate and divide synchronously. Synchronous divisions are typical of most naturally occurring and artificially created multinucleate cells (Johnson and Rao, 1971). Nuclear bridges may connect nuclei at interphase (Fowke *et al.*, 1975a), but their significance in the subsequent division process has not been demonstrated. Mitosis and cytokinesis are closely comparable to the same processes in uninucleate soybean protoplasts (Fowke *et al.*, 1974b, 1975a). Nuclei are partitioned by cell plates, but it has not been determined whether viable cell clusters ultimately are formed by this process. In other experimental systems with cultured multinucleate protoplasts, division has not been observed, e.g., *Ammi visnaga* (Fowke *et al.*, 1974a) and tobacco (Morel *et al.*, 1973).

VI. Protoplast Fusion

One of the most exciting applications of plant protoplasts has been the production of hybrids by induced fusion between freshly isolated protoplasts (Fig. 11). Fused protoplasts retain their totipotency and can regenerate hybrid plants in culture (Table III). There has been a major interest in somatic hybridization studies aimed at the production of hybrid plants from species that cannot be crossed sexually. This goal has been achieved with the recent production of hybrid plants from intergeneric fusions (Melchers *et al.*, 1978; Krumbiegel and Schieder, 1979; Dudits *et al.*, 1979; Gleba and Hoffmann, 1979). Since this aspect of protoplast fusion has been reviewed thoroughly (Constabel, 1976; Scowcroft, 1977; Bajaj, 1977; Cocking, 1978a,b; Gamborg *et al.*, 1974, 1978), it will not be dealt with further in this chapter.

Plant protoplast fusion has two further important applications to the study of plant cells. First, protoplast fusion offers a unique opportunity to study membrane interactions that occur during the actual fusion process (Fig. 12). Second, fusion products can be cultured and provide an excellent system for studying the fate of cell organelles in mixed cytoplasm (Fig. 13). We will consider these applications in some detail.

Protoplasts can be agglutinated and subsequently induced to fuse by treating them with a number of different solutions, e.g., sodium nitrate (Power *et al.*,

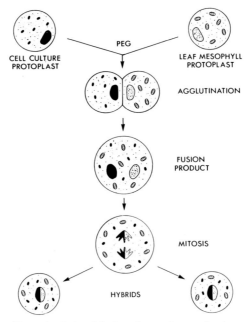

FIG. 11.   Diagram illustrating induced fusion of protoplasts and culture of fusion products.

1970), alkaline calcium solution (Keller and Melchers, 1973), and polyethylene glycol (PEG) (Kao and Michayluk, 1974; Eriksson *et al.*, 1974). More recently both polyvinylalcohol (Nagata, 1978) and liposomes composed of a synthetic, positively charged phospholipid (Nagata *et al.*, 1979) have been shown to induce protoplast fusion. Lectins (Glimelius *et al.*, 1974) and antibodies (Hartmann *et al.*, 1973; Burgess and Flemming, 1974b) will also agglutinate protoplasts but will not fuse them. The most successful and reproducible procedure for protoplast fusion involves the use of PEG. Protoplasts are first agglutinated with high concentrations (28–30% w/v) of high molecular weight PEG (e.g., PEG 1540) and then fused during the gradual elution of the PEG (Fig. 11 and Kao, 1975). Following washing with culture medium, the fusion products can be cultured in petri dishes. Cell wall regeneration, cell division, and, in some cases, complete plant regeneration will occur.

A. MEMBRANE INTERACTIONS

PEG-induced fusion is nonspecific and thus fusions can be induced between protoplasts from widely separated plant species. In fact, fusions have been achieved between algal and higher plant protoplasts (Fowke *et al.*, 1979) and even between animal cells and plant protoplasts (Jones *et al.*, 1976; Dudits *et al.*, 1976b; Willis *et al.*, 1977; Davey *et al.*, 1978; Ward *et al.*, 1979). This

TABLE III

REGENERATION OF PLANTS FROM FUSED PROTOPLASTS

Plant species fused	References
Arabidopsis thaliana + Brassica campestris	Gleba and Hoffman (1979)
Datura innoxia + D. innoxia	Schieder (1977)
Datura innoxia + D. discolor	Schieder (1978)
Datura innoxia + D. stramonium	Schieder (1978)
Datura innoxia + Atropa belladonna	Krumbiegel and Schieder (1979)
Daucus carota + D. capillifolius	Dudits *et al.* (1977)
Daucus carota + Aegopodium podagraria	Dudits *et al.* (1979)
Nicotiana glauca + N. langsdorffii	Carlson *et al.* (1972); Smith *et al.* (1976); Chen *et al.* (1977)
Nicotiana sylvestris + N. knightiana	Maliga *et al.* (1977)
Nicotiana tabacum + N. tabacum	Melchers and Labib (1974); Glimelius *et al.* (1978b); Gleba (1979)
Nicotiana tabacum + N. glauca	Evans *et al.* (1979)
Nicotiana tabacum + N. knightiana	Maliga *et al.* (1978)
Nicotiana tabacum + N. rustica	Nagao (1978)
Nicotiana tabacum + N. sylvestris	Melchers (1977); Zelcer *et al.* (1978)
Petunia hybrida + P. parodii	Power *et al.* (1976b); Cocking *et al.* (1977)
Petunia hybrida + P. axillaris	Izhar and Power (1979)
Solanum tuberosum + Lycopersicum esculentum	Melchers *et al.* (1978)

technique should therefore be very useful for studying interactions of a broad spectrum of membrane systems. Unfortunately during the past few years, very little attention has been focused on the nature of these membrane interactions.

Agglutination and subsequent fusion of higher plant protoplasts has been studied with both light and electron microscopy. Agglutination occurs by tight contact between adjacent plasma membranes. In some cases, the membrane contact is uniform, whereas in others, it occurs in localized regions (Fig. 12 and Burgess and Fleming, 1974b; Wallin *et al.,* 1974; Fowke *et al.,* 1975b). Robenek and Peveling (1978) suggest that agglutination is mediated by lipid droplets secreted by the protoplasts, but other studies do not support their results. The actual fusion event has not been recorded by TEM. Conventional chemical fixation may not be rapid enough to capture the fusion of membranes (Poste and Allison, 1973). An alternative approach that has proved very useful for studying membrane fusion in animal cells is the use of freeze-etching techniques (Orci *et al.,* 1977; Pinto da Silva and Nogueira, 1977). These authors demonstrate the movement of intramembranous particles away from the zone of agglutination and subsequent fusion. A similar mobility of intramembrane particles may occur in the plasma membranes of fusing plant protoplasts (Robenek and Peveling, 1978). Further freeze-etch work is necessary to clarify the membrane changes occurring during protoplast fusion.

The fusion of adjacent plasma membranes results in cytoplasmic continuity

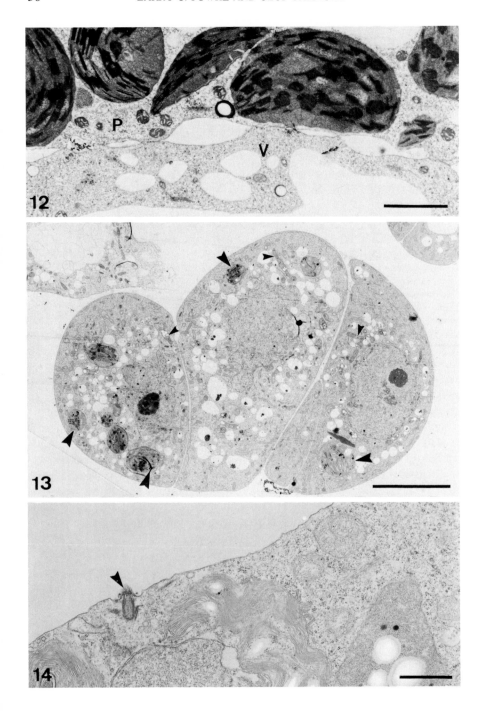

between the two protoplasts. The fate of the agglutinated plasma membranes during the fusion process is uncertain. It has been suggested that the membranes may either vesiculate and disperse in the cytoplasm or possibly disintegrate (Takebe *et al.*, 1973; Burgess and Flemming, 1974b; Fowke *et al.*, 1975b). It should be possible to determine the fate of the protoplast plasma membranes by attaching labels that can be detected in the electron microscope to the surface of protoplasts (e.g., see Table II). Burgess *et al.* (1977a) have initiated work in this direction by demonstrating that protoplasts labeled with colloidal gold are capable of agglutination. It should also be possible to study the fate of fusing membranes at the light microscopic level with fluorescent probes (e.g., Frye and Ediden, 1970; Keller *et al.*, 1977; Galbraith and Galbraith, 1979).

The fusion of small algal protoplasts with higher plant protoplasts may provide a model system for investigating membrane interactions. Fowke *et al.* (1979) demonstrated the fusion of cell wall-less mutants of *Chlamydomonas reinhardii* with carrot protoplasts (Figs. 14 and 15). The fusion resulted in the release of algal organelles into the carrot cytoplasm. This system may be particularly useful for a number of reasons. Cell wall-less mutants of *Chlamydomonas* (e.g., CW15) are easily cultured and therefore provide a ready source of protoplasts. Secondly, the basal bodies associated with the *Chlamydomonas* plasma membrane serve as a useful marker for membrane fusion at the electron microscopic level and possibly also at the light microscopic level with fluorescent techniques (Connolly and Kalnins, 1978). Moreover a variety of mutants have been characterized in *Chlamydomonas* and some may prove valuable for studying membrane properties.

B. ORGANELLES IN CULTURED FUSION PRODUCTS

Fusion products will regenerate cell walls and undergo division when cultured (Figs. 11 and 13). Information is gradually accumulating regarding the fate of cell organelles in such cultured fusion products. Emphasis has been focused on the fate of nuclei and plastids and information regarding other cell organelles in fusion products is rather fragmentary. Since a wide range of protoplast combinations has been studied, the results provide some insight into the problem of incompatability at the cellular level.

FIG. 12. Electron micrograph showing agglutination of pea leaf (P) and *Vicia* cell culture (V) protoplasts. Bar, 2 μm.

FIG. 13. Electron micrograph showing a multicellular fusion product derived from fusion of pea leaf and soybean cell culture protoplasts. Chloroplasts (large arrows) and leucoplasts (small arrows) are present in all cells. Bar, 10 μm.

FIG. 14. Electron micrograph showing a *Chlamydomonas* basal body (arrow) at the surface of a carrot protoplast. See Fig. 15 also. Bar, 1 μm.

Fᴵɢ. 15.   Diagram illustrating fusion of *Chlamydomonas* (Ch) and carrot (C) protoplasts. The algal plasma membrane is incorporated into the carrot plasma membrane and intact algal organelles are released into the carrot cytoplasm. (Adapted from Fowke *et al.*, 1979.)

1. *Nuclei*

Nuclear structure and distribution in fusion products has been traced through early divisions and in some combinations through a callus phase to complete hybrid plants. The fate of individual nuclei varies considerably with different combinations of plants. Nuclei may remain separate in fusion products and may undergo division usually synchronously to produce chimeral calluses (Kao *et al.*, 1974; Constabel, 1976; Gosch and Reinert, 1978). The very large heterokaryocytes with many nuclei generally deteriorate and die (Constabel *et al.*, 1975). Alternatively, nuclei in fusion products may undergo fusion. Nuclear fusion during interphase has been described with both light microscopy (Constabel *et al.*, 1975; Dudits *et al.*, 1976c; Kao, 1977; Binding and Nehls, 1978) and electron microscopy (Fowke *et al.*, 1977). Recent results suggest that fusion during interphase may be an artifact due to treatment with Driselase and may be typical of nonviable fusion products (Constabel, 1978; Kao, 1978). As is the case with animal cell fusions, viable hybrid cells seem to result from nuclear fusions during the process of mitosis (Fig. 11 and Kao *et al.*, 1974; Constabel *et al.*, 1975; Gosch and Reinert, 1978).

Nuclear cytology has been examined in detail in only a few hybrid cell lines. During the first few cell divisions chromatin of the two parents can be distinguished even during interphase (Fowke *et al.*, 1977; Constabel *et al.*, 1977; Constabel, 1978). During prolonged culture, the distribution of chromatin is quite variable. In some hybrid cell lines, little or no chromosome loss is apparent [e.g., *Arabidopsis* + *Brassica* (Gleba and Hoffmann, 1978)], whereas in others only a few chromosomes from one parent are retained [e.g., soybean + tobacco (Kao, 1977)]. Similarly, considerable variation in the type and number of chromosome abnormalities has been reported (Kao, 1977; Binding and Nehls, 1978).

2. *Plastids*

The fate of plastids in mixed cytoplasm has been determined primarily by examining fusion products derived from suspension culture protoplasts (containing leucoplasts) and leaf mesophyll protoplasts (containing chloroplasts) (Figs.

11, 12, and 13). Fusion products have been cultured for short periods of time (ca. 1–2 weeks) and their plastids and those of cocultured parental protoplasts examined with conventional TEM.

The number, distribution, and structure of leucoplasts in all cultured fusion products examined remained essentially unchanged (Fowke *et al.*, 1976, 1977). The fact that a reduction in number of leucoplasts per cell was not observed following cell division suggests that leucoplasts may divide in cultured fusion products. Leucoplasts are therefore apparently not affected by the hybrid cytoplasm.

In contrast, the chloroplasts in cultured fusion products exhibit a variety of changes. In fusion products derived from plants of different genera [e.g., soybean + sweet clover (Fowke *et al.*, 1976), soybean + pea (Fowke *et al.*, 1977)], structural abnormalities are often observed in the chloroplasts. These range from altered internal membranes to complete plastid degeneration and may reflect some basic incompatability at the cellular level. Degeneration of chloroplasts from one parent has also been reported in intergeneric fusions involving the filamentous green algae *Zynema* and *Spirogyra* (Ohiwa, 1978). It is interesting to note that when protoplasts of even more widely separated species are fused (e.g., *Chlamydomonas* and carrot) most of the algal organelles deteriorate in the cultured fusion products (Fowke *et al.*, 1979). On the other hand, with intrageneric fusions [*Vicia hajastana* + *V. narbonensis* or *Vicia hajastana* + *V. hajastana* (Rennie *et al.*, 1980)], chloroplasts in cultured fusion products seem to dedifferentiate rather than degenerate. Mature chloroplasts may dedifferentiate as a result of being transferred to a cytoplasm capable of active cell division. In the intergeneric and intrageneric fusion products examined to date, it appears that chloroplasts are not able to divide and are thus diluted with each cell division. Long-term studies are necessary to determine whether chloroplasts, particularly those which have dedifferentiated, eventually adapt to the mixed cytoplasm and resume division.

3. *Vacuoles*

Evidence is accumulating that suggests that vacuoles fuse during protoplast fusion. When protoplasts containing anthocyanins in their vacuoles are fused with unpigmented protoplasts, all vacuoles in the resulting fusion product apparently contain pigment (Smith, 1977; Constabel, personal communication). Recently Constabel *et al.* (1979) provided further evidence for vacuole fusion. When periwinkle protoplasts containing alkaloids in their vacuoles were fused with *Haplopappus* protoplasts containing polyphenols in their vacuoles, large precipitates were observed in the vacuoles of fusion products. Fusion of vacuoles is inferred since phenolics in aqueous solution are known to precipitate alkaloids and vice versa (Constabel *et al.*, 1979).

4. *Other Organelles*

It is usually impossible to recognize parental mitochondria in cultured fusion products. However, in fusion products derived from protoplasts of carrot and *Vicia,* Davey and Cocking (1979) were able to distinguish the two parental mitochondria on the basis of density of internal matrix. They did not comment on whether mitochondria fused or not. Biochemical and genetic studies of mitochondrial DNA, enzymes, etc. have been completed with animal cell hybrids (Ringertz and Savage, 1976) but not with plant fusion products.

The fate of other organelles such as ribosomes, endoplasmic reticulum, Golgi bodies, and microtubules in cultured plant fusion products is not known.

VII. Uptake Studies

The removal of the cell wall from plant cells exposes the plasma membrane and permits the uptake of a wide range of substances by the protoplasts. Uptake by plant protoplasts has been useful in a number of ways. First, it provides an opportunity to study directly the process of endocytosis at the plant plasma membrane. Second, the research into plant virus uptake and replication has been greatly aided. A variety of viruses can be taken up into plant protoplasts and will undergo replication. Since it is possible to infect a large proportion of protoplasts in a population with virus, this system is excellent for studying the basic mechanism of virus replication. Finally, uptake of a wide range of macromolecules, organelles, and cells of microorganisms has been investigated in an effort to introduce foreign genetic material into plant cells. These attempts have met with variable success. Most studies have failed to clearly demonstrate uptake into plant protoplasts either from a structural or functional standpoint. This section of the review will first examine the progress in understanding endocytosis at the protoplast surface, deal briefly with virus uptake and conclude by looking at the developments in the field of organelle and microbial cell uptake directed at genetically modifying plant protoplasts. Our emphasis will be on the structural aspects of uptake and we will try to assess the weaknesses and strengths in this area. Since very little is known about the structural basis for uptake of DNA by plant protoplasts, this topic will not be dealt with (see reviews by Cocking, 1977; Ohyama *et al.,* 1978).

A. ENDOCYTOSIS

Uptake into plant protoplasts occurs primarily by a process of endocytosis, which involves invagination of the plasma membrane (Fig. 16A) as opposed to membrane fusion (Fig. 16B). Without the use of the electron microscope, it is often impossible to distinguish between these two possibilities. Table IV lists a

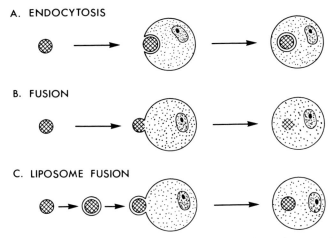

A. ENDOCYTOSIS

B. FUSION

C. LIPOSOME FUSION

FIG. 16.   Diagram illustrating uptake by plant protoplasts. (A) Endocytosis by invagination of the plasma membrane. (B) Membrane fusion. (C) Uptake by enclosure within a liposome.

variety of cells, organelles, and particles that are reported to have been taken up by the process of endocytosis as determined by electron microscopic observations. There is no general agreement regarding the mechanism of uptake of chloroplasts (see following) and viruses (see Burgess *et al.*, 1973).

Studies of uptake of polystyrene spheres have provided some insight into the mechanism of endocytosis in plant protoplasts (Mayo and Cocking, 1969; Willison *et al.*, 1971; Grout *et al.*, 1972; Suzuki *et al.*, 1977). Endocytosis of

TABLE IV

ENDOCYTOSIS BY HIGHER PLANT PROTOPLASTS AS REVEALED BY ELECTRON MICROSCOPY

Component	Origin of protoplasts	References
Ferritin	Tobacco mesophyll cells	Power and Cocking (1970)
	Tomato fruit cells	Cocking (1970)
Polystyrene spheres	Tomato fruit cells	Mayo and Cocking (1969); Willison *et al.* (1971); Grout *et al.* (1972)
	Tobacco mesophyll cells	Suzuki *et al.* (1977)
Viruses	Tomato fruit cells	Cocking (1970)
	Tobacco mesophyll cells	Otsuki *et al.* (1972); Honda *et al.* (1975)
Bacteria	Pea mesophyll cells	Davey and Cocking (1972); Davey (1977)
Chloroplasts	Cultured *Parthenocissus* cells	Davey *et al.* (1976)
Blue green algae	Cultured *Parthenocissus* cells	Davey and Power (1975)
Yeast cells and protoplasts	Cultured *Parthenocissus* cells	Davey and Power (1975)

polystyrene spheres occurs in two distinct phases, adhesion to the plasma membrane and uptake by invagination of the plasma membrane. The attachment phase is apparently not energy dependent but probably is due to electrostatic forces (Grout *et al.*, 1972; Suzuki *et al.*, 1977). The actual movement of spheres into the protoplast as individuals or as clusters occurs very rapidly and is dependent on energy produced by oxidative pathways (Willison *et al.*, 1971; Suzuki *et al.*, 1977). The fate of spheres within the protoplast is uncertain although Suzuki *et al.* (1977) suggest their release into the cytoplasm. The uptake mechanism described bears a number of similarities to endocytosis in animal cells. The presence of a distinct attachment phase, the response to colchicine, and the energy requirements of the process have all be described in animal systems (Silverstein *et al.*, 1977).

Recently Ueda *et al.* (1978) have suggested that protoplasts are capable of phagocytosing one another. The light micrographs from cine film that they present to support their idea of PEG-induced phagocytosis could, however, be interpreted as protoplast fusion. Furthermore, PEG treatments of up to 4 hours were used in this study. Such treatments could cause a wide range of membrane fusions, thus making their results very difficult to assess.

Protoplasts seem to be ideal for studies of endocytosis in plants. It must be remembered, however, that protoplasts represent a rather artificial system for such studies. Protoplasts are maintained in a high osmoticum and lack a cell wall, which normally filters out particles of the size used in most studies of endocytosis. Despite these difficulties, the available evidence favors the existence of a plasma membrane-associated system for energy-mediated uptake of materials into plant protoplasts.

B. VIRUS UPTAKE

The uptake of viruses into plant protoplasts has led to new basic information regarding the replication of viruses in plant cells. Previous studies of virus infection using intact leaf tissues were hampered by low infection rates and slow spread of virus, which resulted in nonsynchronous viral replication. Plant protoplasts offer a number of major advantages when compared to leaf tissue. The most important advantages include high frequency of infection, a high proportion of infected cells, and synchrony of infection and replication. The infection of tobacco protoplasts with tobacco mosaic virus (TMV) has been studied the most thoroughly, and information is now available concerning viral uptake, synthesis of viral RNA and coat protein, and assembly of virus particles (see review by Takebe, 1977). Similar information is accumulating from studies with other viruses and protoplasts. Takebe (1977) concludes that the process of virus replication is basically the same in leaves and protoplasts. He does caution, however, that because of the completely different environment of the leaf cells and isolated protoplasts they may exhibit differences in some of the features of virus replication.

C. Uptake of Organelles and Microorganisms

Organelles and microbial cells have been introduced into plant protoplasts with the hopes of genetically modifying the host protoplasts. In general, these studies have suffered from a lack of basic information regarding the viability of organelles and microorganisms before and after uptake, as well as information concerning the actual uptake process. In many cases, it is impossible to know exactly what has been taken into the cytoplasm of the plant protoplast. This problem highlights the need for critical ultrastructural studies to first determine the mechanism of uptake and second to establish the fate of material taken into protoplasts.

1. *Organelles*

The prospect of introducing cell organelles into naked protoplasts is intriguing. If feasible, this approach might considerably facilitate research on the development and genetics of cell organelles, as well as improve our understanding of nuclear–cytoplasmic interactions and regulation (see reviews by Potrykus and Lörz, 1976; Giles, 1977). Research in this area is limited and little progress has been made. Current work is concerned mainly with the incorporation of isolated chloroplasts and nuclei into higher plant protoplasts.

The uptake of chloroplasts into protoplasts has been observed in studies utilizing both the light microscope (Potrykus, 1973; Bonnett and Eriksson, 1974) and electron microscope (Davey *et al.*, 1976; Bonnett, 1976). Fine structure studies are essential to determine the nature of the uptake process and the structural integrity of chloroplasts following fusion.

There have not been any convincing reports of uptake of *intact* chloroplasts by higher plant protoplasts. The work of Davey *et al.* (1976) illustrates some of the difficulties encountered in this type of research. The isolation of intact chloroplasts proved very difficult and chloroplasts tended to degenerate rapidly. Populations of freshly isolated chloroplasts ready for fusion studies contained intact, swollen, and totally disrupted chloroplasts. Furthermore, the PEG-mediated fusion proved rather complicated. In some cases, the chloroplast envelope apparently fused with the plasma membrane, thus releasing organelle contents into the protoplast cytoplasm (Fig. 16B). Other chloroplasts seemed to be taken up by endocytosis (Fig. 16A).

Bonnett (1976) has also provided electron microscopic evidence for the uptake of chloroplasts into higher plant protoplasts, but again the mechanism is not clear. His electron micrograph depicts an algal chloroplast within the cytoplasm of a carrot protoplast. The chloroplast, however, is lacking its outer envelope, which suggests that membrane fusion as in Fig. 16B may have occurred.

Even less information is available concerning the uptake of nuclei into isolated plant protoplasts. Evidence of nuclear uptake is derived entirely from light microscope observations (Potrykus and Hoffman, 1973; Binding, 1976; Lörz and Potrykus, 1978), and thus, nothing is known about the mechanism of uptake or the fate of nuclei in the host protoplast. The fact that Lörz and Potrykus (1978) obtained the most efficient uptake with PEG suggests uptake may result from a fusion event. If so, the nuclei would not be expected to remain intact following uptake.

Wallin *et al.* (1978) have suggested that a better approach to the problem of nuclear uptake may be the use of miniprotoplasts consisting of a nucleus surrounded by a thin layer of cytoplasm and enclosed within a plasma membrane. Miniprotoplasts capable of cell division have been produced from a number of plants by exposing protoplasts to cytochalasin B during high speed centrifugation. These miniprotoplasts have been used successfully in fusion experiments to produce hybrid plants (Wallin *et al.*, 1979). Miniprotoplasts have also been obtained from fruit tissue and used in fusion experiments (Binding and Kollmann, 1976). Since miniprotoplasts contain most if not all types of organelles in their cytoplasm (Binding and Kollman, 1976; Wallin *et al.*, 1979), they cannot be considered vehicles for transplanting only nuclei. Further technical improvements are necessary to substantially reduce the cytoplasmic component of miniprotoplasts.

2. *Microorganisms*

Attempts to introduce microorganisms into higher plant protoplasts have been motivated by the perhaps overly optimistic desire to transfer functional nitrogenase enzyme to nonlegume plants. Both nitrogen-fixing bacteria [*Rhizobium* (Davey and Cocking, 1972; Davey 1977)] and blue-green algae [*Glaeocapsa* (Burgoon and Bottino, 1977) and *Anabaena* (Meeks *et al.*, 1978)] have been introduced into protoplasts of higher plants. The protoplasts failed to survive and endosymbiosis was not established.

Uptake of bacteria seems to involve a type of endocytosis. During isolation of either pea or cowpea protoplasts with enzymes, *Rhizobium* bacteria appear to be taken up by enclosure within vesicles derived from the plasma membrane. Uptake of blue-green algae by higher plant protoplasts also seems to involve endocytosis (Davey and Power, 1975).

Information regarding uptake of other microorganisms is very limited. Whereas both yeast cells and protoplasts appear to be taken up into *Parthenocissus* protoplasts by endocytosis (Davey and Power, 1975), protoplasts of the single cell green alga *Chlamydomonas* actually fuse with carrot protoplasts and endocytosis has not been observed (Figs. 14 and 15 and Fowke *et al.*, 1979).

3. *Uptake via Liposomes*

It is evident from the foregoing that serious problems exist in attempting to transfer intact organelles or microorganisms into plant protoplasts. Uptake may occur by endocytosis, thus isolating these structures in plasma membrane-derived vesicles, or they may be totally disrupted if uptake results from membrane fusion (Fig. 16A and B). An alternative approach that should avoid these problems is to use liposomes (lipid vesicles) for uptake studies. Organelles and microorganisms might be enclosed within liposomes, which in turn could be fused with the protoplast to release them intact into the protoplast cytoplasm (Fig. 16C).

Considerable success has been achieved in the uptake of materials into animal cells by liposomes (Poste *et al.*, 1976; Tyrrell *et al.*, 1976) and recently the technology has been applied to studies with plant protoplasts. Some success has been achieved with the uptake of fluorescent dyes (Cassells, 1978), RNA (Matthews *et al.*, 1979), and chloroplasts (Giles, 1978). Liposomes may prove to be of general applicability to a broad range of uptake studies.

VIII. Organelle Isolation

The isolation of intact cell organelles permits detailed studies of both their structure and physiology. Conventional methods for organelle isolation involve the rupture of plant cells by mechanical means (grinding, cutting, etc.). Such techniques tend to damage a large proportion of the cell organelles. Plant protoplasts have recently been used for organelle isolation with very encouraging results. Protoplasts can be isolated from many plant tissues and lysed gently by various methods (Fig. 17) to release the cell contents. Organelles isolated in this manner appear to be less damaged than those prepared by conventional techniques. The following discussion is not intended to be a rigorous coverage of the field of organelle isolation (see review by Quail, 1979) but is designed to illustrate the potential of plant protoplasts for isolating cell organelles. The isolation of chloroplasts, nuclei, and vacuoles will be briefly examined.

A. Chloroplasts

Chloroplasts have been isolated from leaves in order to study their structure and to gain a better understanding of a number of basic physiological processes

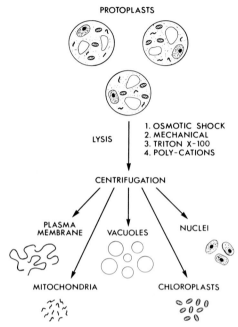

FIG. 17.   Diagram illustrating isolation of cell organelles from plant protoplasts.

(e.g., photosynthesis, photorespiration, nucleic acid, and protein synthesis). Protoplasts are proving very useful for these studies. Chloroplasts have been isolated from leaf protoplasts of both monocots and dicots [e.g., corn (Horváth *et al.,* 1978), wheat (Rathnam and Edwards, 1976; Edwards *et al.,* 1978; Leegood and Walker, 1979), spinach (Nishimura *et al.,* 1976), and tulip (Wagner and Siegelman, 1975)]. The isolation process usually involves gentle rupture of protoplasts by passing them through a hypodermic syringe. Purification is either by sucrose density gradient methods, which yield very clean preparations with rather low CO_2 fixing ability (Nishimura *et al.,* 1976), or by differential centrifugation to produce preparations capable of good CO_2 fixation rates (Rathnam and Edwards, 1976; Horváth *et al.,* 1978). The latter preparations likely are contaminated with other organelles (Quail, 1979).

Most studies involving chloroplast isolation are strictly biochemical, and they lack ultrastructural data to verify the structural integrity of the isolated plastids. Horváth *et al.* (1978) included electron micrographs of isolated corn chloroplasts, but the magnifications were not high enough to assess the quality of their preparations. There is certainly a need for careful ultrastructural analyses to complement the biochemical results.

B. Nuclei

Protoplasts also seem to be useful for rapid isolation of large quantities of plant nuclei (Ohyama *et al.*, 1977; Hughes *et al.*, 1977; Lörz and Potrykus, 1978; Zuily-Fodil *et al.*, 1978; Blaschek, 1979). The technique for isolating nuclei usually involves disruption of protoplasts with Triton X-100 followed by separation of nuclei by differential centrifugation. The resulting preparations are contaminated with cellular debris to varying degrees. The functional integrity of the isolated nuclei has been determined by examining their ability to synthesize RNA and protein (Ohyama *et al.*, 1977; Zuily-Fodil *et al.*, 1978). The structural integrity of nuclei has been assessed by TEM in only one study. Hughes *et al.* (1977) examined the ultrastructure of nuclei from barley and tobacco protoplasts and observed the loss of at least the outer nuclear membrane. Membrane loss likely is due to the use of detergents in the isolation procedure.

C. Vacuoles

Vacuoles of higher plants are rather poorly characterized due to a general lack of methods for large scale isolation and purification of intact and physiologically active vacuoles. Since 1975, techniques have been developed for large scale isolation of vacuoles from mature plant cells. Leigh and Branton (1976) devised a slicing machine to isolate vacuoles from beet root. This technique should be useful with other firm plant tissues. The most generally applicable method, however, involves the use of protoplasts as a source of vacuoles. Protoplasts can be burst gently by various methods (Fig. 17) to release intact vacuoles. This technique has proved very successful for isolating vacuoles from different plant tissues [e.g., endosperm (Nishimura and Beevers, 1978, 1979), leaves (Wagner and Siegelman, 1975; Buser and Matile 1977), petals (Wagner and Siegelman, 1975; Boller and Kende, 1979), and stems, pedicels, filaments, and styles (Wagner and Siegelman, 1975)] as well as from cultured cells (Boller and Kende, 1979). Vacuole purity has in some cases been assessed by the absence of markers for other organelles (Nishimura and Beevers, 1978; Boller and Kende, 1979), but ultrastructural examination of vacuole preparations is almost nonexistent (Leigh and Branton, 1976).

The large scale isolation of vacuoles should be very important in furthering our rather meager knowledge of the function of vacuoles in plant cells. The vacuole is a major component of mature plant cells often occupying up to 90% of the cell volume. Three major functions have been ascribed to the vacuole, (1) osmoregulation, (2) storage, and (3) digestion (Gunning and Steer, 1975; Matile, 1978). Our basic understanding of these functions should be greatly facilitated by detailed studies of the content of vacuoles, as well as studies of the structure and transport characteristics of the tonoplast. Recent research with isolated vacuoles has already contributed to our knowledge of the storage (Buser and Matile, 1977; Wagner, 1979) and digestive functions (Butcher *et al.,* 1977; Nishimura and Beevers, 1978, 1979; Boller and Kende, 1979) of vacuoles in higher plants. Evidence is accumulating to support the idea that the vacuole has a lysosomal function in plants (Matile, 1978). Hydrolytic enzymes have been localized in plant vacuoles (Nishimura and Beevers, 1978; Boller and Kende, 1979), and recently Nishimura and Beevers (1979) reported the hydrolysis of protein in vacuoles isolated from castor bean endosperm tissue. There is some disagreement about the distribution of specific enzymes in plant cells and thus the lysosomal concept of the vacuole remains somewhat controversial (see review by Quail, 1979).

Very little is known about the structure or physiology of the tonoplast of higher plants. Our understanding of this membrane could be enhanced by examining tonoplast preparations derived from isolated, purified vacuoles. Studies of vacuoles isolated from protoplasts of *Tulipa* and *Hippeastrum* indicate the presence of a Mg-dependent ATPase associated with the tonoplast (Lin *et al.,* 1977). Specific permeases for membrane transport seem to be a feature of tonoplasts (Matile, 1978), and vacuole isolation techniques should assist in their detection and characterization.

IX. Studies of Plant Cell Physiology

Plant protoplasts may be useful tools for studying the physiology of plant cells. A detailed discussion of this topic is beyond the scope of this chapter. However, we would like to present very briefly a few examples of the type of research in progress in this area.

Protoplasts can be used to study such basic cellular processes as photosynthesis and photorespiration. Nishimura and Akazawa (1975) have demonstrated that spinach leaf protoplasts are just as active photosynthetically as intact leaf tissue. They used these protoplasts to examine the biosynthesis of ribulose-1,5-biphosphate carboxylase (Nishimura and Akazawa, 1978). The process of photorespiration has also been examined in isolated spinach protoplasts (Nishimura *et al.,* 1975; Inoué *et al.,* 1978). Inoué *et al.* (1978) demonstrated a functional glycolate pathway in protoplasts, and their results suggest that the photorespiration occurring in isolated protoplasts and leaf tissue is similar. Protoplasts are particularly useful for this type of study because they avoid the problems associated with the isolation and purification of active chloroplasts from plant cells. Furthermore, the use of protoplasts rather than isolated chloroplasts maintains organelle interrelationships that are typical of intact cells. Peroxisomes, for example, are important in photorespiration (Inoué *et al.,* 1978) and their association with chloroplasts should be retained in order to study the process of photo-respiration.

Protoplasts have also been used to examine membrane transport properties of plant cells without the complicating factors due to the presence of a cell wall (e.g., adsorption to cell wall components). The results to date have been variable. Some studies suggest a marked alteration in membrane transport ability as a result of protoplast isolation. These include membrane damage with resulting leakiness during ion uptake (Taylor and Hall, 1976) and reduced levels of amino acid and sugar uptake (Guy et al., 1978; Ruesink, 1978). Ruesink has attributed the reduction in leucine uptake by *Convolvulus* protoplasts to water stress. Recent studies indicate that transport of ions across the protoplast plasma membrane may not be affected by protoplast isolation procedures. Protoplasts derived from suspension cultures of tobacco did not show any significant differences in ion transport properties when compared to whole cells (Mettler and Leonard, 1979a,b; Briskin and Leonard, 1979).

The results of Zeiger and Hepler (1976, 1977) suggest that protoplasts may also be used to study the physiology of differentiated, highly specialized plant cells. They prepared protoplasts from mature guard cells of both onion and tobacco epidermal peels. Guard cell protoplasts isolated from onion swell when illuminated by blue light. Zeiger and Hepler (1977) argue that the swelling, which is K^+-dependent, is a manifestation of normal guard cell behavior. Further studies of guard cell protoplasts may be helpful in clarifying the mechanism of guard cell action.

X. Summary

Research with plant protoplasts has contributed significantly to our understanding of the structure and function of plant cells. Perhaps the most important contribution has been to our knowledge of the plant plasma membrane. Studies of the plasma membrane have until recently been rather limited due to its inaccessibility in plant cells. The development of techniques to remove plant cell walls has made it much easier to study the plasma membrane *in situ* and has also facilitated attempts to isolate the plasma membrane. As a result, a broad spectrum of information is accumulating regarding the nature of the plant plasma membrane. The use of specific stains, lectins, fluorescent antibodies, and various TEM techniques is providing information regarding the chemistry and organization of the plasma membrane and associated cell organelles. Preliminary information is also available concerning the interaction of plasma membranes during protoplast fusion, uptake mechanisms at the cell surface, and the relationship of the plasma membrane to the process of cell wall formation by regenerating protoplasts.

In addition to providing information regarding the plasma membrane, plant protoplasts have various other important applications to the study of plant cells. Investigations of cultured protoplasts are providing fundamental information about different aspects of cell wall formation in plants (e.g., microfibril chemistry, structure, and deposition). Fusion studies are contributing to our understanding of organelle interactions in mixed cytoplasms and of organelle growth and development, as well as the production of hybrid plants by somatic hybridization. Research with protoplasts may also facilitate genetic modification of plant cells via uptake of nucleic acids or cell organelles although progress in this field has been slow. The uptake of some viruses into plant protoplasts has been achieved and has contributed significantly to the understanding of virus replication in plant cells. Finally protoplasts and organelles isolated from protoplasts have been utilized with some success to investigate physiological processes in plant cells.

During the past few years, scientists from many disciplines (e.g., cell biology, biochemistry, genetics, physiology) have used protoplasts to study plant cells. Most of these studies are in the initial stages and we look forward to new and exciting discoveries in the near future.

ACKNOWLEDGMENTS

We wish to express our appreciation to Fred Constabel, Pieter Van der Valk, Gerd Weber, and Francis Williamson for their contribution to the research from our laboratories and Pat Rennie, Jim Kirkpatrick, and Jerry Shyluk for excellent technical assistance. Thanks also to those colleagues who kindly supplied preprints of unpublished data and to Joan Ryan and Dennis Dyck for their help in preparing the manuscript. Financial support from the National Research Council of Canada is gratefully acknowledged.

REFERENCES

Abermann, R., Salpeter, M. M., and Bachmann, L. (1972). *In* "Principles and Techniques of Electron Microscopy" (M. A. Hayat, ed.), Vol. 2, pp. 197–217. Van Nostrand-Reinhold, New York.

Asamizu, T., Tanaka, K., Takebe, I., and Nishi, A. (1977). *Physiol. Plant.* **40,** 215–218.

Bajaj, Y. P. S. (1977). *In* "Applied and Fundamental Aspects of Plant Cell, Tissue, and Organ Culture" (J. Reinert and Y. P. S. Bajaj, eds.), pp. 467–496. Springer-Verlag, Berlin and New York.

Bajer, A. S., and Molè-Bajer, J. (1972). *Int. Rev. Cytol.* Suppl. No. 3.

Binding, H. (1974). *Z. Pflanzenphysiol.* **74,** 327–356.

Binding, H. (1976). *Molec. Gen. Genet.* **144,** 171–175.

Binding, H., and Kollmann, R. (1976). *In* "Cell Genetics in Higher Plants" (D. Dudits, G. L. Farkas and P. Maliga, eds.), pp. 191–205. Hungarian Academy of Science, Budapest.

Binding, H., and Nehls, R. (1977). *Z. Pflanzenphysiol.* **85,** 279–280.
Binding, H., and Nehls, R. (1978). *Molec. Gen. Genet.* **164,** 137–143.
Binding, H., Nehls, R., Schieder, O., Sopory, S. K., and Wenzel, G. (1978). *Physiol. Plant.* **43,** 52–54.
Blaschek, W. (1979). *Plant Sci. Lett.* **15,** 139–149.
Boller, T., and Kende, H. (1979). *Plant Physiol.* **63,** 1123–1132.
Bonnett, H. T. (1976). *Planta (Berlin)* **131,** 229–233.
Bonnett, H. T., and Eriksson, T. (1974). *Planta (Berlin)* **120,** 71–79.
Boss, W. F., and Ruesink, A. W. (1979). *Plant Physiol.* **64,** 1005–1011.
Briskin, D. P., and Leonard, R. T. (1979). *Plant Physiol.* **64,** 959–962.
Bui-dang-ha, D., and MacKenzie, I. A. (1973). *Protoplasma* **78,** 215–221.
Burgess, J., and Fleming, E. N. (1974a). *J. Cell Sci.* **14,** 439–449.
Burgess, J., and Fleming, E. N. (1974b). *Planta (Berlin)* **118,** 183–193.
Burgess, J., and Linstead, P. J. (1976a). *Planta (Berlin)* **130,** 73–79.
Burgess, J., and Linstead, P. J. (1976b). *Planta (Berlin)* **131,** 173–178.
Burgess, J., and Linstead, P. J. (1977a). *Planta (Berlin)* **133,** 267–273.
Burgess, J., and Linstead, P. J. (1977b). *Planta (Berlin)* **136,** 253–259.
Burgess, J., and Linstead, P. J. (1979). *Planta (Berlin)* **146,** 203–210.
Burgess, J., Motoyoshi, F., and Fleming, E. N. (1973). *Planta (Berlin)* **112,** 323–332.
Burgess, J., Linstead, P. J., and Fisher, V. E. L. (1977a). *Micron* **8,** 113–122.
Burgess, J., Linstead, P. J., and Harnden, J. M. (1977b). *Micron* **8,** 181–191.
Burgess, J., Linstead, P. J., and Bonsall, V. E. (1978). *Planta (Berlin)* **139,** 85–91.
Burgoon, A. C., and Bottino, P. J. (1977). *In* "Genetic Engineering for Nitrogen Fixation" (A. Hollaender, R. H. Burris, P. R. Day, R. W. F. Hardy, D. R. Helsinki, M. R. Lamborg, L. Owens, and R. C. Valentine, eds.), pp. 213–229. Plenum, New York.
Buser, C., and Matile, P. (1977). *Z. Pflanzenphysiol.* **82,** 462–466.
Butcher, H. C., Wagner, G. J., and Siegelman, H. W. (1977). *Plant Physiol.* **59,** 1098–1103.
Butenko, R. G., Kuchko, A. A., Vitenko, V. A., and Anetisov, V. A. (1977). *Sov. Plant Physiol. (Moscow)* **24,** 541–546 (Transl.).
Carlson, P. S., Smith, H. H., and Dearing, R. D. (1972). *Proc. Natl. Acad. Sci. U.S.A.* **69,** 2292–2294.
Cassells, A. C. (1978). *Nature (London)* **275,** 760.
Chen, K., Wildman, S. G., and Smith, H. H. (1977). *Proc. Natl. Acad. Sci. U.S.A.* **74,** 5109–5112.
Chin, J. C., and Scott, K. J. (1979). *Ann. Bot.* **43,** 33–44.
Chrispeels, M. J. (1976). *Annu. Rev. Plant Physiol.* **27,** 19–38.
Cocking, E. C. (1970). *Int. Rev. Cytol.* **28,** 89–124.
Cocking, E. C. (1972). *Annu. Rev. Plant Physiol.* **23,** 29–50.
Cocking, E. C. (1974). *In* "Dynamic Aspects of Plant Ultrastructure" (A. W. Robards, ed.), pp. 310–330. McGraw-Hill, New York.
Cocking, E. C. (1977). *Int. Rev. Cytol.* **48,** 323–343.
Cocking, E. C. (1978a). *Proc. Symp. Plant Tissue Cult., Peking, 1978,* pp. 255–263. Science Press, Peking.
Cocking, E. C. (1978b). *Proc. Int. Congr. Plant Tissue Cell Cult., 4th, Calgary, Alberta* pp. 151–158.
Cocking, E. C., George, D., Price-Jones, M. J., and Power, J. B. (1977). *Plant Sci. Lett.* **10,** 7–12.
Connolly, J. A., and Kalnins, V. I. (1978). *J. Cell Biol.* **79,** 526–532.
Constabel, F. (1975). *In* "Plant Tissue Culture Methods" (O. L. Gamborg and L. R. Wetter, eds.), pp. 11–21. National Research Council of Canada, Saskatoon.
Constabel, F. (1976). *In Vitro* **12,** 743–748.
Constabel, F. (1978). *Proc. Int. Congr. Plant Tissue Cell Cult., 4th, Calgary, Alberta* pp. 141–149.

Constabel, F., Dudits, D., Gamborg, O. L., and Kao, K. N. (1975). *Can. J. Bot.* **53**, 2092–2095.

Constabel, F., Weber, G., and Kirkpatrick, J. W. (1977). *C. R. Acad. Sci. Paris D* **285**, 319–322.

Constabel, F., Koblitz, H., Kirkpatrick, J. W., and Rambold, S. (1980). *Can. J. Bot.* **58**, 1032–1034.

Davey, M. R. (1977). *In* "Applied and Fundamental Aspects of Plant Cell, Tissue, and Organ Culture" (J. Reinert and Y. P. S. Bajaj, eds.), pp. 551–562. Springer-Verlag, Berlin and New York.

Davey, M. R., and Cocking, E. C. (1972). *Nature (London)* **239**, 455–456.

Davey, M. R., and Cocking, E. C. (1980). *In* "Recent Advances in Biological Nitrogen Fixation" (N. S. S. Rao, ed.). (in press).

Davey, M. R., and Mathias, R. J. (1979). *Protoplasma* **100**, 85–99.

Davey, M. R., and Power, J. B. (1975). *Plant Sci. Lett.* **5**, 269–274.

Davey, M. R., Frearson, E. M., Withers, L. A., and Power, J. B. (1974). *Plant Sci. Lett.* **2**, 23–27.

Davey, M. R., Frearson, E. M., and Power, J. B. (1976). *Plant Sci. Lett.* **7**, 7–16.

Davey, M. R., Clothier, R. H., Balls, M., and Cocking, E. C. (1978). *Protoplasma* **96**, 157–172.

Dorion, N., Chupeau, Y., and Bourgin, J. P. (1975). *Plant Sci. Lett.* **5**, 325–331.

Dudits, D., Kao, K. N., Constabel, F., and Gamborg, O. L. (1976a). *Can. J. Bot.* **54**, 1063–1067.

Dudits, D., Rasko, I., Hadlaczky, G., and Lima-de-faria, A. (1976b). *Hereditas* **82**, 121–124.

Dudits, D., Kao, K. N., Constabel, F., and Gamborg, O. L. (1976c). *Can. J. Genet. Cytol.* **18**, 263–269.

Dudits, D., Hadlaczky, G., Levi, E., Fejer, O., Haydu, Z. S., and Lázár, G. (1977). *Theoret. Appl. Genet.* **51**, 127–132.

Dudits, D., Hadlaczky, G., Bajszár, G. Y., Koncz, C., Lázár, G., and Horváth, G. (1979). *Plant Sci. Lett.* **15**, 101–112.

Durand, J., Potrykus, I., and Donn, G. (1973). *Z. Pflanzenphysiol.* **69**, 26–34.

Edwards, G. E., Robinson, S. P., Tyler, N. J. C., and Walker, D. A. (1978). *Plant Physiol.* **62**, 313–319.

Eriksson, T., and Jonasson, K. (1969). *Planta (Berlin)* **89**, 85–89.

Eriksson, T., Bonnett, H., Glimelius, K., and Wallin, A. (1974). *Proc. Int. Congr. Plant Tissue Cell Cult., 3rd, Leicester, 1974* pp. 213–231.

Eriksson, T., Glimelius, K., and Wallin, A. (1978). *Proc. Int. Congr. Plant Tissue Cell Cult., 4th, Calgary, Alberta* pp. 131–139.

Evans, D. A., Wetter, L. R., and Gamborg, O. L. (1980). *Physiol. Plant.* **48**, 225–230.

Facciotti, D., and Pilet, P. (1979). *Plant Sci. Lett.* **15**, 1–6.

Fowke, L. C. (1975). *In* "Plant Tissue Culture Methods" (O. L. Gamborg and L. R. Wetter, eds.), pp. 55–59. National Research Council of Canada, Saskatoon.

Fowke, L. C. (1978). *Proc. Int. Congr. Plant Tissue Cell Cult., 4th, Calgary, Alberta* pp. 223–233.

Fowke, L. C., and Setterfield, G. (1969). *Can. J. Bot.* **47**, 1873–1877.

Fowke, L. C., Bech-Hansen, C. W., Gamborg, O. L., and Shyluk, J. P. (1973). *Am. J. Bot.* **60**, 304–312.

Fowke, L. C., Bech-Hansen, C. W., and Gamborg, O. L. (1974a). *Protoplasma* **79**, 235–248.

Fowke, L. C., Bech-Hansen, C. W., Constabel, F., and Gamborg, O. L. (1974b). *Protoplasma* **81**, 189–203.

Fowke, L. C., Bech-Hansen, C. W., Gamborg, O. L., and Constabel, F. (1975a). *J. Cell Sci.* **18**, 491–507.

Fowke, L. C., Rennie, P. J., Kirkpatrick, J. W., and Constabel, F. (1975b). *Can. J. Bot.* **53**, 272–278.

Fowke, L. C., Rennie, P. J., Kirkpatrick, J. W., and Constabel, F. (1976). *Planta (Berlin)* **130**, 39–45.

Fowke, L. C., Constabel, F., and Gamborg, O. L. (1977). *Planta (Berlin)* **135**, 257–266.

Fowke, L. C., Gresshoff, P. M., and Marchant, H. J. (1979). *Planta (Berlin)* **144**, 341-347.

Frearson, E. M., Power, J. B., and Cocking, E. C. (1973). *Dev. Biol.* **35**, 130-137.

Frye, L. D., and Ediden, M. (1970). *J. Cell Sci.* **7**, 319-335.

Furner, I. J., King, J., and Gamborg, O. L. (1978). *Plant Sci. Lett.* **11**, 169-176.

Galbraith, D. W., and Galbraith, J. E. C. (1979). *Z. Pflanzenphysiol.* **93**, 149-158.

Gamborg, O. L. (1975). *In* "Plant Tissue Culture Methods" (O. L. Gamborg and L. R. Wetter, eds.), pp. 1-10. National Research Council of Canada, Saskatoon.

Gamborg, O. L. (1976). *In* "Cell Genetics in Higher Plants" (D. Dudits, G. L. Farkas and P. Maliga, eds.), pp. 107-127. Hungarian Academy of Science, Budapest.

Gamborg, O. L., and Miller, R. A. (1973). *Can. J. Bot.* **51**, 1795-1799.

Gamborg, O. L., Constabel, F., Fowke, L., Kao, K. N., Ohyama, K., Kartha, K., and Pelcher, L. (1974). *Can. J. Genet. Cytol.* **16**, 737-750.

Gamborg, O. L., Kartha, K. K., Ohyama, K., and Fowke, L. (1978). *Proc. Symp. Plant Tissue Cult., Peking, 1978,* pp. 265-278. Science Press, Peking.

Gamborg, O. L., Shyluk, J. P., Fowke, L. C., Wetter, L. R., and Evans, D. (1980). *Z. Pflanzenphysiol.* **95**, 255-264.

Giddings, T. H., Brower, D. L., and Staehelin, L. A. (1980). *J. Cell Biol.* **84**, 327-339.

Gigot, C., Kopp, M., Schmitt, C., and Milne, R. G. (1975). *Protoplasma* **84**, 31-41.

Giles, K. L. (1977). *In* "Plant Cell, Tissue, and Organ Culture" (J. Reinert and Y. P. S. Bajaj, eds.), pp. 536-550. Springer-Verlag, Berlin and New York.

Giles, K. L. (1978). *Proc. Int. Congr. Plant Tissue Cell Cult., 4th, Calgary, Alberta* pp. 67-74.

Gleba, Y. Y. (1979). *In* "Plant Cell and Tissue Culture" (W. R. Sharp, P. O. Larsen, E. F. Paddock and V. Raghavan, eds.), pp. 371-398. Ohio State Univ. Press, Columbus.

Gleba, Y. Y., and Hoffmann, F. (1978). *Molec. Gen. Genet.* **165**, 257-264.

Gleba, Y. Y., and Hoffmann, F. (1979). *Naturwissenschaften* **11**, 547-554.

Gleba, Y. Y., Shvydkaya, L. G., Butenko, R. G., and Sytnik, K. M. (1974). *Sov. Plant Physiol.* **21**, 486-492.

Glimelius, K., Wallin, A., and Eriksson, T. (1974). *Physiol. Plant.* **31**, 225-230.

Glimelius, K., Wallin, A., and Eriksson, T. (1978a). *Protoplasma* **97**, 291-300.

Glimelius, K., Eriksson, T., Grafe, R., and Müller, A. J. (1978b). *Physiol. Plant.* **44**, 273-277.

Gosch, G., and Reinert, J. (1978). *Protoplasma* **96**, 23-38.

Gosch, G., Bajaj, Y. P. S., and Reinert, J. (1975). *Protoplasma* **86**, 405-410.

Grambow, H. J., Kao, K. N., Miller, R. A., and Gamborg, O. L. (1972). *Planta (Berlin)* **103**, 348-355.

Grout, B. W. W. (1975). *Planta (Berlin)* **123**, 275-282.

Grout, B. W. W., Willison, J. H. M., and Cocking, E. C. (1972). *Bioenergetics* **4**, 585-602.

Grun, P., and Chu, L.J. (1978). *Am. J. Bot.* **65**, 538-543.

Gunning, B. E. S., and Steer, M. W. (1975). "Ultrastructure and the Biology of Plant Cells." Arnold, London.

Gunning, B. E. S., Steer, M. W., and Cochrane, M. P. (1968). *J. Cell Sci.* **3**, 445-456.

Gunning, B. E. S., Hardham, A. R., and Hughes, J. E. (1978). *Planta (Berlin)* **143**, 145-160.

Guy, M., Reinhold, L., and Laties, G. G. (1978). *Plant Physiol.* **61**, 593-596.

Hanke, D. E., and Northcote, D. H. (1974). *J. Cell Sci.* **14**, 29-50.

Hartmann, J. X., Kao, K. N., Gamborg, O. L., and Miller, R. A. (1973). *Planta (Berlin)* **112**, 45-56.

Hayward, C., and Power, J. B. (1975). *Plant Sci. Lett.* **4**, 407-410.

Heath, I. B. (1974). *J. Theor. Biol.* **48**, 445-449.

Hellmann, S., and Reinert, J. (1971). *Protoplasma* **72**, 479-484.

Henderson, W. J., and Griffiths, K. (1972). *In* "Principles and Techniques of Electron Microscopy" (M. A. Hayat, ed.), Vol. 2, pp. 151-193. Van Nostrand-Reinhold, New York.

Hepler, P. K., and Jackson, W. T. (1968). *J. Cell Biol.* **38**, 437–446.

Hepler, P. K., and Palevitz, B. A. (1974). *Annu. Rev. Plant Physiol.* **25**, 309–362.

Herth, W., and Meyer, Y. (1977). *Rev. Biol. Cell.* **30**, 33–40.

Herth, W., and Meyer, Y. (1978). *Planta (Berlin)* **142**, 11–21.

Honda, Y., Kajita, S., Matsui, C., Otsuki, Y. and Takebe, I. (1975). *Phytopathol. Z.* **84**, 66–74.

Horváth, G., Droppa, M., Mustárdy, L. A., and Faludi-Dániel, Á. (1978). *Planta (Berlin)* **141**, 239–244.

Hughes, B. G., White, F. G., and Smith, M. A. (1976). *Protoplasma* **90**, 399–405.

Hughes, B. G., Hess, W. M., and Smith, M. A. (1977). *Protoplasma* **93**, 267–274.

Hughes, J., and McCully, M. E. (1975). *Stain Technol.* **50**, 319–329.

Inoue, K., Nishimura, M., and Akazawa, T. (1978). *Plant Cell Physiol.* **19**, 317–325.

Izhar, S., and Power, J. B. (1979). *Plant Sci. Lett.* **14**, 49–55.

Johnson, R. T., and Rao, P. N. (1971). *Biol. Rev.* **46**, 97–155.

Jones, C. W., Mastrangelo, I. A., Smith, H. H., and Liu, H. Z. (1976). *Science* **193**, 401–403.

Kao, K. N. (1975). *In* "Plant Tissue Culture Methods" (O. L. Gamborg and L. R. Wetter, eds.), pp. 22–27. National Research Council of Canada, Saskatoon.

Kao, K. N. (1977). *Molec. Gen. Genet.* **150**, 225–230.

Kao, K. N. (1978). *Proc. Symp. Plant Tissue Cult., Peking, 1978*, pp. 331–339. Science Press, Peking.

Kao, K. N., and Michayluk, M. R. (1974). *Planta (Berlin)* **115**, 355–367.

Kao, K. N., and Michayluk, M. R. (1980). *Z. Pflanzenphysiol.* **96**, 135–141.

Kao, K. N., Gamborg, O. L., Miller, R. A., and Keller, W. A. (1971). *Nature (London) New Biol.* **232**, 124.

Kao, K. N., Gamborg, O. L., Michayluk, M. R., and Keller, W. A. (1973). *Colloq. Int. C.N.R.S.* **212**, 207–213.

Kao, K. N., Constabel, F., Michayluk, M. R., and Gamborg, O. L. (1974). *Planta (Berlin)* **120**, 215–227.

Kartha, K. K., Michayluk, M. R., Kao, K. N., Gamborg, O. L., and Constabel, F. (1974). *Plant Sci. Lett.* **3**, 265–271.

Keller, P. M., Person, S., and Snipes, W. (1977). *J. Cell Sci.* **28**, 167–177.

Keller, W. A., and Melchers, G. (1973). *Z. Naturforsch.* **28c**, 737–741.

Klein, A. S., and Delmer, D. P. (1979). *Plant Physiol.* **63**, 51 (Suppl.).

Knox, R. B., and Clarke, A. E. (1978). *In* "Electron Microscopy and Cytochemistry of Plant Cells" (J. L. Hall, ed.), pp. 149–185. Elsevier, Amsterdam.

Koblitz, H. (1978). *Biochem. Physiol. Pflanzen* **172**, 213–222.

Krumbiegel, G., and Schieder, D. (1979). *Planta (Berlin)* **145**, 371–375.

Landgren, C. R. (1976). *Am. J. Bot.* **63**, 473–480.

Larkin, P. J. (1978). *Plant Physiol.* **61**, 626–629.

Leegood, R. C., and Walker, D. A. (1979). *Plant Physiol.* **63**, 1212–1214.

Leigh, R. A., and Branton, D. (1976). *Plant Physiol.* **58**, 656–662.

Lin, W., Wagner, G. J., Siegelman, H. W., and Hind, G. (1977). *Biochim. Biophys. Acta* **465**, 110–117.

Lloyd, C. W., Slabas, A. R., Powell, A. J., MacDonald, G., and Badley, R. A. (1979). *Nature (London)* **279**, 239–241.

Lörz, H., and Potrykus, I. (1978). *Theor. Appl. Genet.* **53**, 251–256.

Lörz, H., and Potrykus, I. (1979). *Experientia* **35**, 313–314.

Lörz, H., Wernicke, W., and Potrykus, I. (1979). *Planta Med.* **36**, 21–29.

Maeda, E., and Hagiwara, T. (1974). *Proc. Crop Sci. Soc. Jpn* **43**, 67–76.

Maliga, P., Lázár, G., Joó, F., H.-Nagy, A., and Menczel, L. (1977). *Molec. Gen. Genet.* **157**, 291–296.

Maliga, P., Kiss, Z. R., Nagy, A. H., and Lázár, G. (1978). *Molec. Gen. Genet.* **163,** 145-151.

Marchant, H. J. (1978). *Exp. Cell Res.* **115,** 25-30.

Marchant, H. J. (1979). *Nature (London)* **278,** 167-168.

Marchant, H. J., and Hines, E. R. (1979). *Planta (Berlin)* **146,** 41-48.

Matile, P. (1978). *Annu. Rev. Plant Physiol.* **29,** 193-213.

Matthews, B., Dray, S., Widholm, J., and Ostro, M. (1979). *Planta (Berlin)* **145,** 37-44.

Mayo, M. A., and Cocking, E. C. (1969). *Protoplasma* **68,** 223-230.

Meeks, J. C., Malmberg, R. L., and Wolk, C. P. (1978). *Planta (Berlin)* **139,** 55-60.

Melchers, G. (1977). *In* "International Cell Biology" (B. R. Brinkley and K. R. Porter, eds.), pp. 207-215. Rockefeller Univ. Press, New York.

Melchers, G., and Labib, G. (1974). *Molec. Gen. Genet.* **135,** 277-294.

Melchers, G., Sacristán, M. D., and Holder, A. A. (1978). *Carlsberg Res. Commun.* **43,** 203-218.

Mettler, I. J., and Leonard, R. T. (1979a). *Plant Physiol.* **63,** 183-190.

Mettler, I. J., and Leonard, R. T. (1979b). *Plant Physiol.* **63,** 191-194.

Meyer, Y., and Abel, W. (1975). *Planta (Berlin)* **123,** 33-40.

Meyer, Y., and Herth, W. (1978). *Planta (Berlin)* **142,** 253-262.

Miller, R. A., Gamborg, O. L., Keller, W. A., and Kao, K. N. (1971). *Can. J. Genet. Cytol.* **13,** 347-353.

Milne, R. G. (1972). *Bot. Gaz.* **133,** 401-404.

Montezinos, D., and Brown, R. M. (1976). *J. Supramolec. Struct.* **5,** 277-290.

Morel, G., Bourgin, J. P., and Chupeau, Y. (1973). *In* "Yeast, Mould and Plant Protoplasts" (J. R. Villanueva *et al.,* eds.), pp. 333-344. Academic Press, New York.

Motoyoshi, F. (1971). *Exp. Cell Res.* **68,** 452-456.

Mueller, S. C., and Brown, R. M. (1980). *J. Cell Biol.* **84,** 315-326.

Mueller, S. C., Brown, R. M., Jr., and Scott, T. K. (1976). *Science* **194,** 949-951.

Nagao, T. (1978). *Jpn. Crop Sci.* **47,** 491-498.

Nagata, T. (1978). *Naturwissenschaften* **65,** 263-264.

Nagata, T., and Takebe, I. (1970). *Planta (Berlin)* **92,** 301-308.

Nagata, T., Eibl, H., and Melchers, G. (1979). *Z. Naturforsch.* **34,** 460-462.

Nehls, R. (1978). *Plant Sci. Lett.* **12,** 183-187.

Newcomb, W. (1978). *Can. J. Bot.* **56,** 483-501.

Nicolson, G. L. (1974). *Int. Rev. Cytol.* **39,** 89-190.

Nishimura, M., and Akazawa, T. (1975). *Plant Physiol.* **55,** 712-716.

Nishimura, M., and Akazawa, T. (1978). *Plant Physiol.* **62,** 97-100.

Nishimura, M., and Beevers, H. (1978). *Plant Physiol.* **62,** 44-48.

Nishimura, M., and Beevers, H. (1979). *Nature (London)* **277,** 412-413.

Nishimura, M., Graham, D., and Akazawa, T. (1975). *Plant Physiol.* **56,** 718-722.

Nishimura, M., Graham, D., and Akazawa, T. (1976). *Plant Physiol.* **58,** 309-314.

Nitsch, J. P., and Ohyama, K. (1971). *C.R. Acad. Sci. Paris D* **273,** 801-804.

Ohiwa, T. (1978). *Protoplasma* **97,** 185-200.

Ohyama, K., Pelcher, L. E., and Horn, D. (1977). *Plant Physiol.* **60,** 179-181.

Ohyama, K., Pelcher, L. E., and Schaefer, A. (1978). *Proc. Int. Congr. Plant Tissue Cell Cult., 4th Calgary, Alberta* pp. 75-84.

Orci, L., Perrelet, A., and Friend, D. S. (1977). *J. Cell Biol.* **75,** 23-30.

Otsuki, Y., Takebe, I., Honda, Y., and Matsui, C. (1972). *Virology* **49,** 188-194.

Pearce, R. S., and Cocking, E. C. (1973). *Protoplasma* **77,** 165-180.

Pearce, R. S., Withers, L. A., and Willison, J. H. M. (1974). *Protoplasma* **82,** 223-236.

Pickett-Heaps, J. D. (1974). *In* "Dynamic Aspects of Plant Ultrastructure" (A. W. Robards, ed.), pp. 129-255. McGraw-Hill, New York.

Pinto da Silva, P., and Nogueira, M. L. (1977). *J. Cell Biol.* **73,** 161-181.

Poste, G., and Allison, A. C. (1973). *Biochim. Biophys. Acta* **300**, 421–465.
Poste, G., Papahadjopoulos, D., and Vail, W. J. (1976). *In* "Methods in Cell Biology" (D. M. Prescott, ed.), Vol. 14, pp. 33–71. Academic Press, New York.
Potrykus, I. (1973). *Z. Pflanzenphysiol.* **70**, 364–366.
Potrykus, I., and Hoffmann, F. (1973). *Z. Pflanzenphysiol.* **69**, 287–289.
Potrykus, I., and Lörz, H. (1976). *In* "Cell Genetics in Higher Plants" (D. Dudits, G. L. Farkas and P. Maliga, eds.), pp. 183–190. Hungarian Academy of Science, Budapest.
Power, J. B., and Cocking, E. C. (1970). *J. Exp. Bot.* **21**, 64–70.
Power, J. B., Cummins, S. E., and Cocking, E. C. (1970). *Nature (London)* **225**, 1016–1018.
Power, J. B., Frearson, E. M., George, D., Evans, P. K., Berry, S. F., Hayward, C., and Cocking, E. C. (1976a). *Plant Sci. Lett.* **7**, 51–55.
Power, J. B., Frearson, E. M., Hayward, C., George, D., Evans, P. K., Berry, S. F., and Cocking, E. C. (1976b). *Nature (London)* **263**, 500–502.
Prat, R., and Poirier-Hamon, S. (1975). *Protoplasma* **86**, 175–187.
Prat, R., and Roland, J.-C. (1970). *C.R. Acad. Sci. Paris D* **271**, 1862–1865.
Prat, R., and Roland, J.-C. (1971). *C.R. Acad. Sci. Paris D* **273**, 165–168.
Prat, R., and Williamson, F. A. (1976). *Soc. Bot. Fr. Coll. Sécrét. Veget.* **123**, 33–45.
Quail, P. H. (1979). *Annu. Rev. Plant Physiol.* **30**, 425–484.
Rathnam, C. K. M., and Edwards, G. E. (1976). *Plant Cell Physiol.* **17**, 177–186.
Raveh, D., and Galun, E. (1975). *Z. Pflanzenphysiol.* **76**, 76–79.
Rennie, P. J., Weber, G., Constabel, F. and Fowke, L. C. (1980). *Protoplasma* (in press).
Ringertz, N. R., and Savage, R. E. (1976). "Cell Hybrids." Academic Press, New York.
Robenek, V. H. (1979). *Biol. Zentralbl.* **98**, 429–447.
Robenek, H., and Peveling, E. (1977). *Planta (Berlin)* **136**, 135–145.
Robenek, H., and Peveling, E. (1978). *Ber. Dtsch. Bot. Ges.* **91**, 351–359.
Robinson, D. G. (1977). *In* "Advances in Botanical Research" (H. W. Woolhouse, ed.), Vol. 5, pp. 89–151. Academic Press, New York.
Roland, J.-C. (1978). *In* "Electron Microscopy and Cytochemistry of Plant Cells" (J. L. Hall, ed.), pp. 1–62. Elsevier, Amsterdam.
Roland, J.-C., and Prat, R. (1973). *In* "Protoplastes et Fusion de Cellules Somatiques Végétales," pp. 243–271. C.N.R.S., Paris.
Ruesink, A. W. (1978). *Physiol. Plant.* **44**, 48–56.
Schieder, O. (1975). *Z. Pflanzenphysiol.* **76**, 462–466.
Schieder, O. (1977). *Planta (Berlin)* **137**, 253–257.
Schieder, O. (1978). *Molec. Gen. Genet.* **162**, 113–119.
Schilde-Rentschler, L. (1977). *Planta (Berlin)* **135**, 177–181.
Scowcroft, W. R. (1977). *In* "Advances in Agronomy" (N. C. Brady, ed.), Vol. 29, pp. 39–81. Academic Press, New York.
Shepard, J. F., and Totten, R. E. (1977). *Plant Physiol.* **60**, 313–316.
Silverstein, S. C., Steinman, R. M., and Cohn, Z. A. (1977). *Annu. Rev. Biochem.* **46**, 669–722.
Singer, S. J., and Nicolson, G. L. (1972). *Science* **175**, 720–731.
Sink, K. C., and Power, J. B. (1977). *Plant Sci. Lett.* **10**, 335–340.
Smith, H. H. (1977). Ann. Tabac, Sect. 2. Bergerac. S.E.I.T.A., pp. 19–30.
Smith, H. H., Kao, K. N., and Combatti, N. C. (1976). *J. Hered.* **67**, 123–128.
Suzuki, M., Takebe, I., Kajita, S., Honda, Y., and Matsui, C. (1977). *Exp. Cell Res.* **105**, 127–135.
Takebe, I. (1977). *In* "Comprehensive Virology" (H. Fraenkel-Conrat and R. R. Wagner, eds.), Vol. 11, pp. 237–283. Plenum, New York.
Takebe, I., Labib, G., and Melchers, G. (1971). *Naturwissenschaften* **58**, 318–320.
Takebe, I., Otsuka, Y., Honda, Y., Nishio, T., and Matsui, C. (1973). *Planta (Berlin)* **113**, 21–27.
Takeuchi, Y., and Komamine, A. (1978). *Planta (Berlin)* **140**, 227–232.

Taylor, A. R. D., and Hall, J. L. (1976). *J. Exp. Bot.* **27,** 383-391.

Taylor, A. R. D., and Hall, J. L. (1978). *Protoplasma* **96,** 113-126.

Taylor, A. R. D., and Hall, J. L. (1979). *Plant Sci. Lett.* **14,** 139-144.

Temmink, J. H. M., Collard, J. G., Spits, H., and Roos, E. (1975). *Exp. Cell Res.* **92,** 307-322.

Thomas, E., Hoffmann, F., Potrykus, I., and Wenzel, G. (1976). *Molec. Gen. Genet.* **145,** 245-247.

Torrey, J. G., and Landgren, C. R. (1977). *In* "La Culture des Tissus et des Cellules des Végétaux" (M. R. Gautheret, ed.), pp. 148-168. Masson, Paris.

Tyrrell, D. A., Heath, T. D., Colley, C. M., and Ryman, B. E. (1976). *Biochim. Biophys. Acta* **457,** 259-302.

Ueda, K., Tan, K., Sato, F., and Yamada, Y. (1978). *Cell Struct. Function* **3,** 25-30.

Vasil, I. K. (1976). *Adv. Agron.* **28,** 119-159.

Wagner, G. J. (1979). *Plant Physiol.* **64,** 88-93.

Wagner, G. J., and Siegelman, H. W. (1975). *Science* **190,** 1298-1299.

Wallin, A., and Eriksson, T. (1973). *Physiol. Plant.* **28,** 33-39.

Wallin, A., Glimelius, K., and Eriksson, T. (1974). *Z. Pflanzenphysiol.* **74,** 64-80.

Wallin, A., Glimelius, K., and Eriksson, T. (1978). *Z. Pflanzenphysiol.* **87,** 333-340.

Wallin, A., Glimelius, K., and Eriksson, T. (1979). *Z. Pflanzenphysiol.* **91,** 89-94.

Ward, M., Davey, M. R., Mathias, R. J., Cocking, E. C., Clothier, R. H., Balls, M., and Lucy, J. A. (1980). *Somatic Cell Genet.* (in press).

Wernicke, W., Lörz, H., and Thomas, E. (1979). *Plant Sci. Lett.* **15,** 239-249.

Williamson, F. A. (1979). *Planta (Berlin)* **144,** 209-215.

Williamson, F. A., Fowke, L. C., Constabel, F. C., and Gamborg, O. L. (1976). *Protoplasma* **89,** 305-316.

Williamson, F. A., Fowke, L. C., Weber, G., Constabel, F., and Gamborg, O. (1977). *Protoplasma* **91,** 213-219.

Willis, G. E., Hartmann, J. X., and de Lamater, E. D. (1977). *Protoplasma* **91,** 1-14.

Willison, J. H. M. (1976). *In* "Microbial and Plant Protoplasts" (J. F. Peberdy, A. H. Rose, H. J. Rogers, and E. C. Cocking, eds.), pp. 283-297. Academic Press, New York.

Willison, J. H. M., and Cocking, E. C. (1972). *Protoplasma* **75,** 397-403.

Willison, J. H. M., and Cocking, E. C. (1975). *Protoplasma* **84,** 147-159.

Willison, J. H. M., and Grout, B. W. W. (1978). *Planta (Berlin)* **140,** 53-58.

Willison, J. H. M., Grout, B. W. W., and Cocking, E. C. (1971). *Bioenergetics* **2,** 371-382.

Withers, L. A., and Cocking, E. C. (1972). *J. Cell Sci.* **11,** 59-75.

Zeiger, E., and Hepler, P. K. (1976). *Plant Physiol.* **58,** 492-498.

Zeiger, E., and Hepler, P. K. (1977). *Science* **196,** 887-889.

Zelcer, A., Aviv, D., and Galun, E. (1978). *Z. Pflanzenphysiol.* **90,** 397-407.

Zuily-Fodil, Y., Passaquet, C., and Esnault, R. (1978). *Physiol. Plant.* **43,** 201-204.

Control of Membrane Morphogenesis in Bacteriophage

GREGORY J. BREWER

Department of Medical Microbiology and Immunology,
Southern Illinois University School of Medicine,
Springfield, Illinois

I. Introduction

A. MORPHOGENESIS IN GENERAL

The most evident consequence of biological development is formation of new shapes and structures for new functions. This morphogenesis is achieved through an alteration in genetic expression. At the organismic level, the approach to understanding this problem has only slowly provided insight. Due to the inherent

complexity at this level, attention has been directed to the investigation of cellular and subcellular levels. Since cellular morphogenesis often depends on reception and processing of an environmental stimulus at the cell membrane, control
of membrane morphogenesis emerges as a critical determinant in cell–cell interaction, reception of signals, and function of organs as well as subcellular
organelles.

Membrane morphogenesis will be defined as the formation of a phospholipid
bilayer whose shape and/or composition are distinct from those previously existing. The emphasis here will be on the phospholipid bilayer as the common
denominator of all biological membranes. I take the perspective of the lipid
bilayer as the matrix into which are inserted specific proteins that imbue characteristic function to the new membrane. In most cases, membrane morphogenesis
arises from previously existing membranes. A ubiquitous example of membrane
morphogenesis is the invagination of the plasma membrane during cell division.
Other examples in prokaryotes include the formation of mesosomes and of in-
tracytoplasmic vesicles in some species (Reaveley and Burge, 1972). At the
subcellular level in eukaryotes, the generation of chloroplast grana and the cristae
of mitochondria are dramatic examples of membrane morphogenesis. Other
examples of membrane morphogenesis include the formation of microvilli and
lamellapodia at the cell surface, the formation of a nuclear membrane after cell
division, and the numerous conversions of other intracellular membranes such as
rough endoplasmic reticulum to smooth endoplasmic reticulum to Golgi membranes. Although generally acknowledged to be stimulated by energy, control of
these differentiations remains poorly understood.

Discussion of all of these examples of membrane morphogenesis is beyond the
scope of this chapter. Rather, with faith in the reductionist principle, membrane
morphogenesis and its possible controlling factors will be addressed at the level
of formation of new viral membranes that occurs upon infection of appropriate
host cells. In general, viruses are used to study membrane morphogenesis because they are simpler than the cells that they infect. Furthermore, the signal for
initiation of membrane morphogenesis must be controlled by a relatively small
viral genome. This fact serves to limit the search for a morphological agent.
Hence, the power of viral genetics can be applied to dissecting the mechanisms
of control of membrane morphogenesis.

B. Modes of Control in Membrane Morphogenesis

1. Evagination by Encapsulation of a Core

Four mechanisms will be advanced for controlling the size, shape, and composition of a new membrane (Fig. 1). In the first mode, a core structure binds to the
membrane (Fig. 1A). The adhesion causes the malleable membrane to be drawn

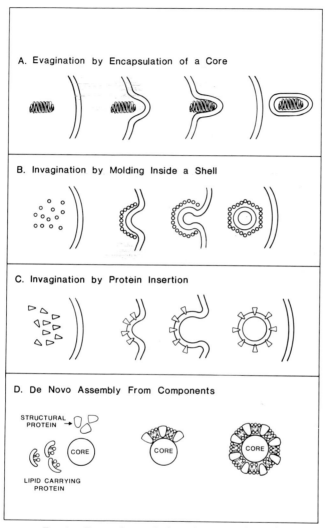

A. Evagination by Encapsulation of a Core

B. Invagination by Molding Inside a Shell

C. Invagination by Protein Insertion

D. De Novo Assembly From Components

STRUCTURAL PROTEIN →

CORE

CORE

CORE

LIPID CARRYING PROTEIN

FIG. 1. Types of control of membrane morphogensis.

onto the core structure. When the core is completely encapsulated, a new membrane is released. The best example of this mechanism is the budding from the plasma membrane of the enveloped animal viruses (see Section II). Prerequisites for this mode of membrane morphogenesis are the preassembly of a core structure and the presence of a recognition site in the membrane. Unless the proteins in the membrane that compose the recognition site have a specific lipid affinity, there is no reason to expect selection from the host membrane of an altered lipid composition in the progeny membrane. In this mode of membrane morphogenesis, membrane size and shape are determined by the size and the shape of the core structure, and the newly formed membrane is released on the opposite side of the membrane from which the core started.

2. *Invagination by Molding inside a Shell*

In a second type of membrane morphogenesis, membrane formation occurs inside a shell (Fig. 1B). Examples of pure protein shells are found in the heads of many bacteriophages (Murialdo and Becker, 1978). According to this mechanism, the protein subunits arrange themselves into an icosahedral shell according to the quasi-equivalence principle of Caspar and Klug (1962). The driving energy for this morphogenesis arises from the binding energy of the protein subunits to each other and to a lesser extent from the binding energy of these subunits to the membrane. If the protein binding to the membrane is specific for a particular lipid, then the new membrane will have a lipid composition different from that of the parent membrane. Here, the size and shape of the new membrane is determined by the size and shape of the icosahedral protein shell. In this mode of membrane morphogenesis, the new membrane is produced on the same side of the parental membrane from which the morphogenesis was initiated. The best example of this mechanism to date is the sequestration of eukaryotic membrane vesicles by the protein clathrin (Keen *et al.*, 1979). This protein forms a shell-like structure around portions of membranes causing them to bud. In the analogy to head formation in bacteriophages T4, P22, λ, and others, the protein shell is formed before the DNA is packaged inside that shell. This requires that a specialized structure exist in the shell through which the DNA can pass as it is packaged (Hendrix, 1978). Here the amount of DNA that is packaged is determined by the size of the shell (Eiserling *et al.*, 1970). This is in contrast to the first mechanism. However, these bacteriophages do not contain a lipid-bilayer membrane.

3. *Invagination by Protein Insertion*

In a third type of membrane morphogenesis, a membrane is caused to invaginate by the insertion of proteins (Fig. 1C). Here proteins synthesized on the cytoplasmic side of a membrane have hydrophobic surfaces that cause them to insert up to halfway into the membrane. According to the bilayer-couple mechanism (Sheetz and Singer, 1974), this insertion causes an expansion of the inner lamella of the bilayer relative to the outer lamella. The change in surface free energy is best accommodated by the bending or invagination of the membrane. The complete vesiculation process produces a new membrane on the same side of the membrane from which the process was initiated. New membranes accumulate inside the cell. The composition of the new membrane is determined by the specific affinity of the protein for specific lipids. If only one type of protein is inserted, the shape would necessarily be spherical to minimize surface free energy. On the other hand, the size of the membrane would be critically determined by the shape of the protein and the number of copies of this protein in the membrane. Here it is easy to conceptualize the protein in the form of a wedge. The angle of the wedge determines the radius of curvature of the forming membrane. Protein insertion would stop when the structure reaches the point of closing on itself. As before, the packaging of a genome would require a specialized structure to allow nucleic acid access to the interior. Alternatively, nucleic acid could be exported to the exterior of the membrane and packaged from the side coincident with membrane morphogenesis.

4. *De Novo Assembly from Components*

A common feature of the previous three modes of membrane morphogenesis is the formation of a new membrane from a preexisting cellular membrane. In a fourth type of control of membrane morphogenesis, membrane assembly is accomplished *de novo* from components (Fig. 1D). This mechanism begins with the synthesis of a core structure. Next, membrane structural proteins bind to the core. These proteins have an affinity for lipids in the form of a bilayer. The formation of the membrane is then initiated by the transfer of lipids from other membranes to these high-affinity, lipid-binding sites. This transfer is mediated by lipid carrier proteins that solubilize the lipids in the cytoplasm and enable their bulk concentration to substantially rise above the critical micelle concentration for free lipids (10^{-12} M). The composition of the new membrane is then determined by the affinity of these carrier proteins for specific lipids and any binding selectivity of the structural proteins. Of course, the size and shape of the new membrane is determined by the size and shape of the core structure. In this mechanism, newly formed membranes accumulate inside the cell.

II. Enveloped Animal Viruses

Several of the concepts for control of membrane morphogenesis have arisen from studies of enveloped animal viruses. This chapter will briefly treat morphogenesis of the enveloped animal viruses in order to give perspective to understanding the morphogenesis of the membranes of bacteriophages.

A. Morphogenesis by Budding

Most of the enveloped animal viruses acquire their membrane by the process of budding through the plasma membrane of their host, i.e., evagination by encapsulation of a core (Fig. 1A). Herpes virus, myxovirus, paramyxovirus, retrovirus, rhabdovirus, and togavirus all mature by the apposition of a core structure to a membrane and subsequent release of a virus particle by budding through that membrane. The membranes of these viruses have one or more glycoproteins inserted in the lipid bilayer of the virus. The envelope of these viruses is at the exterior of the virion. With only minor differences, the lipid composition of the virus represents that of the membrane from which it was formed. In the simplest case of the alpha virus, Sindbis, a single nucleocapsid protein–RNA complex recognizes the Sindbis glycoproteins E1 and E2 in the plasma membrane. The fact that host proteins are excluded from this region of the membrane suggests an interaction of the viral glycoproteins with each other. While the majority of the glycoprotein is exposed on the exterior of the cell and virus, a portion of it extends into the interior where it can be recognized by the nucleocapsid (Garoff and Soderlund, 1978). At restrictive temperature, mutant *ts103* of Sindbis virus dramatically portrays the budding event (Fig. 2) (Strauss *et al.*, 1977). Here, nucleocapsids appear to be bound to regions of the plasma

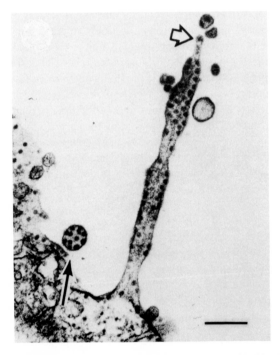

Fɪɢ. 2.   Membrane morphogenesis of an enveloped animal virus, Sindbis. Encapsulation of Sindbis virus nucleocapsids is seen to occur by evagination from the cytoplasmic membrane of the host chick embryo fibroblast grown in tissue culture. In this cell infected by mutant *ts103* of Sindbis, a normal 70-nm virus particle is seen (open arrow) along with particles that contain multiple cores (solid arrow). Bar represents 200 nm. (From Strauss *et al.*, 1977.)

membrane but fail to bud normally as individual nucleocapsids surrounded by an envelope. Instead, infectious particles are released in which multiple nucleocapsids are surrounded by a larger envelope. Most likely, the interaction of the *ts103* nucleocapsid with viral glycoproteins in the plasma membrane is weaker than normal.

In the other enveloped animal viruses that mature by budding, there is an additional protein called M that appears to serve as a linker between the nucleocapsid protein and the viral glycoprotein in the membrane. In contrast to the high specificity of the nucleocapsid of Sindbis, in the rhabdovirus, vesicular stomatitis virus (VSV), the nucleocapsid + M protein is able to form pseudotype viruses; the glycoprotein from retroviruses substitutes for that of VSV (Witte and Baltimore, 1977; Lodish and Weiss, 1979). An interesting mutant of VSV, *ts045*, forms noninfectious virus particles that lack the viral glycoprotein (Schnitzer *et al.*, 1979). This suggests that the viral glycoprotein is not necessary for budding of VSV from the plasma membrane. However, it may be compensated by the presence of host glycoproteins in the envelope of VSV. Studies with other temperature-sensitive mutants of VSV have led to the model of viral budding in which a soluble nucleocapsid and a soluble M protein coalesce at the plasma membrane with G protein or other substitute glycoproteins to form the viral bud (Knipe *et al.*, 1977). The budding process appears most dependent on the M protein, not the N protein (Schnitzer and Lodish, 1979). The size and shape of the envelope of animal viruses appears to be determined by the size and shape of the nucleocapsid. In VSV, this results in a bullet-shaped particle, whereas the other enveloped animal viruses that bud are nearly spherical. The recent finding of the posttranslational attachment of 1–2 moles of fatty acid per mole of VSV glycoprotein (Schmidt and Schlesinger, 1979) or Sindbis glycoprotein (Schmidt *et al.*, 1979) may have significance for viral membrane morphogenesis. At this time, the role is unclear.

In summary, control of membrane morphogenesis in the enveloped animal viruses appears to reside in the binding interaction of the nucleocapsid and/or the M protein to the plasma membrane containing various glycoproteins. The envelope of the virus is acquired by the plasma membrane being drawn onto the surface of the nucleocapsid.

B. MEMBRANE MORPHOGENESIS BY *de Novo* ASSEMBLY

The best studied member of the poxvirus group, vaccinia, acquires its envelope in the cytoplasm at discrete viroplasmic foci that are removed from proximity with other cellular membranes (Dales and Mosbach, 1968). The most simple mechanism for explaining this membrane morphogenesis invokes a phosholipid exchange protein that catalyzes the transfer of phospholipid from microsomes to the assembling vaccinia envelopes (Stern and Dales, 1974) (Fig. 1D). The polar head group selectivity of such a protein could explain the significantly smaller proportion of phosphatidylethanolamine found in the virus than found in host cells. Both preformed and newly synthesized phospholipids are utilized in assembling the vaccinia membrane (Stern and Dales, 1974). When DNA synthesis is inhibited by hydroxyurea, envelope assembly still proceeds (Rosenkranz et al., 1966). Although viral maturation occurs after removal of the block or temperature shift in an analogous mutant (Stern et al., 1977), the preformed envelopes are not utilized and are therefore defective. Nevertheless, their assembly demonstrates the lipid-condensing role of at least one of the vaccinia proteins onto a core that lacks DNA. The identification of this protein is likely to be difficult since there are more than 100 polypeptides in the virion (Essani and Dales, 1979).

III. Enveloped Bacteriophages

Because of their simplicity, ease of isolation in large quantity, and potential for genetic manipulation, the enveloped bacteriophages are well-suited for studies aimed toward understanding a variety of factors important in control of membrane morphogenesis. Table I introduces the enveloped bacteriophages (PM2, ϕ6, and the PR4 group), which are contributing to our understanding of membrane morphogenesis. There are other bacteriophages that contain lipid, but their arrangement into a membrane, much less their contribution to understanding membrane morphogenesis, has not been determined. These include AP50, which infects *Bacillus anthracis* (Nagy et al., 1976); DS6A, which infects *Mycobacterium tuberculosis* H37RV (Bowman et al., 1973); Dp1, which infects *Diplococcus pneumoniae* (Lopez et al., 1977); and ϕNS11, which infects *Bacillus acidocadarias* TA6 (Sakaki et al., 1977). As discussed by Mindich (1978), other bacteriophages that are sensitive to chloroform or ether may not necessarily contain lipid (e.g., CP1, Cihlar et al., 1978). Recent comprehensive reviews have appeared on PM2 and ϕ6 (Franklin, 1974, 1978; Franklin et al., 1976, 1978; Mindich, 1978). For this reason, this chapter will be organized around and limited to those aspects of the structure and development of these phages that pertain to control of membrane morphogenesis.

PM2, ϕ6, and PR4 group of viruses replicate inside their respective bacterial hosts and are released by cell lysis. Presumably the cell wall of the host bacterium precludes an external budding mechanism for membrane morphogenesis. Considering these phages, the question of membrane morphogenesis can be refined to, "What is the information present in the phage genome that says, 'Make new membranes of the desired size, shape, and lipid

TABLE I
Some Properties of Membrane-Containing Phages[a]

Bacteriophage	Host	Genome (daltons)	Diameter (nm)	CsCl bouyant density	Lipid composition (by ^{32}P) (%)					
					Host			Virus		
					PG	PE	DPG	PG	PE	DPG
PM2	*Alteromonas espejiana*	dsDNA circular (6.3×10^6)	60	1.28	24	76	—	63	37	—
φ6	*Pseudomonas phaseolicola* HB10Y	dsRNA 3 segments (10.4×10^6)	82	1.27	29	56	15	57	35	8
PR3, PR4, PR5, PRD1	*Escherichia coli* and *Pseudomonas aeruginosa* and others carrying plasmids P, N, or W compatibility groups	dsDNA linear (7.4×10^6)	65	1.26	16	74	10	44	43	13

[a] For references, see text. Abbreviations: PG, phosphatidylglycerol; PE, phosphatidylethanolamine; DPG, diphosphatidylglycerol, cardiolipin; ds, double-stranded DNA or RNA.

composition'"? The mechanisms for control of membrane morphogenesis that are delineated in Fig. 1 can be used as models against which to test the known characteristics of these viruses. The models also serve to direct experimentation toward understanding the critical details of a mechanism and for eliminating other mechanisms. At this stage of research, three areas of investigation contribute to our understanding of the control of membrane morphogenesis in bacteriophage.

A. CHARACTERISTICS OF THE VIRAL MEMBRANE

After an initial screening of phage for sensitivity to chloroform or ether, demonstration of the presence of a membrane is based on morphological evidence of thin sections in the electron microscope, lipid compositional analysis, and more detailed structural relationships of the membrane to the other proteins in the virus. In marked contrast to the enveloped animal viruses, a common feature of the enveloped bacteriophages is the dramatic difference in lipid composition of the virus from that of the host (Table I). In particular, the phosphatidylglycerol concentrations in $\phi6$ and PM2 are double those of the hosts in which they mature. The altered lipid composition of the viruses raises the questions, "What aspect of the mechanism of membrane morphogenesis is responsible for this dramatic change in lipid composition? Is there a protein with specifically high affinity for phosphatidylglycerol?"

B. RELATIONSHIP TO THE HOST MEMBRANE

As depicted in Fig. 1, very different mechanisms for the control of membrane morphogenesis must prevail depending on the relationship of maturation of the viral envelope to the existing host membrane. Initial observations are designed to answer the questions, "Does the virus mature at the periphery of the cell or at the center? Are the lipids of the virus obtained from previously existing lipids in the host membrane or are they acquired from those newly synthesized?"

C. PACKAGING THE GENOME

Basic knowledge of the order of assembly of the viral components contributes to an understanding of control of membrane morphogenesis. Does membrane formation precede packaging of the nucleic acid? Or, is a nucleocapsid formed before the viral envelope?

IV. Bacteriophage PM2

A. CHARACTERISTICS OF THE MEMBRANE

Bacteriophage PM2 was originally isolated from the coastal waters of Chile by Romilio Espejo and Eliana Canelo. In 1968, they reported the purification of the virus and the presence of at least 10% of the dry weight of the virion as lipid. Careful qualitative analysis by Braunstein and Franklin (1971) revealed a dramatic difference in the lipid composition of the virus from that of the host. While the host membrane contains predominantly phosphatidylethanolamine, the major lipid of the virus is phosphatidylglycerol (Table I). This finding contributed the first information on control of membrane morphogenesis of PM2. There must be specific control of the selection of lipids from the cell for the membrane of PM2. In forming the viral membrane, lipids are not randomly utilized from the preexisting cellular membrane. Quantitative analysis indicates the lipid content of the virion to be 13% of the total by weight (Camerini-Otero and Franklin, 1972). This amount of lipid is sufficient to account for only 50–70% of a bilayer membrane the size of PM2. Thus, if there were a membrane bilayer in PM2, a significant portion of that membrane must be occupied by protein.

1. Evidence for a Membrane

In morphological terms, one of the best ways to identify a membrane is by its characteristic ''railroad-track'' appearance in stained thin sections. This is due to the coprecipitation of the electron-dense osmium and lead ions at the polar surfaces of the bilayer. In thin sections of the virus inside infected cells, such a structure was clearly observed somewhat to the interior of the virus (Silbert *et al.*, 1969). Observation in the electron microscope is perhaps the simplest method for detecting a membrane bilayer structure. The presence of extractable lipids in a virus does not necessarily imply that they are organized in a bilayer; they could be part of a lipoprotein complex such as is present in the serum lipoproteins.

The most definitive evidence for a lipid bilayer membrane in bacteriophage PM2 comes from studies of low angle X-ray diffraction (Harrison *et al.*, 1971). Fourier synthesis of the corresponding transform of the small angle X-ray scattering indicates the presence of a deep minimum at 220 Å with peaks at 200 and 240 Å. These dimensions are precisely those expected for the usual 40-Å-thick membrane with polar head groups at the surface causing high scattering of electrons and fatty-acyl chains in the interior causing low scattering of electrons. Two other observations are noteworthy. The surface area of the outer aspect of the membrane represents 60%, whereas the inner membrane surface is only 40% of the total membrane surface area. Also, the absolute mean electron density at the center of the bilayer, 0.30 e/Å^3, is significantly higher than that for the central region of pure lipid bilayer membranes, 0.17 e/Å^3. This suggests that significant protein (up to 30%) may contribute to the rise in electron density at the center. However, another possible interpretation of these data is that the rise in electron density is due to the spherical averaging of a number of CH_2 groups into the central region of the bilayer. Recently, neutron scattering has lent further support to the existence of a lipid bilayer membrane in PM2 (Schneider *et al.*, 1978).

2. *Relationship of the Membrane to Other Viral Components*

A model for the structure of bacteriophage PM2 is shown in Fig. 3. The four major structural proteins of the virus are listed. As outlined in the introduction, one or more of these could participate in controlling the morphogenesis of the viral membrane. Therefore their disposition relative to the viral membrane is critical. An understanding of these relationships has arisen from (a) analyses by chemical tagging and limited degradation, (b) partial dissociation of the virion and (c) reconstitution of virus structure from the components, (d) physical characterization, and (e) chemical characterization of individual components.

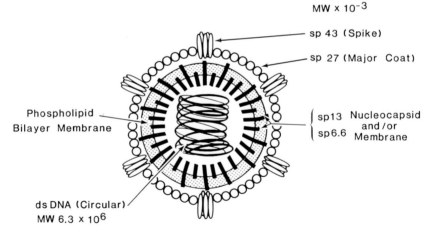

600 Å DIAMETER ICOSAHEDRON

4 MAJOR STRUCTURAL PROTEINS
MW x 10^{-3}

sp 43 (Spike)

sp 27 (Major Coat)

Phospholipid
Bilayer Membrane

sp13 Nucleocapsid
and/or
sp6.6 Membrane

ds DNA (Circular)
MW 6.3 x 10^6

Coding Capacity 315,000 Daltons   Major Structural Proteins = 90,000

8-12 Genes

FIG. 3. Schematic structure of bacteriophage PM2 showing the DNA, lipid bilayer membrane, and four structural proteins. Drawn to scale with a membrane 40 Å thick. The localization of sp13 and sp6.6 is hypothetical.

   The first step in understanding the assembly of a virus is to identify the
components that are assembled. The four major structural proteins of PM2 have
been resolved by SDS–polyacrylamide gel electrophoresis. They are identified
by their molecular weight, e.g., sp43 is a structural protein of molecular weight
43,000 (see Fig. 3). A beginning to the understanding of the disposition of these
proteins is obtained from a knowledge of the number of copies of each protein
present in a virion. Along with a comparison of nomenclatures of Brewer and
Franklin, two independent analyses are compared in Table II. Brewer and Singer
(1974) base their data on purification of viruses that were uniformly labeled in
the presence of [^{14}C]glucose, whereas the data of Franklin et al. (1976) were
based on uniform labeling with [^{14}C]acetate (Datta et al., 1971b). It is surprising
that these numbers are so different since the difference in labeling techniques
should not affect the results. Perhaps one source of error in the low value for the
number of copies of sp6.6 is the dialysis against SDS to which the sample was
subjected prior to electrophoresis (Datta et al., 1971b). At least some sp6.6 can
pass through the pores of the dialysis tubing that has a cutoff of around 12,000
daltons (Brewer, unpublished observation). In support of this contention, Datta
et al. present other data based on [^{3}H]amino acid incorporation into PM2 proteins
in which the percentage of sp6.6 was found to be 14% instead of 6% (possibly
dialyzed for a shorter period). Nevertheless, the most abundant protein is sp27,
the major coat protein of the virus.
   a. *Limited Digestion and Chemical Tagging.*   Limited proteolysis with tryp-
sin revealed that sp27 and sp43 are exposed at the surface of the virion (Brewer
and Singer, 1974). Treatment of the virion with bromelain selectively digests
sp43 (Schafer et al., 1974) as judged by electrophoresis (Hinnen et al., 1974).
Following bromelain treatment, observation of negatively stained preparations in
the electron microscope shows the absence of spikes at the vertices. The loss of

TABLE II
ABUNDANCE OF STRUCTURAL PROTEINS OF PM2

		Copies/virion[a]	
Protein	Molecular weight (Hinnen et al., 1976)	Brewer and Singer (1974)	Datta et al. (1971b) Franklin et al. (1976)
sp43 I	43,000	50	80
sp27 II	27,000	640	820
sp13 III	12,500	320	500
sp6.6 IV	6,600	785	300

   [a] Revised numbers based on the product of PM2 molecular weight, 4.5×10^7, fractional weight of
total protein, 0.72 (Camerini-Otero and Franklin, 1975), and the individual protein percentage of
total protein divided by the individual protein molecular weight.

these spikes was also associated with loss of viral infectivity, indicating their probable importance for attachment to the host (Hinnen *et al.*, 1974). Other labeling studies with [^{14}C]glycine ethylester and transglutaminase or ^{125}I and lactoperoxidase also indicate the exterior exposure of sp43 and sp27 (Brewer and Singer, 1974). sp27 could also be labeled with [^{35}S]sulfanilic acid diazonium salt without loss of viral infectivity (Hinnen *et al.*, 1974). The labeling of other proteins by these techniques is ambiguous since additional labeling occurs only after loss of viral infectivity and possible disruption or rearrangement of the proteins in the virion. Positive controls with disrupted virus indicate that sp13 and sp6.6 can be labeled with these reagents. These data suggest that sp13 and sp6.6 are not exposed at the surface of the virus but their disposition relative to the membrane remains uncertain.

Treatment of virus particles with 0.5% gluteraldehyde or 5 mM dimethyl-suberimidate results in the cross-linking of sp6.6 to sp13 and each of these proteins to phosphatidylethanolamine in the membrane (Schafer *et al.*, 1975). Examination of treated virions in the electron microscope reveals no gross alteration in morphology. These cross-linking studies unequivocally demonstrate the proximity of sp13, sp6.6, and phosphatidylethanolamine in the membrane.

b. *Dissociation.*   Another approach to the relationship of proteins to the viral membrane is by partial dissociation of the viral structure. This has been achieved by multiple cycles of freezing and thawing the virus and in another case by treatment with increasing concentrations of urea. These results must be interpreted cautiously since the unnatural conditions may result in rearrangements of the native structure. A virus subfraction can be isolated after freeze–thaw disruption (Brewer and Singer, 1974). This fraction contains all of the viral membrane, all of the viral DNA, and the viral structural proteins sp13 and sp6.6. These results suggest that sp43 and sp27 can be dissociated from the viral membrane under moderate physical stress, whereas sp6.6 and sp13 are likely to be integral membrane proteins. The DNA is not inside the virus subparticle since it is susceptible to digestion with DNase. Failure to release sp6.6 after digestion of the DNA suggests that this protein is not primarily associated with the DNA, but is associated with the membrane.

By carefully raising the urea concentration in a solution of virus, discrete viral substructures are generated that have been isolated and characterized extensively (Hinnen *et al.*, 1974; Schafer *et al.*, 1978). In 2 M urea, two viral substructures are stable. Seventy percent of the starting virions have become spikeless virions and 7% have become lipid-containing nucleocapsids. The spikeless virions are missing sp43 from their vertices. The lipid-containing nucleocapsids are additionally missing the majority of sp27. Between urea concentrations of 4 and 8.5 M, icosahedral nucleocapsids can be isolated that lack sp43 and lipids and the majority of sp27. Treatment with DNase destroys the structure. The importance of hydrophobic interactions to the integrity of this structure is indicated by its dissolution upon decreasing the NaCl concentration below 0.1 M. Is this lipid-free nucleocapsid a real component of the native virion? Although it is unlikely that an icosahedral structure would be generated if it were not a critical determinant in the native structure, the diameter of the nucleocapsid is 44 nm as determined from electron micrographs. This is significantly larger than the 40-nm diameter of the inner face of the membrane as determined by X-ray diffraction. Some sp13 and/or sp6.6 must penetrate the membrane. Thus there appears to be a real relationship between sp13, sp6.6, and DNA of the virus, but the topological arrangement of these components relative to the membrane remains unclear. In particular, questions regarding the *in vivo* preassembly of a nucleocapsid are discussed in Section IV,C.

c. *Reconstitution.* Another approach to understanding structural organization is to begin with the components and under specialized conditions reconstitute the structure. It is a remarkable accomplishment that Schafer and Franklin (1975a, 1978a) have reconstituted virus-like particles from the nine major components of PM2. Their strategy was basically to reverse the dissociation that they had performed with increasing concentrations of urea. In the first step, PM2 DNA was mixed with sp6.6 at low ionic strength. To the complex that formed was added sp13 in 4 M urea. Based on input DNA, the yield was 52% icosahedral nucleocapsids. These nucleocapsids were identical to nucleocapsids isolated from whole virus with the exception that the reconstituted nucleocapsid did not possess the residual sp27 nor could the residual sp27 be bound to these nucleocapsids. Next, the two lipids, phosphatidylethanolamine and phosphatidylglycerol, were sonicated in 4 M urea and added to the nucleocapsids along with sp27. After lowering the urea concentration to 1 M by dialysis, reconstituted virus-like structures could be isolated with a yield of 26%. To these particles was added sp43 followed by dialysis to remove the urea. Amazingly, viral infectivity was obtained in this last step at the level of 320 PFU/μg. No infectivity was detected at any of the prior steps. Although the recovery of virus-like particles was 14% based on input DNA, the specific activity of native virus is 5 x 10^9 PFU/μg, 10^7-fold higher than that obtained in the *in vitro* reconstitution. The physical properties of these virus-like particles were very similar to those of the native virus. In the electron microscope, they were indistinguishable. One major difference was that the reconstituted particles contained twice as much phospholipid per mole of DNA as the native virus.

Does this remarkable reconstitution of bacteriophage PM2 reflect the natural assembly of the virus inside an infected cell? There is lipid specificity in the proteins used for reconstitution. Regardless of the input ratio of phosphatidylethanolamine and phosphatidylglycerol, the ratio of these lipids in the reconstituted particles is identical to that of the native virus (a 2:1 ratio, respectively). Another aspect of the specificity of assembly is that the stoichiometries of each of the structural proteins is the same in the reconstituted particles as in the native PM2. I interpret this as strong evidence for self-assembly of the virus from its components based on specific lipid–protein and protein–protein interactions. On the other side of the question, a cell does not have high concentrations of urea that it can adjust. The virus has only 10^{-7} times the activity of the native virus and has twice as much lipid as as the native virus. Obviously, virus-like particles have been reconstituted, but only a miniscule portion of them are infectious bacteriophage PM2. Since the reconstituted particles have twice as much lipid as the native virus, I suggest that the crucial element missing in the reconstituted particles is protein insertion into the membrane. The possible importance of lipid–protein interactions in the structure of the virus is obscured by the use of urea to isolate a nucleocapsid core from native virus or to reconstitute a nucleocapsid from two proteins and DNA. Possibly it is these very lipid–protein interactions that control the membrane morphogenesis *in vivo* and render the resultant particles infectious (see model in Section IV,D). The *in vivo* relationships of the viral membrane to the host membrane and packaging of the genome are discussed in Sections IV,B and C.

d. *Physical Studies.*   Electron spin resonance spectroscopy has provided important information about the structure of the membrane of PM2. Scandella *et al.* (1974) report a spin-label order parameter for a nitroxide fatty acid in PM2 that is considerably higher than that for egg-yolk lecithin. It is necessary to increase the temperature of PM2 to 60°C in order to observe motion in the fatty-acyl chains in the virus equivalent to those of egg-yolk lecithin at 25°C. Calculation of the hyperfine splitting constant for spin-labeled PM2 indicates strong interaction between protein and acyl chains, especially in the outer regions of the bilayer. Thus when spin labeled virus are treated with 0.1% glutaraldehyde (Schafer *et al.*, 1975), motion of the acyl chains in the polar head group region is restricted, whereas the motion toward the interior of the PM2 membrane is unaffected. Unfortunately no direct comparisons have been made using a single spin probe for the virus membrane in comparison to lipid extracted from that membrane or the host membrane. However, it seems likely that the relationship found for Sindbis virus (Sefton and Gaffney, 1974) and for influenza (Landsberger *et al.*, 1973) holds for PM2: the fatty-acyl chains in the virus are more rigid than those in the cell, both of which are more ordered than extracted lipids. Extensive lipid–protein interactions are thought to immobilize the viral lipids.

e. *Enzymatic and Other Chemical Characteristics of Individual Components.* Enzymatic and other specific properties have been attributed to specific proteins of PM2.

i. *Sp 13 Is an Endolysin.*   A layer of peptidoglycan surrounds the host of PM2 between the inner and outer membranes. In order for PM2 to infect the host, it must pass through this layer. It must also be degraded at the end of virus maturation when the cell lyses to release the internal phage. An endolysin that may accomplish one or both of these functions is induced in the infected host and is also associated with the purified virion (Tsukagoshi *et al.*, 1977). The ability to digest a purified cell wall preparation in an *in vitro* assay was clearly associated with purified sp13 and not the other three viral structural proteins. However, the specific activity of this purified protein was not increased over that of the intact virus disrupted by osmotic shock. Perhaps the purification of sp13 from an acetic acid extract and column chromatography in guanidine removed the lipid environment necessary for maximal activity. It would be interesting to add lipids back to sp13 and check for an increase in specific activity. It is noteworthy that the calcium dependence for activity of the enzyme precisely coincides with the calcium dependence for production of infectious virus (Snipes *et al.*, 1974).

Studies with temperature-sensitive mutants of bacteriophage PM2 further suggest a relationship of sp13 to viral membrane formation (Brewer, 1978b). Cells infected under restrictive conditions with mutant *ts27* of PM2 show only occasional evidence of viral membrane formation. When the proteins in these cells are examined by polyacrylamide gel electrophoresis, the amount of sp13 present in a membrane sediment is greatly reduced in comparison to cells infected with wild-type virus. This occurs despite production of normal amounts of sp43 and sp27 that remain in the supernatant. Taken together with similar evidence for three other mutant groups that produce very few or only incomplete viral membranes (*ts7, ts8, ts17*), these mutants suggest that sp13 binding to the host membrane may be a crucial event in control of membrane morphogenesis in PM2 (see model in Section IV,D). This hypothesis needs to be tested by appropriate shift-down and pulse–chase experiments, as well as attempts to achieve membrane morphogenesis in cell-free extracts.

The fact that no mutants of PM2 were obtained that produce large amounts of intracellular virus yet fail to lyse the cells (Brewer, 1978a) suggests that the endolysin activity that is needed for cellular lysis is a required structural component of the virus. Excess sp13 that is not assembled into the virus may accumulate in the host cytoplasmic membrane to eventually digest the peptidoglycan of the cell. In providing accessibility to the peptidoglycan, an exterior localization of sp13 in the cytoplasmic membrane would be equivalent in the virus to an interior localization in a model of invagination (Fig. 1C).

ii. Sp6.6 Is a Polymerase. Enzymatic activity is associated with another PM2 structural protein, sp6.6. Originally reported as a DNA-dependent RNA polymerase associated with detergent-treated virions (Datta and Franklin, 1972), further analysis indicated a less specific polymerase activity associated with sp6.6 (Schafer and Franklin, 1975b). The polynucleotide-pyrophosphorylating activity will polymerize into acid-percipitable material either ribo- or deoxyribonucleoside triphosphates. Although PM2 and other DNAs preferentially stimulate the incorporation of ribonucleotides and RNAs preferentially stimulate the incorporation of deoxynucleoside triphosphates, the nucleic acid serves neither as a template nor as a primer for the activity. Therefore, it seems likely that this peculiar activity needs to be fully optimized or requires some host factors for intelligent catalysis. *In vivo,* this activity may be used to block host DNA replication (Franklin *et al.,* 1969) or to catalyze viral DNA replication or transcription. Studies with mutant *ts27* of PM2 (Brewer, 1978a) suggest that sp6.6 may function in viral DNA replication. In cells infected with *ts27* at a restrictive temperature, there is a reduction in total DNA synthesis, a reduction in association of sp6.6 with a membrane sediment, and no assembly of viral structures. Since *ts27* is also defective in sp13, as cited earlier, an interdependence of sp13 and sp6.6 is suggested.

   iii. Sp6.6 and Sp13 Are Hydrophobic Proteins.   The hydrophobic nature of sp6.6 and sp13 is suggested by their solubility in the chloroform phase of a chloroform–methanol water two-phase system (Brewer and Singer, 1974). Using an aqueous phase buffer at pH 7.4, the only viral protein present in the lower phase is sp6.6. If the pH is raised to 9, then both sp13 and sp6.6 can be partitioned into the lower phase. With 56% apolar amino acids, purified sp6.6 is among the most hydrophobic proteins characterized (Brewer and Singer, 1974). It is disturbing that the amino acid composition of sp6.6 reported by Hinnen *et al.* (1976) is so different from that reported by Brewer and Singer (1974). Although the total number of amino acids reported was similar (60 and 63, respectively), there is agreement only on the number of aspartic acids/ asparagines, serines, lysines, arginines, and the absence of histidine. Brewer and Singer found a blocked N-terminus. The purity of the protein was demonstrated by several criteria including the generation of a new N-terminal isoleucine following cyanogen-bromide cleavage at the single methionine. The purity of Hinnen's preparations is attested by their ability to find a single N-terminal methionine and their ability to partially sequence the protein by automated Edman degradation. The most striking feature of this partial sequence is clustering of basic residues in the first ten positions followed by a sequence of at least 16 hydrophobic amino acids. Since the pitch of a protein α-helix is 3.6 residues per turn with an axial 1.5 Å per residue, sp6.6 could easily traverse the membrane one or even two times. The preponderance of basic residues in sp6.6 could interact electrostatically with anionic phosphatidylglycerol in the membrane.

   iv. Sp6.6 Binding to DNA.   Led by the isolation of a nucleocapsid of PM2 and the reconstitution of virus-like particles, Marcoli *et al.* (1979) have studied the interaction between DNA and sp6.6 of bacteriophage PM2. Under appropriate conditions, an interaction of sp6.6 and DNA is demonstrated by cosedimentation, filter binding, and electron microscopy. Binding is independent of DNA structure or sequence. Saturation of the PM2 genome occurs with approximately 26,000 monomers of sp6.6. This is many times the amount of sp6.6 found in each virion. The binding is optimal at 50 m*M* NaCl and no binding occurs at 0.4 *M* NaCl. Although the authors cite the studies of Cavalieri *et al.* (1976) for the dissociation at high ionic strengths of protein 5 from bacteriophage fd and protein 32 from bacteriophage T4 DNAs, the marine habitat of PM2 and its host requires that protein–DNA interactions occur at high ionic strength *in vivo.* Therefore the evidence for specific interaction of sp6.6 with DNA at low ionic strength is no better than the demonstration of binding of cytochrome c to DNA as it is commonly used to spread and visualize DNA for electron microscopy.

   v. Possible Interaction of Sp27 and Phosphatidylglycerol.   Evidence for lipid–protein interaction has been sought by reconstitution (Schafer and Franklin, 1975a). Removal of urea from a solution containing sp6.6, sp13, DNA, phospholipids, and serum albumin results in the binding of phospholipids to a protein–DNA core complex. The proportions of the lipids in the complex are directly related to the ratios of phospholipids at the start. However, when sp27 is present in this incubation mixture, there is a preferential selection of phosphatidylglycerol, independent of the proportions present at the start. Furthermore, sp27 is able to cause the partitioning of phosphatidylglycerol out of chloroform into the aqueous phase of a chloroform–methanol–water mixture. The other structural proteins of PM2 do not have this capability. Thus, an interaction between sp27 and phosphatidylglycerol is likely to occur in the virion. However, as discussed earlier, the relevance of these *in vitro* studies to *in vivo* assembly is not at all certain. The crucial question remains unanswered, "Does the binding of phosphatidylglycerol to sp27 aid its enrichment in the virus during membrane assembly *in vivo,* or does a preassembled membrane containing phosphatidylglycerol bind the coat protein sp27?" Evidence for preassembly of the viral membrane comes from cells infected under restrictive conditions with mutant *ts5* of PM2 (Brewer, 1976). Sp27 made by this mutant fails to associate with what appear to be normal viral membranes. This finding suggests that formation of a viral membrane does not require an outer scaffolding of coat protein (Fig. 1B). Further evidence against a specific role of sp27 in determining the phosphatidylglycerol content of the virus comes from *in vivo* studies by Tsukagoshi *et al.* (1975a). Growth of PM2 in a fatty acid auxotroph host supplied with palmitelaidate (trans 16:1) results in virus whose phosphatidylglycerol to phosphatidylethanolamine ratio is changed to 0.8 from the wild-type ratio of 1.5.

B. Relationship of the Viral Membrane to the Host Membrane

   The acquisition of the phospholipids in the membrane of bacteriophage PM2 depends on the host membrane, at least for the site of lipid synthesis. The relationship of the viral membrane to the host cytoplasmic membrane has been studied morphologically by electron microscopy of infected cells, biochemically by noting the presence of viral proteins in the host cytoplasmic membrane, the kinetics of lipid synthesis during the course of infection, and investigation of lipid asymmetry in the host and the viral membranes.

1. *Electron Microscopy of Infected Cells: Maturation at the Periphery*

Examination in the electron microscope of thin sections of *Alteromonas espejiana* infected with bacteriophage PM2 strikingly shows maturation of virions at the inner face of the cytoplasmic membrane (Cota-Robles *et al.*, 1968; Dahlberg and Franklin, 1970) (Fig. 4). This finding provokes the questions, "In the control of membrane morphogenesis of the virus, what is the role played by the host membrane? Does viral membrane formation proceed by an invagination of the host membrane?" The micrographs of Cota-Robles *et al.* do show mature virions apparently attached to the cytoplasmic membrane. However, precursors to this state have not been reported. This is not surprising since once initiated, the entire membrane formation may occur in less than a minute and intermediate structures are likely to be unstable. Therefore the probability of observing one of these intermediates in section after fixation, dehydration, and embedding is likely to be very small. Nevertheless, mature-appearing virions whose membrane appears to branch off of the host membrane are observed (Brewer, unpublished observations). In most cases, visualization of the membrane is unclear due to ribosomes in the background and the plane of sectioning passing tangentially across the membrane. In a few cases, the contrast is sufficient to observe continuity between putative viral membranes branching from the host cytoplasmic membrane (Fig. 5). The appearance of these structures have stimulated the model described in Section IV,D (Fig. 8).

Additional morphological support for replication of PM2 at the membrane of the host comes from an autoradiographic study (Brewer, 1978c). Viruses labeled *in vivo* with [^{3}H]thymidine were purified and used to infect unlabeled host cells. In the middle of the infectious cycle, cells were fixed and prepared for electron microscope autoradiography. Grains arising from the emission of tritium in the infecting DNA are found to be localized at the bacterial membrane and not distributed evenly throughout the cell. This finding helps to explain low levels of genetic recombination and complementation found between mutants of PM2 (Brewer, 1978a). If the infecting DNA remains bound to the host membrane, the DNA of multiply infected cells would not have the opportunity to exchange.

2. *Viral Proteins in the Host Membrane*

The stimulation of viral membrane morphogenesis may be controlled by the insertion of proteins into the host membrane. Additional viral proteins may be bound at the periphery of the membrane. Fifteen new proteins have been identified by two-dimensional gel electrophoresis of cells infected with PM2 (Table III). The sums of their molecular weights account for 98% of the coding capacity

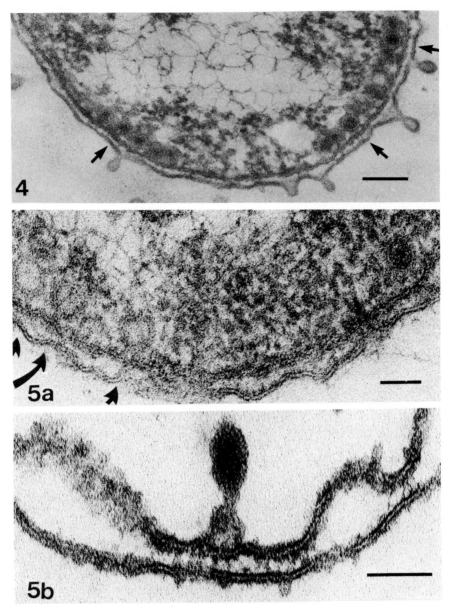

F<small>IG</small>. 4. Morphogenesis of bacteriophage PM2 at the inner face of the cytoplasmic membrane of *A. espejiana*, 45 minutes after infection. Arrows point to mature virions. Bar represents 100 nm. (From Brewer, 1976.)

F<small>IG</small>. 5. Morphogenesis of putative viral membranes from the cytoplasmic membrane of the host. Cells were fixed 45 minutes (a) and 70 minutes (b) after infection. Arrows point to putative viral membranes. Bar represents 50 nm.

TABLE III

PROTEINS WHOSE SYNTHESIS IS STIMULATED BY PM2[a]

MW ($\times 10^{-3}$)	pI	Max rate (cpm $\times$ 10^{-2})	Start (minutes)
p50	4.3	191	29
sp43	5.4,I[b]	73	15
sp43	5.0,I[b]	105	15
p33	4.8	38	8
sp27	6.8	1122	23
sp27	6.4	139	23
sp27	6.2	36	28
p25	7.0	56	15
p22	6.2	68	24
p22	5.8	81	23
p19	4.6	184	15
p15	5.4	77	22
p13.4	5.8	ND	ND
sp13	I[b]	5	26
p12.5	5.3	ND	ND
p9.5	6.6	9	28
p8.6	5.2,I[b]	ND	ND
p8.4	5.2	36	11
sp6.6	I[b]	44	25

[a] Average apparent molecular weight (MW); protein (p); structural protein (sp); average apparent isoelectric point (pI); maximum rate of incorporation of [^{35}S]methionine (max rate); apparent start of synthesis (start); not determined (ND).

[b] I, Insoluble; did not enter isoelectric focusing gel but migrated along border through second-dimension gel.

of PM2 (Brewer, 1980). Based on kinetic analysis, the synthesis of p50, p22, p15, p9.5, and the viral structural proteins is delayed. This suggests transcriptional control in PM2 as with other bacteriophages. SDS–polyacrylamide gel electrophoresis of membranes isolated from infected cells gives evidence for new polypeptides whose association with the membrane is dependent on infection by PM2 (Datta *et al.*, 1971a). In this system, resolution from the large number of host proteins is insufficient to identify these proteins with confidence. Higher resolution analysis indicates at least nine viral proteins associated with purified membranes from the infected host (Fig. 6). Thus, the stronger bands at p19, p15, sp13, and sp6.6 are candidates for controlling membrane morphogenesis in PM2. Further characterization of this membrane fraction and the organization of viral proteins therein are needed in order to approach the problem of control of membrane morphogenesis of PM2.

A more definitive approach to the identification of viral membrane proteins is

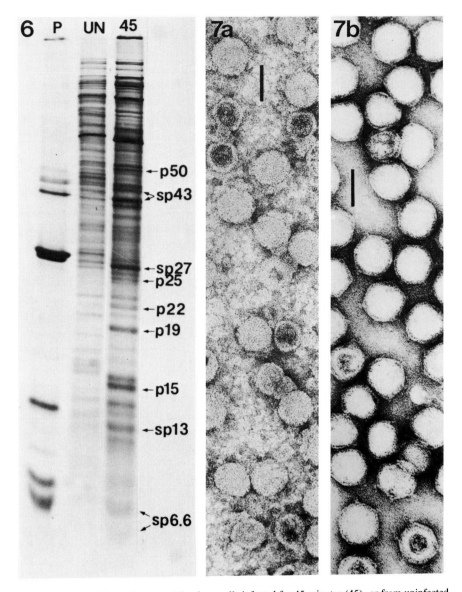

FIG. 6. Bacterial membrane proteins from cells infected for 45 minutes (45), or from uninfected cells (UN). Membranes were prepared from cells (Ito *et al.*, 1977) labeled with [^{35}S]methionine. Membranes were purified by zonal centrifugation and isopycnic gradient centrifugation before resolution of the proteins by sodium dodecyl sulfate–polyacrylamide gel electrophoresis and autoradiography as shown here. Pure PM2 was run for reference purposes (P). Proteins specific for PM2 infection are noted.

FIG. 7. Membrane vesicles (a) produced by infection with *ts1* mutant of bacteriophage PM2. Wild-type PM2 is shown for comparison (b). Preparation is negatively stained with uranyl acetate. Bar represents 60 nm. (From Brewer, 1979.)

to direct their synthesis by PM2 DNA *in vitro* in a coupled transcription–translation system. PM2 DNA directs the *in vitro* synthesis of the enzymatic activities of sp6.6 and sp13 (Schafer and Franklin, 1978b). The claim of synthesis of 11 proteins is equivocal since the sum of their molecular weights (444,300) is beyond the coding capacity of the PM2 genome (315,000), and a number of them are peaks of only one gel slice. Grobovsky and Brewer (unpublished observations) observe similar *in vitro* protein synthesis directed by PM2 DNA. In addition to the four viral structural proteins, two other proteins are seen. Here too, a species that migrates in the region of sp6.6 is the most abundant protein. Of interest to membrane morphogenesis is the 3-fold stimulation of synthesis of this protein when host cytoplasmic membranes are added. Further work is needed to establish the identity of this protein as sp6.6, to identify a possible leader sequence (e.g., Blobel and Dobberstein, 1975; Inouye *et al.,* 1977) as the cause of a shift in migration on gels and to show insertion into the added membranes.

3. *Lipid Synthesis*

The enrichment of phosphatidylglycerol in PM2 from that of the host could be caused by changes in phospholipid synthesis induced by infection. However, in infected cells prior to cell lysis, numerous studies find no effect of infection on host phospholipid metabolism (Tsukagoshi and Franklin, 1974; Diedrich and Cota-Robles, 1976; Ruettinger and Brewer, 1978). The fatty acyl, phosphate, glycerol, and ethanolamine residues of the host lipids are stable during the period of viral morphogenesis. Furthermore, pulse–chase studies show that the viral lipids are acquired at random from newly synthesized and preexisting host lipids. One-third of the viral lipids are derived from lipids synthesized before infection and two-thirds from lipids synthesized during virus replication (Tsukagoshi and Franklin, 1974). Early in the course of infection, there is a rise in activity of phospholipase A against phosphatidylethanolamine (Tsukagoshi *et al.,* 1975b). However, the failure to observe the correlate change in phosphatidylethanolamine in infected cells suggests that the conditions for assay of the enzyme *in vitro* are different from those that exist *in vivo*. Taken together, these studies on lipid synthesis do not support models of active control of lipid synthesis in cells infected by PM2. Rather, there must be specific selection of phosphatidylglycerol from the host membrane into the viral membrane.

4. *Lipid Asymmetry*

In each mechanism of membrane morphogenesis listed in Fig. 1, localization of specific lipids to different sides of the host membrane, membrane lipid asymmetry, could dramatically affect the distribution of lipid in the resulting virus. For example, if phosphatidylglycerol were enriched on the inner face of the host cytoplasmic membrane, then by Fig. 1C, invagination transposes the inner face of the cytoplasmic membrane to be the outer face of the viral membrane, which consequently would be enriched in phosphatidylglycerol. The chemical labeling approach to the determination of membrane asymmetry is based on the accessibility of the outer lamella to reaction with a membrane-impermeable reagent. As a positive control, other conditions are sought in which both the inner and the outer lamellae of the bilayer are fully modified. Intact bacteriophage PM2 is treated with 0.75 M LiCl in order to make the outer protein shell of PM2 permeable to [^{35}S]sulfanylic acid diazonium salt (Schafer *et al.*, 1974). Under these conditions, the amino groups of sp43 and sp27 react with the diazonium salt, whereas those of sp13 and sp6.6 do not. Failure to label these latter proteins suggests that under these conditions the reagent does not reach the inner lamella of the membrane. All four of these structural proteins are labeled at higher concentrations of LiCl or in 0.05% Triton X-100 nonionic detergent. At 0.75 M LiCl, the majority of the phosphatidylglycerol is converted by the diazonium salt to an aldehyde, which is identified by reduction with tritiated borohydride. Under these conditions, only a small percentage of the phosphatidylethanolamine is derivatized. In the presence of 0.05% Triton X-100, all the rest of the phosphatidylethanolamine and a little more phosphatidylglycerol is modified. These results suggest an assymmetric lipid distribution in the virus with a majority of phosphatidylglycerol in the outer lamella. Quantitation of the results are not provided. However, from the data it can be estimated that 60% of the phosphatidylglycerol was modified. Without the kinetics of the reaction, it is uncertain that the reaction went to completion. In addition, it seems unlikely that conditions were found in which the reaction absolutely modified only one-half of the bilayer and none of the reactant penetrated to the other side. Furthermore, if only 60% of the phosphatidylglycerol is in the outer lamella, then the rest of that lamella must be occupied by phosphatidylethanolamine. Yet, under the appropriate conditions, less than 15% of the phosphatidylethanolamine reacted as if present in the outer lamella. Despite the failure of the bookkeeping, the qualitative result seems valid: that the majority of the phosphatidylglycerol occupies the outer lamella of the viral membrane.

The asymmetry of phosphatidylethanolamine in the host membrane can also be determined by reaction with 2,4,6-trinitrobenzene sulfonic acid (TNBS) (Rothman and Kennedy, 1977a). By monitoring the kinetics of reaction of TNBS with PM2, Brewer and Goto (1980) find two kinetic components corresponding to labeling of the outer lamella and transport of the TNBS across the membrane to label lipid in the inner labella. From these data, it can be concluded that 20–25% of the membrane lipid is phosphatidylethanolamine on the inside of the virus (Table IV). Taking into account the lipid composition of the virus and the greater amount of lipid in the outer lamella than the inner lamella due to its small radius, 67–75% of the phosphatidylglycerol is inferred to be in the outer lamella of the virus. These data confrim the asymmetry of phosphatidylglycerol in the

TABLE IV

DISTRIBUTION OF PHOSPHOLIPID BY TYPE AND SIDE OF MEMBRANE FOR PM2 AND A. *espejiana*[a]

	Total (%)	Inside (%)	Outside (%)
A. Membrane of PM2			
PE	37	20–25(53–67)	17–12(47–33)
PG	63	21–16(33–25)	42–47(67–75)
Total	100	41	59
B. Bacterial cytoplasmic membrane			
PE	76	38(50)	38(50)
PG	24	12(50)	12(50)
Total	100	50	50

[a] Based on extrapolation to zero time of kinetic component of reaction of TNBS with phosphatidylethanolamine (PE) in the inner lamella of the membrane. Listed as percentage of total lipids (PE + PG) and (percentage of each lipid class) (Brewer and Goto, 1980). Data for the bacterial membranes are ± 10% of the values.

virus, predominantly on the outside, as reported by Schafer *et al.* (1974). However, the asymmetry of this lipid is only 8–16% different from the surface area asymmetry between the inner and outer lamellae of the virus.

The possible contribution of a host membrane asymmetry to that of the virus could provide information on the mode of morphogenesis of the viral membrane (Fig. 1). In order to perform labeling studies on the cytoplasmic membrane of the host, direct access to that membrane is achieved by quantitatively removing the outer membrane of the host by a series of washes with salt and sucrose (Diedrich and Cota-Robles, 1974). The biphasic kinetics of reaction of TNBS with phosphatidylethanolamine correspond to labeling exclusively this lipid in the outer lamella followed by penetration of reagent and labeling in the inner lamella (Brewer and Goto, 1980). The outer lamella of the cytoplasmic membrane of the bacterium contains 50 ± 5% of the phosphatidylethanolamine in the membrane (Table IV). Thus, the distributions of phosphatidylethanolamine and phosphatidylglycerol in the membrane of the host are not asymmetric; the lipids are equally distributed between the inner and outer lamellae. Again, it appears that a simple passive mechanism of viral membrane morphogenesis is not supported. Preferential segregation of phosphatidylglycerol from the host membrane by a viral protein ramains an attractive mechanism for enriching this lipid in the outer lamella of the virion (see model, Section IV,D).

At first sight, the reconstitution of bacteriophage PM2 might be taken as evidence that viral assembly can occur in the absence of another membrane such as that of the host. However, the specific activity of infectious particles of 1 in 10^7, the lack of access in the cell to varying concentrations of urea, the excess quantity of lipid in the reconstituted particles, the insertion of viral proteins into the host membrane, and a function for the two-thirds of the viral genome that codes for nonstructural viral proteins all suggest a role for the host membrane in control of membrane morphogenesis of PM2.

C. PACKAGING THE GENOME

The relationship of the assembly of a new viral membrane to that of a nucleocapsid core is a critical parameter in deciding which of the mechanisms of membrane morphogenesis apply to bacteriophage PM2 (Fig. 1).

1. *Is the Core Made before the Membrane of PM2?*

If maturation of PM2 occurs by an assembly-line process, then by one mechanism, the assembly of a core of DNA and protein is needed before it can be encapsulated by a membrane. Evidence for such a nucleocapsid was described earlier (Section IV,A). The fact that a nucleocapsid core can be isolated from a virus by stepwise dissociation of the structure with urea suggests that a core could be important for assembly of the virus. Additional evidence for a nucleocapsid core comes from the reconstitution of such a structure from DNA and two other viral structural proteins. Is this the process that occurs *in vivo*? If nucleocapsids are performed before being enveloped, then mutants of the virus should be found that make nucleocapsids but fail to coat them with a membrane. Although mutants have been isolated representing 12 phenotypic and complementation groups, when examined in the electron microscope, none of these show extensive production of nucleocapsid cores (Brewer, 1978a). Under restrictive conditions, one mutant, *ts27*, produces a few electron-dense structures that could be nucleocapsid cores. But, even these cores have structures resembling a membrane envelope that is difficult to see because of the poor contrast adjacent to the electron-dense core. However, these are likely to be aberrant products since temperature shift studies show *ts27* to be an early mutant. Shift from permissive to restrictive conditions early in infection results in substantial virus production. One would expect this shift to inhibit encapsulation that occurs late in infection.

2. *Is the Membrane Made before Packaging the DNA?*

Evidence for a preformed viral membrane shell is compelling despite the topological problem of DNA packaging. Early studies by Dahlberg and Franklin (1970) document the appearance of virus-sized, empty-appearing membrane vesicles inside cells infected with PM2. As a function of time after infection, the number of these vesicles reaches a peak at 30 minutes and thereafter begins to decline. In a reciprocal fashion, the number of mature virions per infected cell rises. These observations support the concept of a conversion of the vesicles to mature virions.

Further support for this pathway comes from the extensive characterization of mutant *ts1* of PM2 (Brewer, 1979). Under restrictive conditions inside infected cells, this mutant produces empty membrane vesicles (Fig. 7). These vesicles can be purified from infected cells and shown to be the proper size, shape, and lipid composition as would be expected for the membrane of PM2. The viral DNA is associated with the particle but is not inside since it is susceptible to DNase. The DNA is in a linear rather than the superhelical circular structure found in wild-type virus (Brewer, unpublished results). There is also a single protein in these vesicles, sp6.6. The possible role of this protein in stimulating membrane morphogenesis is discussed in the model in Section IV,D. Shifting cells infected with *ts1* from restrictive to permissive temperature results in the rapid maturation of infectious virus (lag time only 2 minutes). This rapid maturation suggests that the previously existing membranes are being utilized to package DNA and that they are true intermediates in the natural assembly process. To further support this contention, infected cells were pulse-labeled with ^{32}P and shifted to permissive temperature. The specific activities of the membranes of isolated virions were the same regardless of whether a chase was imposed with the shift (Brewer, unpublished observations).

Knowledge of the primary defect in the *ts1* mutant is needed. The mutation could be in a gene required for packaging the viral DNA. Yet it remains possible that the DNA is packaged and that, through purification of the *ts1* vesicles, it is released because of a failure to stabilize the encapsulated genome. Another possibility is that the binding of capsid protein triggers the packaging of the DNA and that this binding is defective in *ts1* under restrictive conditions. In support of this possibility, the isoelectric point of sp27 from *ts1* is different from that of wild-type sp27 (Brewer, manuscript in preparation). Since sp6.6 is the only protein present in *ts1* vesicles, it is likely to play an important role in stimulating membrane morphogenesis (see model in Section IV,D).

D. A Model for PM2 Membrane Morphogenesis

1. *Forces That Maintain Membrane Bilayer Integrity*

In order to change the shape of a membrane, there must be a perturbation to the equilibrium that maintains the lipid bilayer in an essentially planar conformation. Before considering this initiation of membrane morphogenesis, the forces that maintain the integrity of the lipid bilayer should be considered. First, there are the lateral cohesive forces between individual phospholipids. These forces allow elasticity in the plane of the membrane. As evidenced by the enthalpy of melting of pure phospholipids, the primary contribution comes from the interaction of the fatty-acyl chains rather than the polar head groups. Secondly, there are transverse forces that maintain the two lamella of the membrane in apposition. In considering the thermodynamics of membrane integrity, the critical factor is the negative entropy of aqueous solution of individual phospholipids that results from the strong self-attraction of water (Tanford, 1979). This factor keeps water out of the bilayer and the lipids out of the aqueous phase. Osmotic pressure is another factor that keeps biological membranes from wrinkling or collapsing on themselves. The transmembrane osmotic pressure of a cell keeps the membrane taut. If the pressure is greater than the surface tension of the membrane, the membrane will rupture.

2. Steps in Morphogenesis

Figure 8 shows a model of PM2 membrane morphogenesis. The entire process, which probably takes only a few seconds, can be broken down into discrete steps. The first step is initiation of the morphogenic event in order to introduce a branch into an otherwise planar lipid bilayer. The second step is propagation of

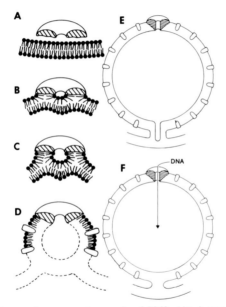

Fig. 8. A model for membrane morphogenesis of PM2. (A) A toroidal amphipathic protein approaches the bilayer membrane from the inside of the cell. (B) Initiation of membrane morphogenesis begins with the interaction of the hydrophobic face of the protein with the acyl chains of the lipids. (C) The distortion in the bilayer and lateral pressure in the membrane contribute to propagation of morphogenesis. (D) Further propagation is stimulated by the insertion of proteins into the newly forming membrane. (E) Limitation of size is achieved by the defined radius imposed by each inserted protein. (F) DNA packaging occurs through the hole in the morphogenic protein and the new membrane pinches off from the host membrane. (Model is drawn to scale with membrane thickness equal to 40 Å; E and F are reduced 41%.)

this branch into a larger structure. Once the increase in size has begun, there must be other factors that control the shape and limit the size. The final step of membrane morphogenesis is termination or pinching off of the new membrane from the cytoplasmic membrane of the host.

a. *Initiation.*   Initiation of membrane morphogenesis is likely to begin with the synthesis of a morphogenic protein. The stimulus for morphogenesis is not likely to be a polysaccharide or a polynucleotide since the penetration of the membrane by either would be entropically unfavorable. Although it is possible to imagine an anionic lipid, for example, that is confined to one half of the bilayer causing a gentle bulge in the membrane due to charge–charge repulsion in the polar head group region, the localization of this effect to one portion of the cell is not likely to be controlled against diffusion. The design of a protein that will accomplish perturbation of the bilayer structure is shown in Fig. 8A. The critical characteristic of this morphogenic protein is that it is amphipathic. One face is hydrophobic and the rest is hydrophilic. The hydrophobic face allows it to partition into the membrane fatty-acyl region. The hydrophilic face keeps a portion exposed to the cytoplasm. The conformation of the protein is depicted as a torus. The utility of a 2-nm aqueous hole will be evident at the end. The width of the hydrophobic face is another critical factor. It is designed to be the width of the bilayer, 4 nm. The hydrophobic face, with a strong affinity for the acyl chains of the phospholipids, causes a disruption of the transverse lipid–lipid interactions. The outer lamella accommodates the distortion by dimpling. Trans-bilayer interactions of the fatty acyl chains are weakened in the region of insertion. These distortions lead to propagation of membrane morphogenesis.

Since it is unlikely that a single polypeptide can fold into this conformation, the aggregation of several subunits is suggested. From these dimensions of one-half of a torus, the volume and hence mass of this protein can be estimated. The volume equals $\pi^2 R r^2$, where r is the radius of the edible portion of the donut and R is the radius from the center of the hole to the center of the edible part (Beyer, 1978). This results in a protein volume of 118 nm^3, which at 1.4 gm/cm^3 is equivalent to 100,000 daltons. If this morphogenic protein were a hexamer, the molecular weight of each subunit would be 16,600. Present at only six copies per virion, this protein would be difficult to distinguish from contaminants in SDS–polyacrylamide gel electrophoresis.

b. *Propagation.*   Steps B and C of Fig. 8 depict the propagation of the morphogenesis on the interior side of the host membrane. Some of the strain in lipid conformation could be relieved by a slight conformational change in the morphogenic protein. This change would also aid in the initial insertion. Since this model depicts the morphogenesis of a new membrane derived from the inner lamella of the host membrane, the morphogenic protein must have no affinity for the outer lamella of the bilayer; it must not penetrate too deeply. Such an interaction would prevent propagation of morphogenesis. It is precisely the disruption of the transverse cohesive interactions of the lipids that allows the new membrane to form.

Why don't all integral membrane proteins stimulate membrane morphogenesis? Many integral membrane proteins examined to date span the bilayer and therefore cannot cause one lamella to separate from the other. The other factor is likely to be the orientation of the lipid–protein interface relative to the plane of the bilayer. If this interface is perpendicular to the plane of the bilayer, then the protein is essentially substituting directly for the lipids and there is little change in the forces bearing on the planar structure. Analysis of the forces prevailing on the region where morphogenesis has been initiated will lead to an understanding of propagation of that morphogenesis. If the lipid–protein interface approaches being parallel to the plane of the bilayer as is shown in Fig. 8B, a region of distortion of the fatty-acyl chains is created. This disordering creates a localized pressure drop that can be partially relieved by a flow of lipids into the region of distortion.

Another force contributing to this lipid flow is the lateral pressure of the lipids in the plane of the membrane due to their active biosynthesis. As newly synthesized lipids are inserted, they tend to expand the inner lamella of the membrane. Lipid–protein interaction may be another factor that contributes to a large internal pressure in the membrane (Conrad and Singer, 1979). The osmotic pressure inside the cell tends to keep the membrane from buckling. Due to the hole in the morphogenic protein, the interior of the newly forming membrane is at the same osmotic pressure as the rest of the cytoplasm. Therefore the lipids move into the region of lower pressure in the membrane and the branch is extended.

The extension of the branch creates a topological problem of supplying lipids to the interior of the newly forming membrane. This is the problem of flip–flop of lipids from one lamella to the other. Rothman and Kennedy (1977b) have measured the half-time for phospholipid flip–flop in the whole membrane of *Bacillus megaterium* and found it to be on the order of 4 minutes. Since the volume of lipids in the mature virus is less than 1/1000 that of the cytoplasmic membrane of the host, the equilibration of sufficient lipids to the inside of the maturing virus membrane could take less than several milliseconds. Although, in the whole cytoplasmic membrane of *A. espejiana,* flip–flop appears to totally equilibrate the phosphatidylethanolamine and phosphatidylglycerol across the membrane, this may not be true for a specialized region of membrane morphogenesis. The zwitterion, phosphatidylethanolamine, would be expected to equilibrate more rapidly than the anion, phosphatidylglycerol. Since the interior of this new membrane is entirely the result of lipid flip–flop, these differences in rate would cause phosphatidylglycerol to accumulate on the outside of the new membrane as found in the mature virus.

On the other hand, if conditions are assumed that permit equilibrium of lipids between the lamellae, then minimum surface free energy occurs with a preferential enrichment of charged lipids in the outer lamella of a curved bilayer (Israelachvili, 1973). The electrostatic free energy in an environment of 0.5 M salt (Debye length 3 Å) is minimized with 60% of the phosphatidylglycerol on the outside of a membrane of radius 240 Å. Considering that 59% of surface area is on the outside, this is not significantly different from that of uniform surface charge density. However, the proportion of phosphatidylglycerol that is found in the outer lamella of PM2, 67% (75%) (Table IV), could arise from a membrane whose composition was determined by equilibrium while at a radius of 145 Å (100 Å). Subsequent addition of protein or elevation of the intracellular Ca^{2+} concentration (Snipes *et al.*, 1974) could block further equilibration as enlargement proceeded.

c. *Shape and Size Limitations.* The predominant factor contributing to membrane shape is the surface free energy. Therefore the new membrane will take the form of a sphere in order to minimize surface free energy (Fig. 8D), other factors not interfering. For example, a long narrow tube is energetically unfavorable because the radius of curvature of the cross section would be very small, whereas that of the longitudinal section would be infinite. Surface tension forces would cause the shape to change until the radii in each plane were equal.

Two factors contribute to the continued growth of the new membrane. The first is the aforementioned lateral pressure in the membrane caused by the continuing synthesis and insertion of new lipid. However, the crucial element of further growth is the proposed insertion of a protein into the new membrane (Fig. 8D and E). This protein would preferentially insert in regions of high fluidity. The partitioning of the majority of its mass in the outer lamella relative to the inner lamella would relieve some of the strain of the small radius of curvature. There is already evidence that some proteins preferentially associate with fluid regions of a membrane (Letellier *et al.*, 1977). If the protein is thought of as a wedge, the angle of the wedge and the number of copies inserted would influence the final radius. Each associated protein would impose the final radius of curvature on its immediate lipid environment. Based on the evidence presented earlier, the viral structural proteins sp6.6 and sp13 are strong candidates for this role. Their higher solubility in more fluid regions of the membrane (e.g., those of small radius) would maintain their localization in this new membrane, opposing diffusion into the rest of the bilayer. Alternatively, the total membrane concentration of these proteins may be high and their localization in regions of morphogenesis would be due to the exclusion of host proteins. The inserted protein may also have an affinity for phosphatidylglycerol that would account for its increased proportion in the viral membrane versus the host membrane. This affinity would be sufficient to prevent the back diffusion and randomization of phosphatidylglycerol out of the region of membrane morphogenesis into the membrane of the host.

Two factors would tend to limit the size of the new membrane. First is the rate of flip–flop, since the outer lamella cannot grow much faster than its partner, the inner lamella. Secondly, when enough copies of the membrane structural protein have been inserted, the cumulative curvature of each unit combines to impose the defined total radius and the structure closes on itself. In this way, numerous copies of new membranes can be made, all with the desired size and shape. The required characteristics are engineered into the membrane structural protein.

d. *Termination.*   At this point (Fig. 8E and F), the nexus constricts and the new membrane is forced to detach from the host membrane. Several other events may occur at this latter stage or just prior to detachment of the new viral membrane. One of these is packaging of the viral DNA. This occurs through the 20 Å hole in the morphogenic cap protein. It could be triggered by a conformational change in this protein as a result of reaching the terminal diameter. Hendrix (1978) has proposed a model for DNA packaging in other phages that may apply to PM2 as well. Meanwhile the buildup of sp27 and sp43 in the cytoplasm adjoining the site of membrane morphogenesis would have reached a critical concentration whereupon these proteins would aggregate on the surface of the new membrane to form the icosahedral shell of the virus with spikes at the vertices.

E. Summary and Prospects

Bacteriophage PM2 matures at the inner face of the cytoplasmic membrane of its host, *A. espejiana*. Morphogenesis does not appear to proceed by evagination by encapsulation of a core, nor by invagination by formation inside a shell. Candidate viral proteins for control of membrane morphogenesis have been identified. Initiation of membrane morphogenesis appears to require formation of a branch in an otherwise planar bilayer. The asymmetric localization of phosphatidylglycerol in the outer lamella of the viral membrane is largely explained by the larger volume of that lamella relative to the inner lamella. The specific enrichment of this lipid in the viral membrane is not yet explained. Enrichment is likely to require extraction of lipid from the host membrane by a protein with specific affinity for phosphatidylglycerol. Size and shape limitation may be controlled by the insertion into the viral membrane of viral structural proteins. The proteins responsible for these processes have not been identified with certainty. The possible insertion of sp6.6 into the viral membrane needs morphological or physicochemical demonstration. The complete amino acid sequences of sp13 and sp6.6 should provide additional insight into their possible interaction with the membrane. Further temperature-shift experiments with mutants of the virus should prove preassembly of the viral membrane before packaging the DNA. Establishment of a suppressor system in the host of PM2 is needed in order to isolate amber mutants of the virus. Analysis of viral mutants is likely to identify which viral gene products are required for specific steps of membrane assembly. The *in situ* demonstration of membrane morphogenesis from host membranes and appropriate cell extracts would dramatically delineate the specific components required for control of membrane morphogenesis in bacteriophage PM2.

V. Bacteriophage ϕ6

A. CHARACTERISTICS OF THE MEMBRANE

Bacteriophage ϕ6 was originally isolated as a possible lipid-containing phage of the plant pathogen *Pseudomonas phaseolicola* HB10Y by Vidaver *et al.* (1973). The buoyant density in CsCl (Table I) and lipid composition (25% by weight) suggested that a second membrane-containing bacteriophage had been discovered. The infectivity of the phage was extremely sensitive to detergents and organic solvents. In a fashion similar to PM2, the lipid composition of the virus was dramatically enriched in phosphatidylglycerol relative to the host lipid composition (Table I). Analysis of the fatty acids of those lipids indicated essentially no difference from those of the host. Progress on the structural proteins of the virus proceeded rapidly with the ingenious construction of a suppressor-bearing host (Mindich *et al.*, 1976a) that could be used in the isolation of suppressible amber mutants of the phage (Sinclair *et al.*, 1976).

1. *Evidence for a Membrane*

The presence of a membrane in ϕ6 was supported in the beginning by the observation in the electron microscope of a pleiomorphic envelope at the exterior of the virus (Vidaver *et al.*, 1973). The bilayer nature of this membrane and its surface exposure is demonstrated by the digestion of the lipid fatty acids at the 2 position with phospholipase A, resulting in a loss of infectivity. The membrane envelope was further described by the electron microscopy of Ellis and Schlegel (1974). At higher resolution in negative stain and in section, the exterior localization of the lipid bilayer is clearly observed (Bamford *et al.*, 1976). More recent lipid analysis based on incorporation of ^{32}P into lipid relative to RNA indicates that there is sufficient lipid to account for only one-half of the bilayer; the lipid composition is 20% by weight (Day and Mindich, 1980).

Electron spin resonance spectra of virus mixed with a spin-labeled hydrocarbon gave evidence for a membrane bilayer in ϕ6 (Sands *et al.*, 1974). As with other enveloped animal viruses (Landsberger *et al.*, 1973; Sefton and Gaffney, 1974), evidence is found for stiffening of the viral membrane relative to that of the host or to that of extracted lipids due most likely to protein–lipid interactions in the viral membrane. Of unclear significance is the relationship of a break around 30°C in the slope of the Arrhenius plot for the spin-label motion parameter and the failure of virus maturation at this temperature at a specific time late in infection. In contrast to the case with PM2 (Cupp *et al.*, 1975), assembly of the membrane of φ6 is not affected by agents such as adamantanone or elevated Ca^{2+} that alter the fluidity of the host membrane (Sands *et al.*, 1975).

2. *Relationship of the Membrane to Other Viral Components*

The proteins of φ6 are enumerated in Table V. Their assignment to the three segments of the double-stranded RNA genome and their structural assignments within the virus have been based on the elegant analysis of the amber mutants of the phage (Sinclair *et al.*, 1975). The relationship of the membrane of φ6 to the viral proteins is depicted in Fig. 9. This model is based on the analysis of intermediates that accumulate in cells infected with amber mutants in each of the viral proteins (Mindich *et al.*, 1976c; Lehman and Mindich, 1979). A procapsid can be formed *in vivo* from P1, P2, P4, and P7 in cells infected with an amber mutant in P8. An unenveloped nucleocapsid can also be formed that contains the previous proteins plus viral RNA and P8 (Mindich and Davidoff-Abelson, 1980). P8 is on the surface of the nucleocapsid as judged by its ability to be iodinated with lactoperoxidase (Van Etten *et al.*, 1976). These structures do not appear to be directly related to the membrane of φ6. However, defects in any of these proteins prevent the assembly of the membrane of φ6. Membrane formation requires the viral nonstructural protein P12, lipids, and the viral structural proteins P6, P9, and P10. The P5/P11 proteins, components of the cell wall lysin (Mindich and Lehman, 1979; Iba *et al.*, 1979), are not required for membrane assembly. However, they are components of the natural viral membrane. The final maturation of virions is completed by the addition of protein P3. These proteins are involved in the processes of viral attachment to the host and penetration. P6 appears to serve as an anchor in the membrane for P3 as judged by variations in the amounts of these proteins depending on the temperature at which maturation occurs (Mindich *et al.*, 1979). Surface iodination with lactoperoxidase indicates the exterior localization of P3 (Van Etten *et al.*, 1976).

Aside from genetic analysis, evidence for the relationship of viral proteins to the envelope of φ6 has come from the differential extraction of the virus with 2% Triton X-100 (Table V; Sinclair *et al.*, 1975; Van Etten *et al.*, 1976). Extraction with detergent releases an intact nucleocapsid while destroying the membrane and releasing P3, P6, P9, and P10. The possible interaction of P5 with the surface phosphate groups of the membrane lipids is suggested by its release when the detergent is buffered with phosphate and its failure to be released when the detergent extraction is buffered with Tris (Mindich and Lehman, 1979).

B. RELATIONSHIP OF THE VIRAL MEMBRANE TO THE HOST MEMBRANE

The relationship of the membrane of φ6 to that of its host is unclear at this time. Certainly the lipids of the virus must be derived from the lipids present in the host cytoplasmic membrane since that is their site of synthesis. However, the

TABLE V
THE PROTEINS OF BACTERIOPHAGE $\phi6$

Name	MW $\times$ 10^{-3a}	Linkage set[b]	Copies/ virion[c]	Regulation[a]	Extraction by detergent[a,d]	Other characteristics
Structural proteins[a]						
P1	93	A	110	Early	Core	RNA polymerase component (Sinclair and Mindich, 1976)
P2	88	A	15	Early	Core	RNA polymerase component (Sinclair and Mindich, 1976)
P3	84	B	73		+	Needed for adsorption (Mindich et al., 1976c), surface iodinatable[d]
P4	37	A?	111	Early	Core	
P5	24	C	89		+	Derived from P11 (Sinclair et al., 1976), component of lysin (Mindich and Lehman, 1979)
P6	21	B	120		±	Anchor for P3 (Mindich et al., 1979)
P7	20	A[e]	92	Early	Core	
P8	10.5	C	1400		Core	Only late protein in core, exposed on the surface of core[d]
P9	8.7	C	2000		+	Needed for membrane assembly (Mindich et al., 1976c)
P10	<6	B[e]	>1200		+	
Nonstructural proteins						
P11	25	C	variable			Precursor for P5 (Sinclair et al., 1976)
P12	20	C				Needed for membrane assembly (Mindich et al., 1976c)

[a] Sinclair et al. (1975).
[b] Mindich et al. (1976b).
[c] Day and Mindich (1980).
[d] Van Etten et al. (1976).
[e] Lehman and Mindich (1979).

accumulation of mature virions in the center of the cell, removed from the proximity of the cytoplasmic membrane (Fig. 10), does not preclude the envelopment of the nucleocapsid at the cytoplasmic membrane.

In contrast to the absence of an effect on cellular lipid synthesis after infection with PM2, cellular lipid synthesis after infection with $\phi6$ is markedly altered (Sands and Lowlicht, 1976). There is a 2-fold increase in synthesis of phos-

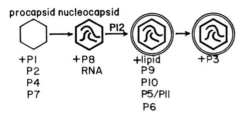

procapsid nucleocapsid

FIG. 9. Assembly of bacteriophage $\phi6$. A procapsid is first formed from P1, P2, P4, and P7. Next, viral RNA and P8 are added to form the nucleocapsid. P12 is needed to promote the envelopment of the nucleocapsid, which includes the addition of P6, P9, P10, and P5/P11. Upon addition of P3, the particle becomes infectious. (Based on Mindich *et al.*, 1976c, and Mindich and Davidoff-Abelson, 1980.)

phatidylglycerol that begins halfway through the $\phi6$ cycle of infection. Pulse–chase studies with ^{32}P indicate that the viral lipids are derived primarily from the lipids that existed in the cell prior to infection. Surprisingly, of the remainder that are derived from those made during infection, 80% are newly synthesized phosphatidylglycerol. This suggests the interesting prospects of (1) viral control of host lipid synthesis and (2) selection of newly synthesized phosphatidylglycerol for the virus from the host membrane. Sands and Lowlicht (1976) propose that the latter may occur by viral protein–lipid interaction pulling lipids out of the host membrane from regions of synthesis of phosphatidylglycerol. A hybrid of models C and D of Fig. 1 is envisioned with portions of the host membrane invaginating to form micelles that migrate to and coat a nucleocapsid in the interior of the cell.

An important role for the nonstructural viral protein P12 of $\phi6$ in stimulating membrane morphogenesis is suggested by the failure of mutants in P12 to assemble the viral membrane (Mindich *et al.*, 1976c). It has been suggested that P12 may be a lipid exchange protein, shuttling lipids from the host membrane to the nucleocapsid (Wirtz, 1974; Fig. 1D). However, no evidence has been obtained to support this idea. Shift experiments with temperature-sensitive mutants of the envelope proteins are needed to delineate control of membrane morphogenesis in $\phi6$. In this regard, it is interesting that a temperature-sensitive mutant of P12 at restrictive conditions traps nucleocapsids at the cytoplasmic membrane (L. Mindich, personal communication). Also, a high percentage of *P. phaseolicola* that are resistant to $\phi6$ exist in a carrier state that continues to produce virus (Cuppels *et al.*, 1979). Morphogenesis of virus in such a mutant, which was isolated independently (D. H. Bamford, personal communication), shows nucleocapsids and virions in contact with the cytoplasmic membrane (Fig. 10). Thus, an adaptation of Fig. 1D might describe membrane morphogenesis in $\phi6$ in which the lipid carrier protein is replaced by the proximity of the nucleocapsid to the cytoplasmic membrane of the host. Lipid–protein interactions might then cause extraction of lipid from the host membrane onto the nucleocapsid.

It will also be interesting to determine whether these potential membrane

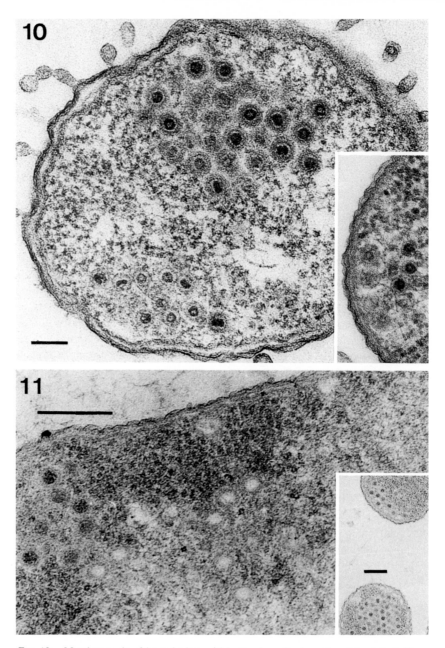

FIG. 10. Morphogenesis of bacteriophage φ6 in *P. phaseolicola* carrier-state mutant. The accumulation of mature virions in the center of the cell is typical of φ6 maturation in a wild-type host. Unique to the mutant is the maturation of putative virion precursors at the inner face of the cytoplasmic membrane (inset). Bar represents 100 nm. Courtesy of Dennis Bamford (unpublished).

FIG. 11. Morphogenesis of bacteriophage PR4 in *E. coli*. Note the appearance of empty mem-

proteins are made as precursors with leader sequences that target their insertion into the host cytoplasmic membrane. An *in vitro* coupled transcription-translation system primed with $\phi6$ RNA synthesizes P6, P9, and P12 at their usual sizes (L. Mindich, personal communication). In this system, P5 is made the same size as P11, suggesting further that P5 is derived from P11. The mobility in SDS electrophoresis for P10 made *in vitro* is slower than that of P10 in the mature virus. This raises the possibility that this protein could be made with a leader sequence (Blobel and Dobberstein, 1975; Inouye *et al.*, 1977). However, other charged amino acid additions or modifications must be ruled out before this conclusion is justified.

C. Packaging the Genome

In certain mutants of $\phi6$ described above, the production of procapsids and nucleocapsids that lack an envelope suggests that the genome is packaged inside a core before the envelope goes on. Pulse–chase studies or shift studies with temperature-sensitive mutants of $\phi6$ are needed for proof. If true, the size and shape of the envelope would be determined by the size and shape of the core.

D. Summary and Prospects

Morphogenesis of the membrane of $\phi6$ is controlled by three viral proteins. One of these is a nonstructural protein of the virus and must therefore serve a catalytic function. The nature of this function and the entire mechanism of membrane morphogenesis of $\phi6$ is not known. However, mechanisms of invagination or evagination of the host membrane appear to be excluded by the acquisition of an envelope only after the assembly of a nucleocapsid core and by the maturation of the virus inside the cell. Prospects for understanding membrane morphogenesis in $\phi6$ are likely to come from integrating genetics with electron microscopy. Further revelation will surely come from analysis of temperature-sensitive mutants of $\phi6$ and *in situ* envelopment of nucleocapsids by isolated membranes together with judicious cell extracts.

VI. Bacteriophages PR3, PR4, PR5, and PRD1

Bacteriophages PR3 and PR4 were originally isolated from sewage by their ability to infect *Pseudomonas aeruginosa* bearing the drug resistance plasmid RP1; they are unable to infect the bacterium without the plasmid (Stanisich,

---

brane vesicles as well as mature-appearing virions and that maturation appears to be localized to the poles of the cell (inset). Bar represents 200 nm. (From Lundstrom *et al.*, 1979.)

1974). PRD1 (Olsen *et al.*, 1974) and PR5 (Wong and Bryan, 1978) were similarly isolated. The phages are infectious in hosts that bear a plasmid that belongs to the P, N, or W compatibility groups (Bradley and Rutherford, 1975) and that probably codes for the phage receptor. These phages are morphologically indistinguishable, with an icosahedral head of diameter 65 nm and a tail 60 nm long. They are serologically related but not identical (Wong and Bryan, 1978; Bradley and Rutherford, 1975). Other properties are listed in Table I. Unfortunately, little work has been directed toward membrane morphogenesis in these phages. Nevertheless, the groundwork necessary for such investigations will be reviewed. PR4 is emerging as the prototype of this group.

PR4 is 12% lipid by weight (Sands, 1976). There is some variation in the lipid composition, dependent on the host in which the virus is grown (Wong and Bryan, 1978). Sands and Cadden (1975) report an unidentified lipid that may be lyso-phosphatidylglycerol and large amounts of a lipid that cochromatographs with phosphatidylserine in one solvent system. These could be artifacts due to autooxidation during the heating and extended time used to extract the lipids. The generalization emerges that the virus is greatly enriched in phosphatidylglycerol relative to the host. Evidence for a lipid bilayer in the virion comes from electron microscopy of PR3 and PR4 (Bradley and Rutherford, 1975; Lundstrom *et al.*, 1979) and PR5 (Wong and Bryan, 1978). The electron-translucent center of the bilayer is located between the core and the capsid, as is the case with PM2. The subsurface localization of the membrane is supproted by its resistance to hydrolysis with phospholipase A (Bradley and Rutherford, 1975). The six (Cadden and Sands, 1977) to fifteen (Lundstrom *et al.*, 1979) proteins of the virion range in molecular weight up to 68,000. The most abundant structural protein has a molecular weight of 39,000 (Lundstrom *et al.*, 1979); it may be the capsid protein. Highly purified preparations, which are more likely to be free of minor contaminants from the host, show seven or eight proteins (J. Cronan, personal communication). There have been no reports to indicate which of these proteins might interact with the membrane. Nonstructural viral proteins have not been identified due to the continued synthesis of host proteins during viral maturation (Cadden and Sands, 1977).

Maturation of PR4 involves the production of intracellular empty membrane vesicles (Fig. 11; Lundstrom *et al.*, 1979). The observation of a putative viral membrane contacting the cytoplasmic membrane raises the possibility of formation from the host membrane similar to the proposed mechanism for PM2. Mature virions begin to appear at the poles of the cell 45 minutes after infection. They do not contact the cytoplasmic membrane. The DNA could be packaged in the nuclear region of the cell followed by encapsidation and migration to the poles of the cell. As isolated, these empty membranes contain all the viral structural proteins found in infectious virions. However, they could be defective in a minor component or they could be breakdown products of mature phage.

Pulse–chase studies are needed to resolve the relationship of the empty membranes to the mature virions.

As with PM2, the reason is unclear for the enrichment of phosphatidylglycerol and possibly phosphatidylserine. It is not due to changes in lipid synthesis (Sands, 1976). One-third of the viral lipids are acquired from those that existed in the host before infection. Host mutants that are blocked in lipid synthesis allow the continued assembly of PR4 (Sands and Auperin, 1977). These observations suggest that lipids for the virion are acquired from the host membrane and are not directly pirated from the lipid biosynthetic machinery. In contrast to $\phi6$ and PM2, PR4 assembly is cold sensitive (Sands and Auperin, 1977). Total patching or phase separation of phosphatidylglycerol in the membrane at low temperatures may block virus assembly. Work with temperature-sensitive and amber mutants of the virus is needed to begin to understand morphogenesis of the membrane of PR4.

VII. Conclusion

The advantage of bacteriophages in understanding control of membrane morphogenesis is beginning to be realized. With their genomes of relatively small size, the search for a morphological agent is restricted to roughly a dozen gene products. Candidate proteins have been identified for PM2 and $\phi6$. Understanding the morphogenesis of the membranes of bacteriophages has begun with knowledge of their structure.

The membranes of PM2 and the PR4 group of bacteriophages are morphologically similar. The lipid bilayer is sandwiched between the DNA and an external coat of protein, the capsid. Evidence suggests that the membrane is assembled before the DNA is packaged and before the capsid is formed. This fact requires that the size, shape, and lipid composition of the membrane be determined by the physicochemical properties of the membrane components. The membrane of PM2 appears to be derived directly from the cytoplasmic membrane of the host. A model of membrane morphogenesis has been considered in which a differentiation of the host membrane is stimulated by a morphogenic protein.

The morphogenesis of $\phi6$ is very different from that of PM2. In $\phi6$, the genetic–morphological evidence clearly indicates preassembly of the nucleocapsid followed by envelopment. There is no capsid. Thus, control of membrane size and shape is determined by the nucleocapsid. The question of control of membrane morphogenesis is reduced to identifying the mechanism by which the lipids are mobilized onto the surface of the nucleocapsid. In a mutant cell, the observation of proximity of the $\phi6$ nucleocapsid to the host cytoplasmic membrane simplifies a mechanism of lipid mobilization based on lipid transfer from the host to the viral membrane. The molecular details remain to be discovered. The membrane proteins of $\phi6$ need to be purified in order to determine their lipid-binding capabilities. Specific mutants in these proteins may produce aberrant membranes whose characterization will reveal their specific function.

In summary, great strides have been made toward understanding membrane morphogenesis in bacteriophage. Investigation in the next few years promises to identify the controlling elements and their specific function. An important question that remains unanswered is how these viruses become so dramatically enriched in the acidic phospholipid, phosphatidylglycerol. The field is ripe for understanding the molecular basis for control of membrane morphogenesis.

REFERENCES

Bamford, D. H., Palva, E. T., and Lounatmaa, K. (1976). *J. Gen. Virol.* **32,** 249.
Beyer, W. H., ed. (1978). CRC Standard Mathematical Tables, 25th edition, p.150. CRC Press, Cleveland, Ohio.
Blobel, G., and Dobberstein, B. (1975). *J. Cell Biol.* **67,** 835.
Bowman, B. U., Newman, H. A. I., Moritz, J. M., and Koehler, R. M. (1973). *Am. Rev. Respir. Dis.* **107,** 42.
Bradley, D. E., and Rutherford, E. L. (1975). *Can. J. Microbiol.* **21,** 152.
Braunstein, S. N., and Franklin, R. M. (1971). *Virology* **43,** 685.
Brewer, G. J. (1976). *J. Supramol. Struct.* **5,** 73.
Brewer, G. J. (1978a). *Molec. Gen. Genet.* **167,** 65.
Brewer, G. J. (1978b). *J. Virol.* **30,** 875.
Brewer, G. J. (1978c). *Virology* **84,** 242.
Brewer, G. J. (1979). *J. Virol.* **30,** 875.
Brewer, G. J. (1980). *J. Virol.* Submitted.
Brewer, G. J., and Goto, R. M. (1980). *J. Biol. Chem.* Submitted.
Brewer, G. J., and Singer, S. J. (1974). *Biochemistry* **13,** 3580.
Cadden, S. P., and Sands, J. A. (1977). *Can. J. Microbiol.* **23,** 1084.
Camerini-Otero, R. D., and Franklin, R. M. (1972). *Virology* **49,** 385.
Camerini-Otero, R. D., and Franklin, R. M. (1975). *Eur. J. Biochem.* **53,** 343.
Caspar, D. L. D., and Klug, A. (1962). *Cold Spring Harbor Symp. Quant. Biol.* **27,** 1.
Cavalieri, S. J., Neet, K. E., and Goldthwait, D. A. (1976). *J. Molec. Biol.* **102,** 697.
Cihlar, R. L., Lessie, T. G., and Holt, S. C. (1978). *Can. J. Microbiol.* **24,** 1404.
Conrad, M. J., and Singer, S. J. (1979). *Proc. Natl. Acad. Sci., U.S.A.* **76,** 5202.
Cota-Robles, E., Espejo, R. T., and Haywood, P. W. (1968). *J. Virol.* **2,** 56.
Cupp, J., Klymkowski, M., Sands, J., Keith, A., and Snipes, W. (1975). *Biochim. Biophys. Acta* **389,** 345.
Cuppels, D. A., Vidaver, A. K., and Van Etten, J. L. (1979). *J. Gen. Virol.* **44,** 493.
Dahlberg, J. E., and Franklin, R. M. (1970). *Virology* **42,** 1073.
Dales, S., and Mosbach, E. H. (1968). *Virology* **35,** 564.
Datta, A., and Franklin, R. M. (1972). *Nature (London) New Biol.* **236,** 131.
Datta, A., Braunstein, S., and Franklin, R. M. (1971a). *Virology* **43,** 696.
Datta, A., Camerini-Otero, R. D., Braunstein, S. N., and Franklin, R. M. (1971b). *Virology* **45,** 232.

Day, L. A., and Mindich, L. (1980). *Virology* **103**, 376.
Diedrich, D. L., and Cota-Robles, E. H. (1974). *J. Bacteriol.* **119**, 1006.
Diedrich, D. L., and Cota-Robles, E. H. (1976). *J. Virol.* **19**, 446.
Eiserling, F. A., Geiduschek, E. P., Epstein, R. H., and Metter, E. J.(1970). *J. Virol.* **6**, 865.
Ellis, L. F., and Schlegel, R. A. (1974). *J. Virol.* **14**, 1547.
Espejo, R. T., and Canelo, E. S. (1968). *Virology* **34**, 738.
Essani, K., and Dales, S. (1979). *Virology* **95**, 385.
Franklin, R. M. (1974). *Curr. Top. Microbiol. Immunol.* **68**, 108.
Franklin, R. M. (1978). *In* "Cell Surface Reviews" (G. Poste and G. L. Nicholson, eds.), Vol. 4, p. 803. Elsevier, Amsterdam.
Franklin, R. M., Salditt, M., and Silbert, J. A. (1969). *Virology* **38**, 627.
Franklin, R. M., Hinnen, R., Schaefer, R., and Tsukagoshi, N. (1976). *Philos. Trans. R. Soc. London B Biol. Sci.* **276**, 63.
Franklin, R. M., Marcoli, R., Satake, H., Schaefer, R., and Schneider, D. (1978). *Med. Microbiol. Immunol.* **164**, 87.
Garoff, H., and Soderland, H. (1978). *J. Molec. Biol.* **124**, 535.
Harrison, S. C., Caspar, D. L. D., Camerini-Otero, R. D., and Franklin, R. M. (1971). *Nature (London) New Biol.* **229**, 197.
Hendrix, R. W. (1978). *Proc. Natl. Acad. Sci. U.S.A.* **75**, 4779.
Hinnen, R., Schafer, R., and Franklin, R. M. (1974). *Eur. J. Biochem.* **50**, 1.
Hinnen, R., Chassin, R., Schafer, R., Franklin, R. M., Hitz, H., and Schafer, D. (1976). *Eur. J. Biochem.* **68**, 139.
Iba, H., Nanno, M., and Okada, Y. (1979). *FEBS Lett.* **103**, 234.
Inouye, S., Wang, S., Sekizawa, J., Halegoua, S., and Inouye, M. (1977). *Proc. Natl. Acad. Sci. U.S.A.* **74**, 1004.
Israelachvili, J. N. (1973). *Biochim. Biophys. Acta* **232**, 659.
Ito, K., Sato, T., and Yura, T. (1977). *Cell* **11**, 551.
Keen, J. H., Willingham, M. C., and Pastan, I. H. (1979). *Cell* **16**, 303.
Knipe, D. M., Baltimore, D., and Lodish, H. F. (1977). *J. Virol.* **21**, 1149.
Landsberger, F. R., Compans, R. W., Choppin, P. W., and Lenard, J. (1973). *Biochemistry* **12**, 4498.
Lehman, J. F., and Mindich, L. (1979). *Virology* **97**, 164.
Letellier, L., Weil, R., and Shechter, E. (1977). *Biochemistry* **16**, 3777.
Lodish, H. F., and Weiss, R. A. (1979). *J. Virol.* **30**, 177.
Lopez, R., Ronda, C., Tomasz, A., and Portoles, A. (1977). *J. Virol.* **24**, 201.
Lundstrom, K. H., Bamford, D. H., Palva, E. T., and Lounatmaa, K. (1979). *J. Gen. Virol.* **43**, 583.
Marcoli, R., Pirrotta, V., and Franklin, R. M. (1979). *J. Molec. Biol.* **131**, 107.
Mindich, L. (1978). *In* "Comparative Virology" (H. Fraenkel-Conrat and R. Wagner, eds.), Vol. 12, p. 271. Plenum, New York.
Mindich, L., and Davidoff-Abelson, R. (1980). *Virology* **103**, 386.
Mindich, L., and Lehman, J. (1979). *J. Virol.* **30**, 489.
Mindich, L., Cohen, J., and Weisburd, M. (1976a). *J. Bacteriol.* **126**, 177.
Mindich, L., Sinclair, J. F., Levine, D., and Cohen, J. (1976b). *Virology* **75**, 218.
Mindich, L., Sinclair, J. F., and Cohen, J. (1976c). *Virology* **75**, 224.
Mindich, L., Lehman, J., and Huang, R. (1979). *Virology* **97**, 171.
Murialdo, H., and Becker, A. (1978). *Microbiol. Rev.* **42**, 529.
Nagy, E., Pragai, B., and Ivanovics, G. (1976). *J. Gen. Virol.* **32**, 129.
Olsen, R. H., Siak, J., and Gray, R. H. (1974). *J. Virol.* **14**, 689.
Reaveley, D. A., and Burge, R. E. (1972). *Adv. Microbiol. Physiol.* **7**, 1.

Rosenkranz, H. S., Rose, H. M., Morgan, C., and Hsu, K. C. (1966). *Virology* **28,** 510.

Rothman, J. E., and Kennedy, E. P. (1977a). *J. Molec. Biol.* **110,** 603.

Rothman, J. E., and Kennedy, E. P. (1977b). *Proc. Natl. Acad. Sci. U.S.A.* **74,** 1821.

Ruettinger, R. T., and Brewer G. J. (1978). *Biochim. Biophys. Acta* **529,** 181.

Sakaki, Y., Yamada, K., Oshima, M., and Oshima, T. (1977). *J. Biochem.* **82,** 1451.

Sands, J. A. (1973). *Biochem. Biophys. Res. Commun.* **55,** 111.

Sands, J. A. (1976). *J. Virol.* **19,** 296.

Sands, J. A., and Auperin, D. (1977). *J. Virol.* **22,** 315.

Sands, J. A., and Cadden, S. P. (1975). *FEBS Lett.* **58,** 43.

Sands, J. A., and Lowlicht, R. A. (1976). *Can. J. Microbiol.* **22,** 154.

Sands, J. A., Cupp, J., Keith, A., and Snipes, W. (1974). *Biochim. Biophys. Acta* **373,** 277.

Sands, J. A., Lowlicht, R. A., Cadden, S. C., and Haneman, J. (1975). *Can. J. Microbiol.* **21,** 1287.

Scandella, C. J., Schindler, H., Franklin, R. M., and Seelig, J. (1974). *Eur. J. Biochem.* **50,** 29.

Schafer, R., and Franklin, R. M. (1975a). *J. Molec. Biol.* **97,** 21.

Schafer, R., and Franklin, R. M. (1975b). *Eur. J. Biochem.* **58,** 81.

Schafer, R., and Franklin, R. M. (1978a). *Eur. J. Biochem.* **92,** 589.

Schafer, R., and Franklin, R. M. (1978b). *FEBS Leff.* **94,** 353.

Schafer, R., Hinnen, R., and Franklin, R. M. (1974). *Eur. J. Biochem.* **50,** 15.

Schafer, R., Huber, V., Franklin, R. M., and Seelig, J. (1975). *Eur. J. Biochem.* **58,** 291.

Schafer, R., Kunzler, P., Lustig, A., and Franklin, R. M. (1978). *Eur. J. Biochem.* **92,** 579.

Schmidt, M. F. G., and Schlesinger, M. J. (1979). *Cell* **17,** 813.

Schmidt, M. F. G., Bracha, M., and Schlesinger, M. J. (1979). *Proc. Natl. Acad. Sci. U.S.A.* **76,** 1687.

Schneider, D., Zulauf, M., Schafer, R., and Franklin, R. M. (1978). *J. Molec. Biol.* **124,** 97.

Schnitzer, T. J., and Lodish, H. F. (1979). *J. Virol.* **29,** 443.

Schnitzer, T. J., Dickson, C., and Weiss, R. A. (1979). *J. Virol.* **29,** 185.

Sefton, B. M., and Gaffney, B. J. (1974). *J. Molec. Biol.* **90,** 343.

Sheetz, M. P., and Singer, S. J. (1974). *Proc. Natl. Acad. Sci. U.S.A.* **71,** 4457.

Silbert, J. A., Salditt, M., and Franklin, R. M. (1969). *Virology* **39,** 666.

Sinclair, J. F., and Mindich, L. (1976). *Virology* **75,** 209.

Sinclair, J. F., Tzagoloff, A., Levine, D., and Mindich, L. (1975). *J. Virol.* **16,** 685.

Sinclair, J. F., Cohen, J., and Mindich, L. (1976). *Virology* **75,** 198.

Snipes, W., Cupp, J., Sands, J. A., Keith, A., and Davis, A. (1974). *Biochim. Biophys. Acta* **339,** 311.

Stanisich, V. A. (1974). *J. Gen. Microbiol.* **84,** 332.

Stern, W., and Dales, S. (1974). *Virology* **62,** 293.

Stern, W., and Pogo, B., and Dales, S. (1977). *Proc. Natl. Acad. Sci. U.S.A.* **74,** 2162.

Strauss, E. G., Birdwell, C. R., Lenches, E. M., Staples, S. E., and Strauss, J. H. (1977). *Virology* **82,** 122.

Tanford, C. (1979). *Proc. Natl. Acad. Sci. U.S.A.* **76,** 4175.

Tsukagoshi, N., and Franklin, R. M. (1974). *Virology* **59,** 408.

Tsukagoshi, N., Peterson, M. H., and Franklin, R. M. (1975a). *Virology* **66,** 206.

Tsukagoshi, N., Peterson, M. H., and Franklin, R. M. (1975b). *Eur. J. Biochem.* **60,** 603.

Tsukagoshi, N., Schafer, R., and Franklin, R. M. (1977). *Eur. J. Biochem.* **77,** 585.

Van Etten, J. L., Lane, L., Gonzalez, C., Partridge, J., and Vidaver, A. (1976). *J. Virol.* **18,** 652.

Vidaver, A. K., Koski, R. K., and van Etten, J. L. (1973). *J. Virol.* **11,** 799.

Wirtz, K. W. A. (1974). *Biochim. Biophys. Acta* **344,** 95.

Witte, O. N., and Baltimore, D. (1977). *Cell* **11,** 505.

Wong, F. H., and Bryan, L. E. (1978). *Can. J. Microbiol.* **24,** 875.

INTERNATIONAL REVIEW OF CYTOLOGY, VOL. 68

Scanning Electron Microscopy of Intracellular Structures

KEIICHI TANAKA

Department of Anatomy,
Tottori University School of Medicine,
Yonago, Tottori-Ken, Japan

I. Introduction

Since the introduction of the scanning electron microscope to biological research within the last decade, this instrument has been employed in the study of biological materials. It has been used mostly for surface observation of cells, tissues, and organs with little attention given to intracellular structures. Observation of intracellular structures by scanning electron microscope (SEM) commenced with the development of many kinds of cracking method such as the cryofracture method (Haggis, 1970; Lim, 1971; Nemanic, 1972), the frozen resin cracking method (Tanaka, 1972), and others (Humphreys *et al.*, 1974; Tanaka *et al.*, 1974; Tokunaga *et al.*, 1974). Unfortunately, it was practically impossible to expose the inner structures of the cell usually hidden in the fixed cytoplasm only by the cracking method. In addition, conventional SEM did not provide sufficient resolution to make an adequate visualization of cell organelles. Thus, progress in SEM studies in this field has been very slow.

On the other hand, with the application of the ion-etching method to biological materials (Lewis *et al.*, 1968; Hodges *et al.*, 1972; Fulker *et al.*, 1973; Frisch *et al.*, 1974), cytoplasmic granules of blood cells were observed (Fujita *et al.*, 1974).

Subsequently the ion-etching method was utilized on the cracked surface of cells to permit adequate observation of the intracellular structures (Tanaka *et al.*, 1976a).

Moderate ion-etching on the cracked cell surface has proven effective in distinguishing membranous structures of cell organelles form the cytoplasmic matrix, which is far less etch resistant. This is because the rate of etching in a glow discharge apparatus is dependent upon the compactness of the structure to be etched.

In addition, a new type of SEM equipped with a field emission gun was developed (Komoda and Saito, 1972) and has a superior performance in comparison with the conventional SEM. Its resolution is 30 Å at point-to-point. With advancements made in both specimen preparation technique and in SEM instruments, studies on intracellular structures by SEM have become possible. However, the ion-etching method has some shortcomings. It often produces artificial cone figures that fail to reflect the intrinsic structures. It was, therefore, necessary to devise a new method that would reveal the intracellular structures without ion-etching.

Haggis and Phipps-Todd (1977) devised their so-called "freeze–fracture, thaw, fix method." In their method, fresh tissues are first infiltrated with cryoprotectant (DMSO), frozen, and fractured in liquid nitrogen. Then the specimens are thawed and fixed simultaneously in a glutaraldehyde solution. They are finally dehydrated with a graded series of alcohol and dried by the critical point drying method. It is the aim of this method to wash out soluble proteins from the fractured surface at the thawing stage. They were able to observe intracellular structures of carrot protoplasts, chicken erythrocytes, and leaf mesophyll by this method. Tanaka *et al.* (1977) also developed a preparation method using a hypotonic solution treatment and have been able to obtain fairly good results, but unfortunately, the treatment caused cells and cell organelles to swell. In fact, mitochondria and others were markedly swollen or ruptured during perfusion with extremely hypotonic buffer solution (30 mOsm) prior to fixation. This method was later improved. To minimize artifacts, perfusion with hypotonic solution was changed into fixation with slightly hypotonic solution (200 mOsm). In addition, the maceration technique with dilute osmic solution is used to remove the cytoplasmic matrix in the new method (Tanaka and Naguro, 1980). By this revised method, the intracellular structures could be disclosed without introducing any significant artifacts. We have called it the osmium–DMSO–osmium method. In the following sections, many SEM micrographs obtained by this method will be presented. It may be wise to briefly describe here the essential procedures of this method.

1. Small blocks of tissues are removed from the animal and fixed in 1% osmium tetroxide solution buffered at pH 7.4 with M/15 phosphate buffer for about 1–2 hours.

2. After rinsing with buffer solution, the specimens are successively immersed in 15, 30, and 50% dimethyl sulfoxide (DMSO) for 30 minutes each.

3. The specimens are frozen on a metal plate chilled with liquid nitrogen and cracked into two pieces by a razor blade and a hammer.

4. The cracked pieces are immediately placed in 50% DMSO.

5. After rinsing with buffer solution, they are transferred to 0.1% osmium tetroxide solution buffered at pH 7.4 with M/15 phosphate solution and left standing for 24–72 hours. This procedure is important in removing the cytoplasmic matrix.

6. The specimens are again fixed in 1% osmium tetroxide solution for 2 hours.

7. They are dehydrated in a graded series of ethanol.

8. After treatment with isoamyl acetate, they are dried by the critical point drying method.

When specimens thus prepared are observed by SEM, intracellular structures can be seen clearly in bold relief (Fig. 1).

As described earlier, it now is possible to observe intracellular structures through the advances made in specimen preparation techniques and SEM equipment. Detailed three-dimensional configuration of cell organelles and their spatial arrangement may be clearly elucidated by SEM in the near future. It therefore might be worthwhile to summarize the findings obtained to date.

II. Cell Organelles

A. Endoplasmic Reticulum

Endoplasmic reticulum has been observed from the early stages of biological SEM study because it could be readily distinguished on the cracked cell surface. The rough endoplasmic reticula (RER) of a dog pancreatic aciner cell were first observed by frozen resin cracking method (Tanaka, 1972).

Later Woods and Ledbetter (1976) observed an endoplasmic reticulum of the corn root tip that had been treated by the ligand-mediated osmium binding technique and fractured in chilled epoxy resin. In these studies, however, the endoplasmic reticulum could not be clearly observed, being hidden in the cytoplasmic matrix.

A few years later, both the rough and smooth endoplasmic reticula were clearly disclosed by the ion-etching method (Tanaka *et al.*, 1976a). In this study, they appeared very stereoscopically, almost like a schematic drawing made by reconstruction of a series of ultrathin sections of TEM. Unfortunately, ribosomes attached to the endoplasmic reticulum, being poorly resistant to ion-etching, would often disappear in such specimens. For better observation of RER, a

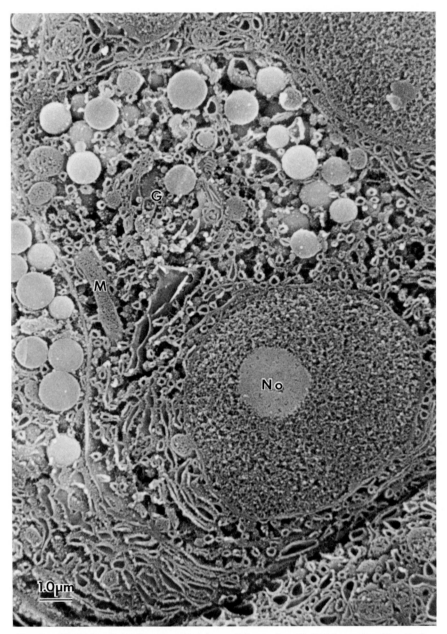

FIG. 1.   Cracked surface of a pancreatic aciner cell from a dog. G, Golgi complex; M, mitochondria; No, nucleolus. ×8700. (Micrograph courtesy of T. Naguro.)

non-ion-etched specimen is recommended for examination. The author recently studied the endoplasmic reticulum using specimens prepared by the osmium–DMSO–osmium method. In these specimens, ribosomes were preserved as well as in non-ion-etched specimens. The following is a detailed description of the endoplasmic reticulum observed in such specimens.

Usually, RER were seen in cisternal or vesicular form. Cisternae of the reticulum were often aggregated to form a lamellar system of flat cavities (Fig. 2). Vesicular endoplasmic reticula of various sizes were firmly connected to each other or secretory granules by threadlike structures, which may be microtubules (Fig. 3). The outer surfaces of the endoplasmic reticula were associated with small granules, that is, ribosomes about 240 Å in diameter. In general, ribosomes observed by SEM are considerably larger than those reported by TEM studies (150 Å). Kirschner et al. (1977) reported ribosomes to be 200–250 Å in diameter in their SEM study on isolated mouse liver nuclei. The difference in size between these observations depends upon the metal coat around the ribosomes of the SEM specimens. According to TEM findings, ribosomes consisted of two subunits of unequal size. These subunits were observed clearly also by SEM (Fig. 4). Kirschner et al. (1977) observed tortuous polysomes in the sample of ribosomes isolated from nuclear envelopes of rat liver cells. The author also found polyribosomes in situ consisting of 4–5 ribosomes linked together in a cracked cell of the dog pancreas (Fig. 4).

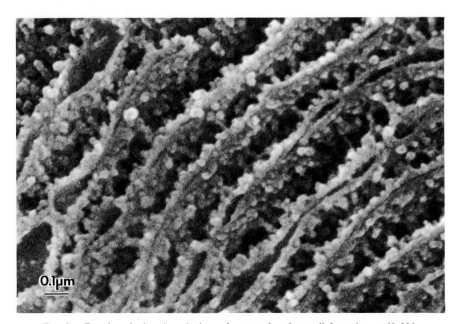

0.1μm

FIG. 2.   Rough endoplasmic reticulum of pancreatic aciner cell from dog. ×60,000.

FIG. 3.   Vesicular endoplasmic reticula. They are connected to each other or to secretory granules by threadlike structures. ×94,000.

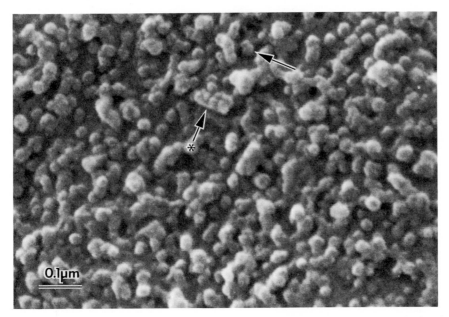

Fig. 4. Outer surface of rough endoplasmic reticulum. Some ribosomes consist of two subunits of unequal size (arrow) and others form tortuous polysomes (asterisk). ×110,000.

On the other hand, the inner surface of the RER is relatively smooth with small granules of various sizes attached to it. These granules are probably protein mass secreted in the cisternal cavity. In addition, small openings on the inner surface of endoplasmic reticulum were seen sporadically (Fig. 5) (Tanaka *et al.*, 1976a). It is still unknown whether these openings are artifacts or not, but generally, no significant artifacts have been observed in SEM specimens. These openings may be a channel of protein produced by a ribosome, but this has not been yet confirmed.

Smooth endoplasmic reticulum differs from RER chiefly in its lack of associated ribosomes. It forms a closely meshed network of branching tubules of rather variable caliber and its outer surface looks coarse with adhesion of minute particles (Fig. 6). As the configuration of the branching tubules can be observed better in the SEM micrographs than TEM micrographs, in the near future the construction of SER and its relation to other organelles may be examined more minutely in stereo-pairs taken by SEM.

B. Golgi Complex

Until recently, there have been hardly any reports on the SEM observations of the Golgi apparatus. It was first observed in a pancreatic acinar cell from a rabbit

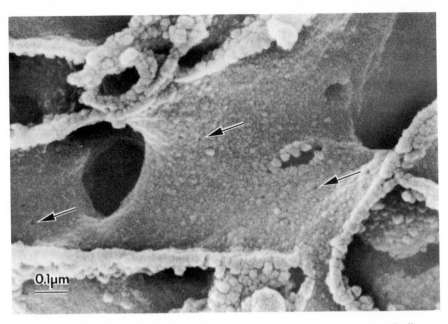

FIG. 5. Inner surface of rough endoplasmic reticulum. Small openings are sporadically seen (arrow). ×87,000. (From Tanaka *et al.*, 1976a.)

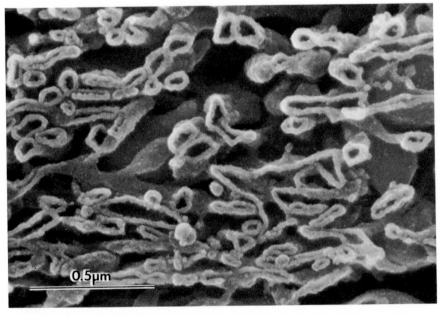

FIG. 6. Smooth endoplasmic reticulum. It forms a network of branching tubules of rather variable caliber. ×66,400. (From Tanaka and Naguro, 1980.)

by the frozen resin cracking method (Tanaka, 1974). In this study, only a part of the Golgi complex was shown, but 2 years later it was disclosed clearly by ion-etching method (Tanaka *et al.*, 1976a).

Sturgess and Moscarello (1976) examined the isolated Golgi apparatus of rat liver cells and observed Golgi cisternae that were continuous with long tubules and a fine tubular network. Since the three-dimensional arrangement of the Golgi elements changed during isolation, their appearance *in situ* could not be confirmed.

Kinose (1979) recently observed the Golgi complex of epididymal cells from a rat by using the osmium–DMSO–osmium method (Tanaka and Naguro, 1980). According to his findings, the Golgi complex is composed of three elements, as in the TEM findings; that is, Golgi cisternae, Golgi vacuoles, and Golgi granules (Fig. 7). Golgi cisternae were piled one upon the other in close parallel array. Five to eight cisternae formed a multilayered array of cisternae (Fig. 8). Around their ends and along their convex outer surface were numerous small vesicles, about 600 Å in diameter, whose surfaces were not smooth but coarse as a result of association with very fine granules (Fig. 9). On the other hand, the innermost cisterna was combined with a network consisting of tubular elements. Budding out on the surface of the network were many vesicles in various sizes containing specific cell products (Fig. 10).

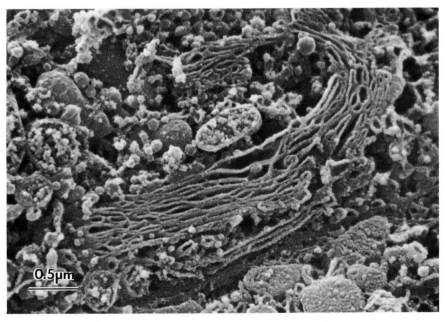

Fig. 7. Golgi complex of a rat edipidymal cell. ×26,000. (From Kinose, 1979.)

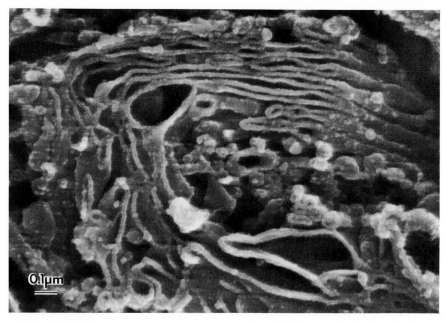

FIG. 8. Golgi cisternae of an epididymal cell from a rat. Several cisternae are piled one upon the other in close parallel array. ×58,000. (From Kinose, 1979.)

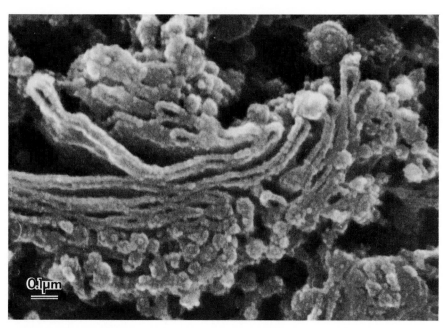

FIG. 9. Close up view of Golgi cisternae. Around their ends and along their convex outer surface, many small vesicles are observed. ×72,000. (From Kinose, 1979.)

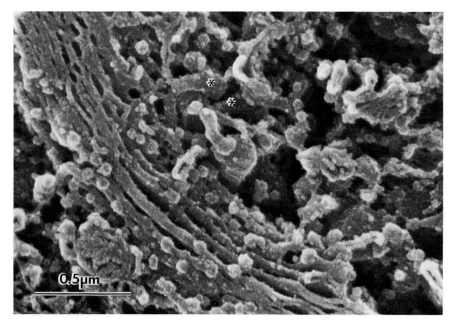

FIG. 10.   Golgi cisternae of a rat epididymal cell. Innermost cisterna is combined with a network that consists of tubular elements (asterisk). ×50,000. (From Tanaka and Naguro, 1980.)

Vacuoles of various sizes were also seen sporadically in the Golgi complex. Some of them involved several granules (multivesicular body, Fig. 7) and others contained an irregularly shaped substance (lysosome, Fig. 29).

The inner surface of the membrane of Golgi cisterna appeared rather smooth, but sometimes it was interrupted with small openings of various sizes (Fig. 11). These openings were encircled with a bank composed of fine granules.

As mentioned earlier, SEM observations have not added much to the present knowledge of Golgi apparatus derived from TEM studies. In the future, however, SEM will become a very useful instrument for three-dimensional analysis of GERL and other elements of the Golgi complex.

C. Mitochondria

Mitochondria can be readily observed by SEM on the cracked surfaces of cells (Lim, 1971; Nemanic, 1972; Panessa and Gennaro, 1972; Humphreys *et al.*, 1974), but it has been difficult to distinguish them (especially the spherical ones) from the other intracellular granules because cristae mitochondriales hidden in the fixed mitochondrial matrix were not exposed clearly until recently. By the ion-etching method, mitochondrial cristae were definitely revealed for the first time (Tanaka *et al.*, 1976a).

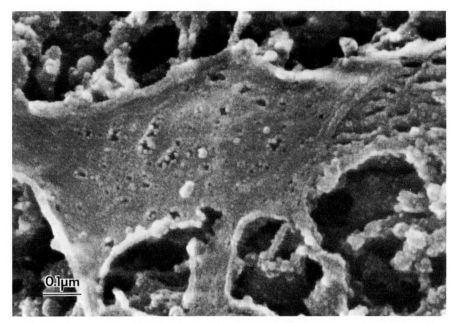

Fig. 11.   Inner surface of a Golgi cisterna. Many openings of various sizes can be seen. ×100,000. (From Kinose, 1979.)

Haggis *et al.* (1976) also observed the mitochondrial cristae of heart muscle cells by means of his freeze–fracture, thaw, fix method. In their study, however, the cristae were found to be forced out of the cracked surface of mitochondria.

Several studies were performed on isolated mitochondria. Andrews and Hackenbrock (1975) examined the surface topography of isolated inner mitochondrial membrane of the rat liver in various states of metabolic activity. Kirschner and Rusli (1976) also studied isolated mitochondria and reported that the inner mitochondrial membrane has a granular surface. Osumi and Torigata (1977) observed the cristae of isolated mitochondria from *Candida tropicalis* and Shimada *et al.* (1978) observed those of the liver or cardiac muscle by means of the resin cracking method. In these studies, however, the detailed construction of mitochondria, especially the details of cristae, was not investigated. Recently Masunaga (1979) in our department studied mitochondria *in situ* by using specimens prepared by the osmium–DMSO–osmium method (Tanaka and Naguro, 1980) and presented a detailed report on cristae mitochondriales.

According to Masunaga, two membranes that enclose mitochondria were sometimes recognized by SEM. The outer membrane was a smooth contoured, continuous limiting membrane, and the inner membrane generally ran parallel to the outer membrane but also formed numerous narrow cristae that projected into the interior of the organelle (Fig. 12).

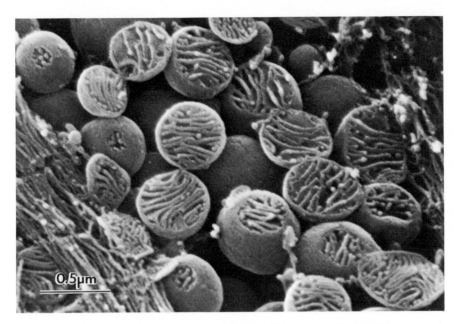

FIG. 12.   Mitochondria of a heart muscle cell from a rabbit. Cristae mitochondriales are clearly seen. ×27,000. (From Masunaga, 1979.)

The mitochondrial membranes formed two compartments, that is, the inner compartment comprised all of the area within the inner membrane and the outer compartment consisted of the narrow cleft between the outer and inner membrane including its inward extensions between leaves of the cristae (intracristal space) (Fig. 13). By SEM observation, the outer compartment usually was not distinguished and consequently the limiting membranes of mitochondria and cristae were seen as a thick plate. The inner compartment, on the other hand, appeared empty, since the mitochondrial matrix was completely removed. In this space, no matrix granules were observed, but rarely fine threadlike structures were found (Fig. 14). Its configuration in the inner compartment suggests that this threadlike structure might be mitochondrial DNA, though no clear evidence for this can be found.

It is well known that the inner mitochondrial membrane has a great number of small particles associated with its inner surface. They are generally called inner membrane particles. This particle consists of a globular head, 90–100 Å in diameter, connected by a slender stem to a baseplate that is incorporated in the inner membrane. By SEM observation, numerous small spherical granules, about 200 Å in diameter, were seen on and in the membranes of cristae mitochondriales (Fig. 15). Kirschner and Rusli (1976), in their study of isolated mitochondria, noted that the inner membrane had a granular surface, and

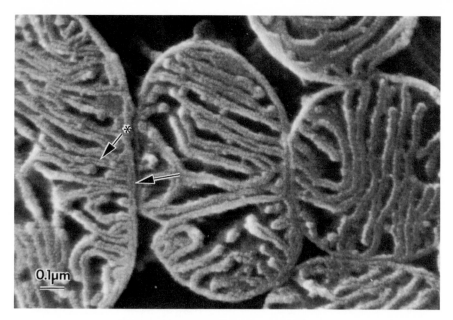

Fig. 13.   Cracked surface of mitochondria of a rabbit heart muscle cell. Outer compartments (arrow) and intracristal spaces (asterisk) can be observed. ×66,000. (From Masunaga, 1979.)

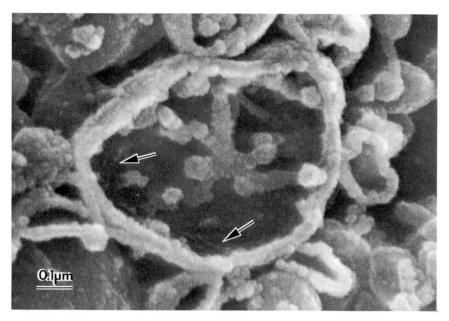

Fig. 14.   Mitochondrion of a suprarenal cell from a rabbit. In the inner compartment, fine threadlike structures are seen (arrow). ×96,000. (Unpublished micrograph, courtesy of Masunaga.)

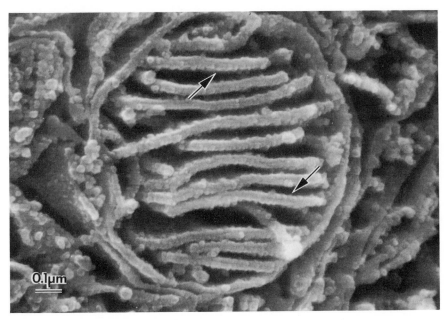

FIG. 15. Mitochondrion of an epididymal cell from rat. Numerous small granules (arrow) are seen on and in the membranes of cristae. ×70,000. (From Tanaka and Naguro, 1980.)

these granules may represent respiratory enzyme complexes similar to the AT-Pase particles visible in the negatively stained inner mitochondrial membrane.

It has not been confirmed whether the particles shown in Fig. 15 are inner membrane particles or not, but this possibility cannot be excluded. The difference in size between the inner membrane particles reported by TEM studies and the particles observed by SEM is probably caused by the metal coat evaporated on the SEM specimens.

Next, the architectural structure of mitochondrial cristae will be discussed. According to Masunaga's findings, there are four types of mitochondrial cristae: (1) The first type extends directly to the inner limiting membrane of the opposite side (Figs. 13 and 15). (2) The second type does not reach the opposite wall (Figs. 13 and 15). (3) The third type fuses together with the neighboring cristae on its way (Fig. 16a). (4) The fourth type turns back, forming a U shape, to the starting side of the mitochondrial wall (Fig. 16b). In addition, there are several perforations of various sizes on the attached portions of the cristae to the inner limiting membrane (Fig. 17a).

According to Daems and Wisse (1966), mitochondrial cristae are attached to the inner limiting membrane of the mitochondrion, not with their entire edge but with their small pedicles. If the portion between two perforations can be regarded as pedicle, there is no difference between both observations.

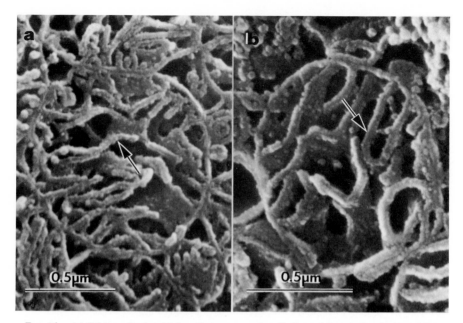

FIG. 16.   (a) Cristae mitochondriales of a heart muscle cell. One of them is fused with neighboring cristae (arrow). ×48,000. (b) Mitochondrion of a rabbit heart muscle cell. One of the cristae mitochondriales turns back (forming a U shape) to the starting side of the mitochondrial wall (arrow). ×56,000. (From Masunaga, 1979.)

Finally, small openings observed on the outer mitochondrial surface will be briefly mentioned. Employing the freeze replica method, Moor and Mühlerthaler (1963) found small openings, 50–100 Å in diameter, on the surface of yeast mitochondria. By SEM, these openings are often recognized, especially in the ion-etched samples (Fig. 17b), but they cannot be found in all kinds of tissues. Further studies must be made to determine what kinds of tissues have mitochondrial openings.

III. Nucleus and Nuclear Envelope

A. Nucleus

The nucleus is a repository of deoxyribonucleic acid, a genetic material. It is an essential organelle present in nearly all cells. Its structure at interphase and the dividing stage has been studied extensively by light microscopy in the past and recently it has been examined by electron microscopy.

Unfortunately, a transmission electron micrograph obtained by ultrathin sec-

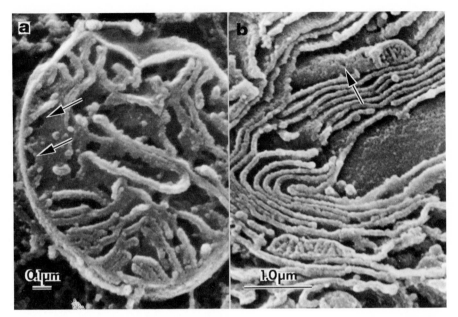

Fig. 17.   (a) Perforations on the attached portion of a crista to the inner limiting membrane of a mitochondria (arrow). ×50,000. (From Masunaga, 1977.) (b) Mitochondria of a pancreatic acinal cell. Small openings are found on the outer surface (arrow). ×18,000. (From Tanaka *et al.*, 1976a.)

tion technique cannot add any significant information to the knowledge of nuclear fine structure derived from the light microscope, because the nucleoplasma presents, in this case, a bewildering array of minute punctate and elongated profiles of varying size and density. By scanning electron microscopy, on the other hand, supercoiled chromatin fibers can be observed successfully in the interphase nuclei of cells that are fixed with potassium permanganate and fractured by frozen resin cracking method (Tanaka and Iino, 1973). According to their finding, these fibers were 350–700 Å in diameter and consist of ~250-Å coiled fibers that were themselves formed of 100–150 Å supercoiled subfibrils (Fig. 18). The fibers sometimes entwined about each other like a tangled rope and some fibers showed a clockwise rotation (dextrose coiling structure) and others a counterclockwise rotation (sinistrose coiling structure) (Fig. 19). Haggis and Bond (1978) also observed chromatin fibers of interphase nucleus with their own "freeze-fracture, thaw, fix method." They attempted to clarify the three-dimensional arrangement of chromatin fibers with stereo-pairs taken by SEM. In comparison with studies of interphase nucleus, many investigations have been made on chromosomes by means of SEM. These studies were made mainly on chromosomes that were spread on a slide glass by the routine method for light microscopy (Neurath *et al.*, 1976; Christenhuss *et al.*, 1967; Pawlowitzki *et al.*,

FIG. 18.   Chromatin fibers in a hepatic interphase nucleus of a rabbit. Diagonal striation with
spacing of 100–150 Å can be seen perpendicular to the diagonal striation with a pitch spacing of about
250 Å (arrow). ×100,000. (From Tanaka and Iino, 1973.)

1968; Tanaka and Iino, 1970; Sweney *et al.*, 1979). In these studies, unfortunately, the fine structure of chromosomes could not be observed because Car-
noy's fixation and air drying method, which were inadequate for the study of
surface fine structures, were used for specimen preparation.

   Iino (1974, 1975) observed, by using his special hypotonic solution, that
human somatic chromosomes were composed of tortuous and entangled fibers of
about 860 Å in diameter (Fig. 20). He also detected around the chromosome
body straight radiating fibers about 300–400 Å in diameter, which had a uniform
cross banding (250 Å). On the other hand, Golomb and Bahr (1971, 1974)
observed a whole mounted metaphase chromosome dried by the critical point
drying method. They described the feature of the chromosome surface as a
"skein of yarn." Daskal *et al.* (1976, 1978), Korf and Diacumakos (1978), and
Mace *et al.* (1978) also examined chromosomes dried by the critical point drying
method. Besides metaphase chromosomes, which were spread on a slide glass or
water surface, prophase chromosomes *in situ* in the nucleus of grasshopper
testiculer cells were observed (Tanaka *et al.*, 1976b). In this SEM micrograph,
three-dimensional arrangement of chromosomes connecting each other with in-
terchromosomal fibers was observed fairly well (Fig. 21).

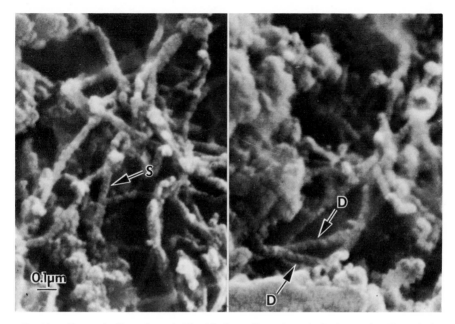

FIG.19.   Chromatin fibers shown in Fig. 18. Some fibers show clockwise rotation (D) and other fibers counterclockwise rotation (S). ×50,000. (From Tanaka and Iino, 1973.)

In addition, giant chromosomes of the salivary gland from *Chironomus tentans* or *C. thummi* were studied by SEM (Trösch and Lindemann, 1973; Holderegger, 1975), but their fine structures such as feature and arrangement of chromatin fibers were not analyzed in detail in these observations. Tanaka *et al.* (1976b) also studied polytene salivary gland chromosomes from a *Drosophila melanogaster* that was treated previously with aurin tricarboxylic acid. These investigators observed many fibers approximately 250 Å in diameter with helix periodicity at an interval of 100 Å (Figs. 22 and 23). Thus the author assumes that SEM is very useful in examining the coiled chromatin fibers and chromosomes.

It would be possible in the future to elucidate by SEM how a chromosome with chromatin fibers is formed and how the chromatin fibers are constructed with DNA thread and protein particles.

B. NUCLEAR ENVELOPE

The outer limit of the nucleus is clearly demarcated by a nuclear envelope that consists of two parallel membranes enclosing a narrow perinuclear space. When cells are cracked by the frozen resin cracking method, nuclear envelopes are

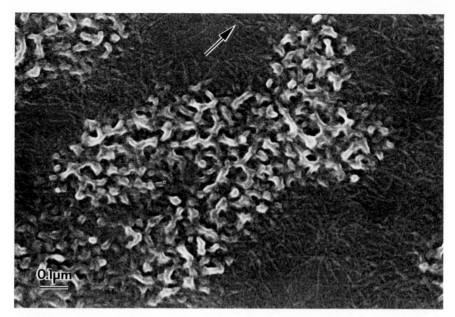

FIG. 20. One of the E group chromosomes of a man. It is composed of fibers about 860 Å. More slender fibers, 300–400 Å in diameter, which have uniform cross-banding, were also detected around the chromosome body (arrow). ×75,000. (From Iino, 1974.)

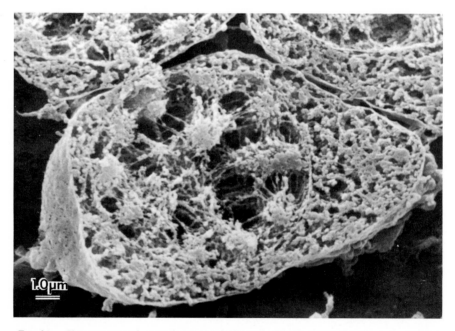

FIG. 21. Chromosomes of a grasshopper testicular cell. ×6500. (From Tanaka *et al.*, 1976b.)

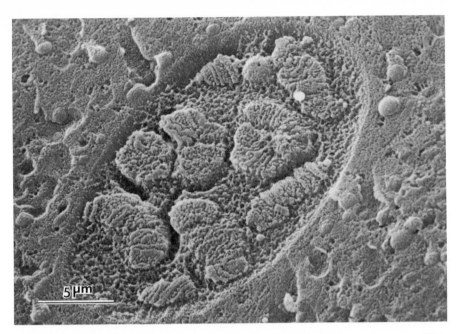

FIG. 22. Giant chromosomes of a salivary gland cell from a *Drosophila Melanogaster*. ×4000. (From Tanaka *et al.*, 1976b.)

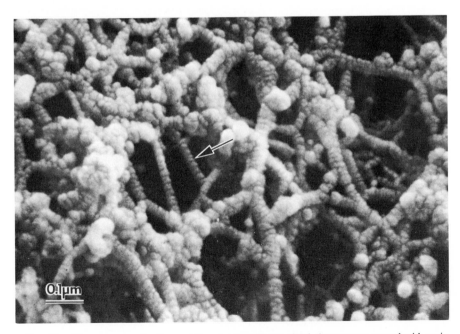

FIG. 23. Fine structure of a *Drosophila* polytene salivary gland chromosome treated with aurin tricarboxylic acid. Chromatin fibers, about 250 Å in diameter, with helix periodicity at intervals of approximately 100 Å can be observed. ×100,000. (From Tanaka *et al.*, 1976b.)

sometimes peeled off in the plane between the inner and outer membrane, and consequently hemispherical nuclei covered by only the inner membrane are exposed on the specimen surface. On the surface of the inner membrane, many nuclear pores with diameter of about 500 Å are seen. They are observed especially clearly on the nuclei treated with the ion-etching method because aggregated protein around the nuclear pore is removed during ion-etching.

On the surface of the outer membrane, numerous ribosomes are attached closely, and among them nuclear pores with annuli of nuclear pore complex are observed sporadically (Fig. 24).

It is well known that the nuclear complex consists of 4–5 subunits each about 850 Å in diameter. We have recently observed by SEM that the subunit was divided into two parts of equal size (Fig. 25). The pore complex was attached to an underlying, peripheral nuclear lamina (Barton *et al.*, 1971; Aaronson and Blobel, 1974; Kirschner *et al.*, 1977), and this nuclear lamina was interrupted by a funnel-shaped intranuclear channel (120–640 Å in diameter) that narrowed toward the pore complex (Schatten and Thoman, 1978). According to the assumption of Schatten and Thoman (1978), the intranuclear channel and pore complex are associated together by some mode, and consequently they are involved in nucleocytoplasmic transport.

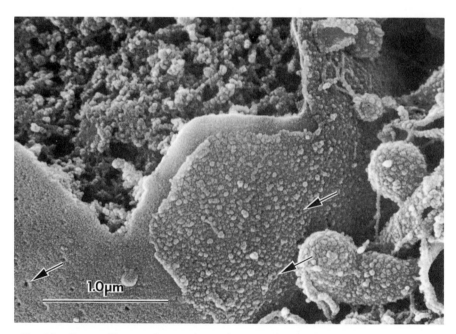

Fig. 24. Outer and inner nuclear membrane of a pancreatic aciner cell. On their surface, nuclear pores are sporadically seen (arrow). ×33,000. (From Tanaka *et al.*, 1976a.)

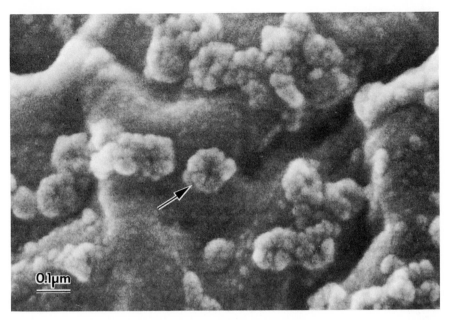

FIG. 25.   Nuclear pore complexes observed on the surface of outer nuclear membrane of rat. They consist of four to five subunits and these subunits are also divided into two parts (arrow). ×90,000.

IV. Secretory Granules and Others

Glandular cells often contain granules that represent intracellular accumulations of the precursors of their secretory products. These granules are readily observed by SEM when the cells have been fractured by some method (Makita and Sandborn, 1971; Tanaka, 1972). Secretory granules were mainly spherical in form and were often connected to each other with threadlike structures that terminated on the surface of secretory granules (Fig. 26) (Tanaka *et al.*, 1976a). The threadlike structures, each about 450 Å in diameter, formed a bundle before their termination but diverged again and were anchored on the surface of the secretory granules. Judging from the thickness of microtubules of cockroach epidermal cells (Nagano and Suzuki, 1975), it seems reasonable to assume that the threadlike structures might be microtubules.

On the other hand, crystalline structures in the secretory granules (e.g., insulin crystals of Langerhans islet B cells from the dog pancreas) can be well observed by the ion-etching method because they are generally etch-resistant and are consequently exposed clearly from the surrounding matrix (Tanaka *et al.*, 1976a). However, a better observation of them can be made in the samples treated by the hypotonic solution method (Fig. 27) (Tanaka *et al.*, 1977) or by the osmium–DMSO–osmium method (Tanaka and Naguro, 1980). In the cyto-

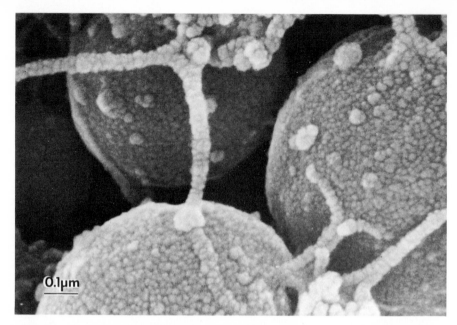

FIG. 26.   Zymogen granules of a dog and microtubules that terminate on the granules. ×90,000. (From Tanaka *et al.*, 1976a.)

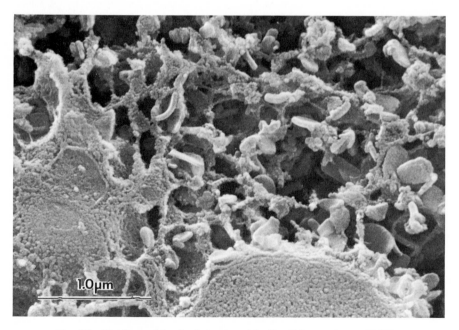

FIG. 27.   Insulin crystals of a Langerhans islet B cell from a dog. ×30,000.

plasm, some of the other granules can be recognized by SEM. For example, one has a circinate membrane (Fig. 28) and another contains an irregularly shaped substance (Fig. 29). Though these two kinds of granules probably are lysosomes, this cannot be verified by histochemical methods.

V. Concluding Remarks

As alluded to earlier, there are only a few works on intracellular structures by SEM in comparison with studies on the cell surface and thus the results obtained do not add much new information to the present knowledge of intracellular structures derived from TEM studies. The reason for this may be that good techniques to expose them have not been developed until recently, and conventional SEM does not have sufficient resolution to adequately visualize cell organelles. A good method for revealing the intracellular structure was, as mentioned before, devised, and the resolution of SEM, especially that of field emission SEM, was further improved (Nagatani and Okura, 1977). It is thus expected that SEM studies on intracellular structures will be made increasingly in the near future.

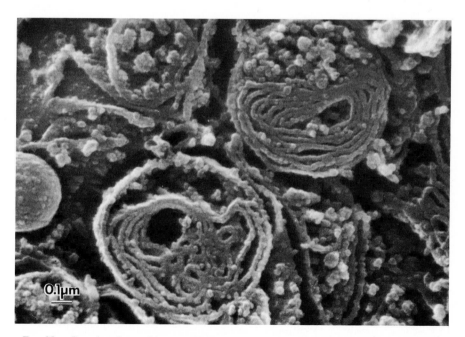

Fig. 28.   Granules observed in an epididymal cell of a rat. They contain circinate membrane. ×50,000.

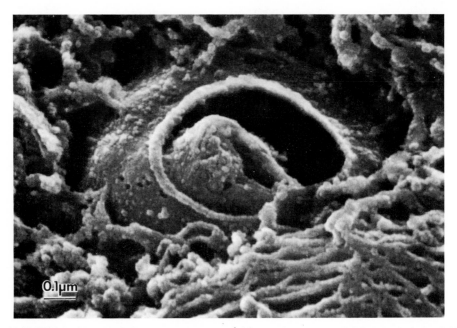

FIG. 29.   A granule observed in an epididymal cell of a rat. It contains substance of an irregular shape. ×80,000. (From Kinose, 1979.)

On the other hand, high resolution SEM of intracellular structures presents a problem. The thickness of the metal coat evaporated on the SEM specimen has a great influence on high resolution SEM. For example, if cell inner particles that are 30 Å in diameter are coated with metal film 100 Å in thickness, their intrinsic configuration can never be observed. Therefore, the metal coating must be thin and homogeneously evaporated as much as possible. Since it is desirable to have no coating at all, it is expected that a good electric conductive method, which completely prevents buildup of electric charges on the specimen surface without metal coating for high resolution SEM, would be developed. When such a method is developed, pictures of the intracellular elements of better quality than those obtained by the freeze replica method will become available.

Finally, future prospects of SEM will be briefly described. It is well known that many kinds of information can be simultaneously obtained by SEM. This is one of its most advantageous characteristics. When a narrow beam of electrons strikes a specimen, many kinds of signals are emitted (e.g., secondary electron, backscattered electron, Auger electron, characteristic X-ray, and cathodoluminescence). Among these, only secondary electron imaging has been used widely for biological research because it provides seemingly three-dimensional pictures of high quality. However, other signals have their own

Fig. 30.   Color SEM micrograph of spermatids of a rat. Orange-colored intracellular granules are clearly seen in the green-colored cytoplasm. At the upper right-hand corner, granules under the cell membrane are clearly observed. ×14,000. (From Tanaka, 1980.)

characteristics. For example, backscattered electron imaging provides atomic number contrast. If specimens stained by silver, uranium, or other heavy metals are examined by the backscattered electron mode, the stained portions such as nucleus or Golgi complex are brightly observed on the cathode ray tube of SEM (Abraham and DeNee, 1973, 1974; Lin and Becker, 1975; Becker and DeBruyn, 1976; Vogel et al., 1976; DeNee et al., 1977; Mitsushima et al., 1979). Application of signals other than secondary electron, especially backscattered electron, for study of intracellular structures, may become very important in the future.

In addition, a method by which different signals are simultaneously shown in a picture will become necessary when two or more signals are widely used for biological research. For this purpose, the author devised a method of preparing colored SEM micrographs using secondary and backscattered electron images (Tanaka, 1980). One of the pictures obtained by this method is given in this review (Fig. 30). In the picture, the granules of rat spermatids stained by heavy metals are shown in orange color, whereas the topography of cracked cell surface is indicated in green color. At the upper right hand corner, granules under the cell membrane can be clearly seen. This color picture provides not only the surface topography and atomic number contrast but also information on the subsurface of the specimen. It is therefore expected that color SEM micrographs, which contain various kinds of information, will be very useful in studies of intracellular structures in the future.

On the other hand, another method for preparing color micrographs has been developed. In this method, different signals of X-ray are presented in different colors directly on the color cathode ray tube of SEM by electronic manipulation (Pawley and Fisher, 1976). Thus the scanning electron micrograph in the future may contain many kinds of information indicated by different colors.

REFERENCES

Aaronson, R. P., and Blobel, G. (1974). *J. Cell Biol.* **62,** 746–754.
Abraham, J. L., and DeNee, P. B. (1973). *Lancet* **19,** 1125.
Abraham, J. L., and DeNee, P. B. (1974). *In* "Scanning Electron Microscopy/1974" (Om. Johari, ed.), pp. 251–258. IITRI, Chicago, Illinois.
Andrews, P. M., and Hackenbrock, Ch. R. (1975). *Exp. Cell Res.* **90,** 127–136.
Barton, A. D., Kisieleski, W. E., Wassermann, F., and Mackevicius, F. (1971). *Z. Zellforsch.* **115,** 299–306.
Becker, R. P., and Debruyn, P. H. (1976). *In* "Scanning Electron Microscopy/1976" (Om. Johari, ed.), Vol. 2, pp. 171–178. IITRI, Chicago, Illinois.
Christenhuss, R., Büchner, Th., and Pfeiffer, R. A. (1967). *Nature (London)* **216,** 379–380.
Daems, W. Th., and Wisse, E. (1966). *J. Ultrastruct. Res.* **16,** 123–140.
Daskal, Y., Mace, M. L., Wray, W., and Busch, H. (1976). *Exp. Cell Res.* **100,** 204–212.
Daskal, Y., Mace, M. L., and Busch, H. (1978). *Exp. Cell Res.* **111,** 472–475.

DeNee, P. B., Frederickson, R. G., and Pope, R. S. (1977). *In* "Scanning Electron Microscopy/ 1977" (Om. Johari, ed.), Vol. 2, pp. 83–92. IITRI, Chicago, Illinois.

Frisch, B., Lewis, S. M., Sherman, D., Stuart, P. R., and Osborn, J. S. (1974). *In* "Scanning Electron Microscopy/1974" (Om. Johari, ed.), pp. 655–664. IITRI, Chicago, Illinois.

Fujita, T., Nagatani, T., and Hattori, A. (1974). *Arch. Histol. Jpn.* **36**, 195–206.

Fulker, M. J., Holland, L., and Hurely, R. E. (1973). *In* "Scanning Electron Microscopy/1973" (Om. Johari, ed.), pp. 379–386. IITRI, Chicago, Illinois.

Golomb, H. M., and Bahr, G. F. (1971). *Science* **171**, 1024–1026.

Golomb, H. M., and Bahr, G. F. (1974). *Exp. Cell Res.* **84**, 79–87.

Haggis, G. H. (1970). *In* "Scanning Electron Microscopy/1970" (Om. Johari, ed.), pp. 99–104. IITRI, Chicago, Illinois.

Haggis, G. H., and Bond, E. F. (1978). *J. Microsc. (Oxford)* **115**, 225–234.

Haggis, G. H., and Phipps-Todd, B. (1977). *J. Microsc. (Oxford)* **111**, 193–201.

Haggis, G. H., Bond, E. F., and Phipps, B. (1976). *In* "Scanning Electron Microscopy/1976" (Om. Johari, ed.), vol. 1, pp. 281–286. IITRI., Chicago, Illinois.

Hodges, G. M., Muir, M. D., Sella, C., and Carteaud, A. J. P. (1972). *J. Microsc. (Oxford)* **95**, 445–451.

Holderegger, Ch. (1975). *In* "Principles and Techniques of Scanning Electron Microscopy" (M. A. Hayat, ed.), pp. 1–16. Van Nostrand-Reinhold, Princeton, New Jersey.

Humphreys, W. J., Spurlock, B. O., and Jhonson, J. S. (1974). *In* "Scanning Electron Microscopy/1974" (Om. Johari, ed.), pp. 275–282. IITRI, Chicago, Illinois.

Iino, A. (1974). *Acta Anat. Nippon* **49**, 337–344.

Iino, A. (1975). *Cytobios* **14**, 39–48.

Kinose, Y. (1979). *J. Yonago Med. Assoc.* **30**, 527–534.

Kirschner, R. H., and Rusli, M. (1976). *In* "Scanning Electron Microscopy/1976" (Om. Johari, ed.), pp. 153–162. IITRI, Chicago, Illinois.

Kirschner, R. H., Rusli, M., and Martin, T. E. (1977). *J. Cell Biol.* **72**, 118–132.

Komoda, T., and Saito, S. (1972). *In* "Scanning Electron Microscopy/1972" (Om. Johari, ed.), pp. 129–136. IITRI, Chicago, Illinois.

Korf, B. R., and Diacumakos, E. G. (1978). *Exp. Cell Res.* **111**, 83–93.

Lewis, S. M., Osborn, J. S., and Stuart, P. R. (1968). *Nature (London)* **220**, 614–616.

Lim, D. J. (1971). *In* "Scanning Electron Microscopy/1971" (Om. Johari, ed.), pp. 257–264. IITRI, Chicago, Illinois.

Lin, P. S. D., and Becker, R. P. (1975). *In* "Scanning Electron Microscopy/1975" (Om. Johari, ed.), pp. 61–70. IITRI, Chicago, Illinois.

Mace, M. L., Daskal, Y., and Wray, W. (1978). *Mutat. Res.* **52**, 199–206.

Makita, T., and Sandborn, E. B. (1971). *Exp. Cell Res.* **67**, 211–214.

Masunaga, Y. (1979). *J. Yonago Med. Assoc.* **30**, 519–526.

Mitsushima, A., Tanaka, K., Kashima, Y., Saito, S., and Nagatani, T. (1979). *J. Electron Microsc.* **28**, 240.

Moor, H., and Mühlerthaler, K. (1963). *J. Cell Biol.* **17**, 609–628.

Nagano, T., and Suzuki, A. (1975). *J. Cell Biol.* **64**, 242–245.

Nagatani, T., and Okura, A. (1977). *In* "Scanning Electron Microscopy/1977" (Om. Johari, ed.), Vol. 1, pp. 695–702. IITRI, Chicago, Illinois.

Nemanic, M. K. (1972). *In* "Scanning Electron Microscopy/1972" (Om. Johari, ed.), pp. 297–304. IITRI, Chicago, Illinois.

Neurath, P. W., Ampola, M. G., and Vetter, H. G. (1976). *Lancet* **2**, 1366–1367.

Osumi, M., and Torigata, S. (1977). *In* "Scanning Electron Microscopy/1977" (Om. Johari, ed.), Vol. 2, pp. 617–622. IITRI, Chicago, Illinois.

Panessa, B. J., and Gennaro, J. F. (1972). *In* "Scanning Electron Microscopy/1972" (Om. Johari, ed.), pp. 327–334. IITRI, Chicago, Illinois.

Pawley, J. B., and Fisher, G. L. (1979). *J. Miscosc, (Oxford)* **110**, 87–101.

Pawlowitzki, I. H., Blaschke, R., and Christenhuss, R. (1968). *Naturwissenschaften* **55**, 63–64.

Schatten, G., and Thoman, M. (1978). *J. Cell Biol.* **77**, 517–535.

Shimada, T., Morizono, T., Yoshimura, T., Murakami, M., and Ogura, R. (1978). *J. Electron Microsc.* **27**, 207–213.

Sturgess, J. M., and Moscarello, M. A. (1976). *In* "Scanning Electron Microscopy/1976" (Om. Johari, ed.), pp. 145–152. IITRI, Chicago, Illinois.

Sweney, L. R., Lam, F. H., and Shapiro, B. L. (1979). *J. Microsc. (Oxford)* **115**, 151–160.

Tanaka, K. (1972). *Naturwissenschaften* **59**, 77.

Tanaka, K. (1974). *In* "Principles and Technique of Scanning Electron Microscopy" (M. A. Hayat, ed.), Vol. 1, pp. 125–134. Van Nostrand-Reinhold, Princeton, New Jersey.

Tanaka, K. (1980). *Scanning* (in press).

Tanaka, K., and Iino, A. (1970). *Arch. Histol. Jpn.* **32**, 203–211.

Tanaka, K., and Iino, A. (1973). *Exp. Cell Res.* **81**, 40–46.

Tanaka, K., and Naguro, T. (1980). *Biomed. Res. Suppl.* (in press).

Tanaka, K., Iino, A., and Naguro, T. (1974). *J. Electron Microsc.* **23**, 313–315.

Tanaka, K., Iino, A., and Naguro, T. (1976a). *Arch. Histol. Jpn.* **39**, 165–175.

Tanaka, K., Iino, A., Naguro, T., and Fukudome, H. (1976b). *J. Electron Microsc.* **25**, 212.

Tanaka, K., Ozawa, K., Naguro, T., and Kashima, Y. (1977). *J. Electron Microsc.* **26**, 251.

Tokunaga, J., Edanaga, M., and Fujita, T. (1974). *Arch. Histol. Jpn.* **37**, 165–182.

Trösch, W., and Lindemann, B. (1973). *Micron* **4**, 370–375.

Vogel, G., Becker, R. P., and Swift, H. (1977). *In* "Scanning Electron Microscopy/1977" (Om. Johari, ed.), Vol. 2, pp. 409–415. IITRI, Chicago, Illinois.

Woods, P. S., and Ledbetter, M. C. (1976). *J. Cell Sci.* **21**, 47–58.

INTERNATIONAL REVIEW OF CYTOLOGY, VOL. 68

The Relevance of the State of Growth and Transformation of Cells to Their Patterns of Metabolite Uptake

RUTH KOREN

Department of Pharmacology,
The Hebrew University,
Hadassah Medical School,
Jerusalem, Israel

I. Introduction

A significant number of recent investigations have been devoted to studies of the interrelationship between the state of growth or transformation of cells in culture and their pattern of nutrient uptake. Many of the investigators in this field were encouraged by a hypothesis concerning the state of malignant growth that was proposed by Holley (1972, 1975) and by Pardee (1975). These investigators proposed that alterations in the cell membrane lead to an increase in transport and, thus, to an increased supply of key nutrients. The increased intracellular concentration of these key nutrients enables malignant cells to proliferate in an

127

uncontrolled manner or at least at a faster rate than their untransformed counter-parts. Now, several years later, in view of the significant amount of information that has accumulated in the field, it has become possible to test the validity of this hypothesis.

The first section of this chapter will be devoted to the presentation of some of the data related to the correlation between uptake rates and the state of growth or transformation of cells. We will try to answer the question of whether trans-formed or malignant cells take up nutrients at a faster rate than untransformed or "normal" cells.

Nutrients enter the cells by mediated, active, or passive transport across the plasma membrane. They are subsequently trapped within the cell by intracellular metabolic reactions, for instance, phosphorylation. In the second part of this chapter, we will present experimental and theoretical criteria that are available for differentiation between the two processes. Our purpose will be to decide whether indeed membrane transport is affected by changes in the state of growth or transformation of the cells or whether the changes in the uptake patterns are due to different rates of the intracellular metabolic reactions. We will review the information regarding the mechanism involved in the stimulation processes.

Finally, we will present some recent studies that are relevant to the general problem of whether the rate of nutrient uptake has a role in the control of cell proliferation.

The data obtained in avian cell cultures are sometimes both quantitatively and qualitatively different from results obtained in mammalian cells. This chapter deals only with data obtained from mammalian cell culture. This selection re-flects the personal experience of the author. For additional reviews with different emphasis, the reader is referred to works by Hatanaka (1974), Berlin and Oliver (1975), Plagemann and Richey (1974), Parnes and Isselbacher (1978), and Kap-lan (1978).

II. Correlation of Uptake Rates with Growth or Transformation States

In Table I, we present some of the data that have accumulated over the last decade or so regarding the correlation between the uptake rate of glucose (as a carbon or energy source), nucleosides (as nucleic acid precursors), and amino acids (as precursors for protein synthesis) and the rate of growth and state of transformation of mammalian cells in culture. A number of points have to be clarified in approaching Table I.

A. CELL GROWTH *in Vitro*

Untransformed cells growing in a monolayer reduce their multiplication rate after the cells become "confluent." This phenomenon has been called "density-dependent inhibition of growth" or "contact inhibition of growth" (Abercrombie, 1962; Stocker and Rubin, 1967). Many mammalian cells also stop growing when the culture medium becomes depleted of growth factors. Such cells are arrested in the G_1/G_0 or A state of the cell cycle (Burns and Tannock, 1970; Smith and Martin, 1973). Both density-inhibited and growth factor-depleted cells are stimulated to synthesize DNA and proliferate when fresh serum (Todaro *et al.*, 1967; Pardee, 1974; Holley, 1975) or purified growth factors (for a recent review, see Rudland and Jimenez de Asua, 1979) are supplied. The phenomena of density inhibition of growth and growth control by serum depletion are closely related. For instance, the maximum density cells reach at confluence is dependent upon the serum content of the growth medium.

B. Untransformed Cells

Untransformed cells discussed in this chapter are of two different kinds. One type comprises the cultures derived from embryo cells. These cells have the advantage of being only a few generations *in vitro,* and may thus possess most of the characteristics of the cell in the living animal. The other type of "untransformed cells" that have been used in the majority of the studies discussed here are the permanent cell lines. These lines have been in culture for many generations. It has been shown that malignant tumors can be induced in mice when the so-called "normal" 3T3 cells are implanted subcutaneously (Boone, 1975; Boone *et al.*, 1976). It thus seems that 3T3 cells and probably other cells from established lines cannot be considered "normal" cells in a strict sense. We have thus preferred to use the word "untransformed" rather than "normal" in the course of this discussion. However, the growth properties of such untransformed cells of established lines are similar to those reported for embryo cells, and hence they can still be used as control systems. Critical evaluation of the data in Table I reveals the fact that at least as far as the uptake processes are concerned, the results obtained for embryo cells and untransformed cells of the established lines are qualitatively similar. Unfortunately, not enough data regarding mammalian embryo cells are available to support a meaningful quantitative comparison. The experimental convenience of working with an established line is probably the reason why most of the available data indeed result from experiments performed on the limited number of cell lines.

C. Transformed Cells

The transformed lines represented in Table I are of different sources and were obtained under different conditions. Some of the cell lines were obtained by transformation *in vitro* with oncogenic virus. Some were obtained by transformation with a carcinogen *in vitro* or *in vivo*. Some cell lines arose by spontaneous tranformation from established untransformed lines or from cultures derived from embryo cells. In a few instances, cell lines were obtained from a malignant tumor. The cells are regarded as transformed by one or more of the established criteria, namely, decreased adhesion to the substrata, altered cellular morphology, ability to grow in soft agar, and ability to induce tumors in suscep- tible hosts. All transformed cell lines discussed here lost to some degree the density- and serum-dependent growth control properties, and, as will be dis- cussed in detail, these lost properties are probably the most relevant to the control of nutrient uptake rate.

In order to stress the general nature of the findings described in Table I, we have included data regarding the activation of uptake processes by lymphocytes from various sources.

D. Experimental Systems

The data presented in Table I were obtained in three different, closely related, experimental systems.

1. Relationship between Nutrient Uptake Rate and Cell Density

The most consistent result that was demonstrated in numerous experiments performed in a period of some 10 years in many laboratories is that the uptake rate of sugars, uridine, and some amino acids decreases significantly as cell density increases or when cells are transferred to a suboptimal medium. This phenomenon, like density inhibition of growth, is lost (or reduced) when cells are transformed either by an oncogenic virus or by a carcinogen. Cells, which for one reason or another revert and regain some measure of density inhibition of growth, regain at the same time "density inhibition of uptake." Of great interest in this respect is the work described by Schultz and Culp (1973) and Bradley and Culp (1974). These authors studied the uptake of 2-deoxyglucose and described a cell line derived from SV40 virus-transformed 3T3 cells (SVT2 clone). These cells have retained the SV40 genome but regained some morphological and growth properties of the original 3T3 cells, including that of density inhibition of proliferation. It should be noted, however, that the maximum drop in uptake rate is ~10-fold in the untransformed cells as compared to an ~5-fold drop in the revertant cells as they reach confluency.

Dubrow et al. (1978) studied the uptake rate of 3-O-methylglucose and of α-aminoisobutyric acid (AIB) into 3T3 cells, SV40-transformed 3T3 cells (SV101), and three revertant cell lines derived from SV101 that regained some manifestations of growth control. The revertant cell lines described in their work differ from the line studied by Culp and his co-workers, since the former re- gained sensitivity either to cell density or to the concentration of serum in the medium. Dubrow et al. showed that cessation of 3T3 cell growth in G_1 (or G_0) under conditions of confluence or serum deprivation was associated with reduced rates of uptake for both compounds. Density and serum dependence of growth and uptake was eliminated in the transformed line SV101. The revertant line FIS V101 retained density regulation of growth and also density regulation of uptake, but in that case neither growth nor uptake was serum dependent. The serum revertants, AfSV7 and LsSV6 have regained both density and serum regulation of growth, but these regulations differ from the original mechanism of 3T3 cells (i.e., entry into a G_1 or G_0 arrest). For these cells, uptake rates were high even under conditions of confluence or serum deprivation. Thus, the authors conclude that uptake rates are not reduced simply as a result of slower cell growth. It seems that growth arrest at a specific point in the cell cycle is necessary in order to slow down the rate of uptake.

At this point, one asks the following questions: Could the phenomenon of density regulation of uptake be due to the fact that cells are synchronized by some as yet unknown mechanism at a certain point in the G_1 phase of the cell cycle? Could the differences of uptake rate observed between growing and quiescent cells by explained by their different distribution in the different phases of the cycle? At the moment, these questions cannot be answered rigorously. The answers can be given only when the same cells are synchronized both by serum depletion and by some different independent method (for instance, double thymidine block or mitotic shaking) and the uptake rate at various stages in the cell cycle is estimated in parallel for the two synchronization methods. The evidence that such an explanation is at least a reasonable one is presented in a study by Sander and Pardee (1972). CHO cells were synchronized by isoleucine deprivation and by mitotic shaking. It was shown that the rate of uptake of AIB was reduced at M and early G_1 periods of the cell cycle. A doubling of the uptake rate occurred when the cells progressed further into G_1. From G_1 onward, the uptake activity (assayed per mg protein) remained constant throughout the cell cycle. The rates of uridine and thymidine uptake also doubled at that stage of the cycle. Measurements of uptake at intervals over a period of several days involves a comparison between cells at different metabolic stages. The question arises as to what is the most appropriate unit of comparison. Should uptake rate be expressed per cell, per mg of cell protein, or possibly as a function of cell volume or the surface area of the membrane? This question was discussed previously by Parnes and Isselbacher (1978) and is especially important when small changes in uptake rate ($\sim$2-fold) are described. Spaggiare et al. (1976) estimated the median volume and protein content of 3T3 and SV40-transformed 3T3 cells (SV40-3T3 cells) as a function of days in culture and found a 2-fold decrease both in the average volume and in the protein content of the cells within a period of 6 days. The effects for untransformed and transformed cells were very similar. Since transformed cells do not arrest in G_1 the decrease in cell volume and protein content is probably not caused by a specific growth arrest in G_1. Most of the data in Table I consist of effects that are considerably larger than a 2-fold decrease and thus are probably not due only to an artifact caused by the decrease in volume or protein content of the cells as they reach confluency.

In conclusion, it is possible to state that the most striking and consistent effect presented in the data of Table I is a significant decrease in the uptake rate of glucose and some of its analogs, nucleosides (uridine and thymidine), and some amino acids, as untransformed cells proliferate in culture and reach confluency. This phenomenon is lost when cells are transformed, and seems to be closely related to the phenomenon of density inhibition of growth. There are some indications that the decrease is specific to cells that are arrested at a specific point in the cell cycle (in G_1 or G_0), and may be due to the fact that rates of uptake are lower at the G_1 phase of cell cycle than at the later phases.

2. *Stimulation of Uptake by Growth-Related Stimuli*

A closely related phenomenon to the one of "contact inhibition of uptake" is the stimulation of nutrient uptake into quiescent cells by growth-promoting stimuli. Addition of fresh serum to quiescent fibroblasts induces a rapid (within minutes) metabolic change and subsequently accelerates the entry into DNA synthesis and cell division. These rapid changes include the increased transmembrane fluxes of ions, which will be discussed in another section of this chapter, and an increase in the uptake rate of nucleosides, glucose analogs, and some amino acids. Since these changes occur within minutes of the addition of serum or other growth-promoting agents, changes of cell volume or cell protein content cannot play an important role. This observation, in addition to the fact that cells of the same culture can be compared before and after the activation, enables one to estimate the changes in an accurate and quantitative manner. A 2-fold increase of uptake rate within minutes of the addition of activator can be considered a significant increase. Again, as discussed earlier for the inhibition of uptake with increasing density, the stimulation of uptake by fresh serum, polypeptide hormones, and even nonspecific activators like calcium phosphate (Barnes and Colowick, 1977) seems to be a widespread phenomenon, occurring in numerous cell lines and upon activation of lymphocytes by mitogenic stimuli. The sensitivity of the uptake process to serum, again, is lost upon transformation of fibroblasts to the malignant state. The uptake of DNA precursors like thymidine and deoxycytidine is not one of the early events that are stimulated upon activation. The increase in uptake occurs only hours after serum addition and upon the cells' entry into the S-phase (DNA synthesis) of the cell cycle. The stimulation of uptake processes is one of the phenomena that accompany the reentry of quiescent cells into the growth cycle and can be considered as the reversal of the inhibition of uptake that occurred when cells were grown to a high density or in the absence of serum. Sugar uptake by revertant, SV40-3T3 cells, which regained contact inhibition of growth, could be stimulated by insulin in a manner similar to that of the parent untransformed line (Bradley and Culp, 1974).

3. Transformation-Specific Increase in Uptake Rates

We have discussed in detail the relation between contact inhibition of growth, the state of quiescence, and the rate of nutrient uptake. It becomes apparent that due to the fact that transformed cells have lost the ability to reduce their proliferation rate at high cell density, these transformed cells, at such high densities, can take up nutrients at a faster rate than their untransformed counterparts. A strict comparison between the uptake rate of transformed and untransformed cells should be made when both cell lines are at their initial, logarithmic phase of growth. Weber et al. (1976) addressed the question of transformation-specific increase in uptake rate in a culture of chick fibroblasts. They tried to define those changes in uptake that can be detected when transformed and untransformed cells are growing at the same rate. They found that changes in hexose uptake persisted under conditions of equal rate of growth. Indeed, a close inspection of Table I leads us to the same conclusion for mammalian cells. Schultz and Culp (1973), Oshiro and Di Paola (1974), Siddiqi and Iypes (1975), and Dubrow et al. (1978) have all taken into account the rate of growth of the untransformed cells and found an enhanced rate of sugar uptake upon transformation. This conclusion has been widely accepted, and indeed, the use of the enhanced rate of sugar uptake as a diagnostic tool for the detection of brain tumors has been suggested by Wassenaar et al. (1975). When dealing with transformation by an oncogenic virus, one has to rule out the possibility that the effect on uptake rate is due to virus infection rather than to transformation itself. Temperature-sensitive viruses were used for this purpose. Moolten et al. (1977) studied the rate of uptake of 2-deoxyglucose into untransformed BHK cells and a BHK cell line that was transformed by a temperature-sensitive virus DMN4B. The infected cells exhibit transformed behavior when grown at 38.5°C and do not display the transformed phenotype at 32°C. When assayed at 32°C, 2-deoxyglucose uptake is significantly lower for the infected DMN4B line as compared to the untransformed BHK line. At 38.5°C, the uptake rate for the two cell lines is comparable. In this study, the rate of sugar uptake is not faster for the transformed line, even at the permissive temperature. May et al. (1973) studied the rate of uptake of 2-deoxyglucose into NRK (normal rat kidney) cells untransformed and transformed by Moloney sarcoma virus and a temperature-sensitive strain of the same virus. The enhanced uptake rate was detected for the cells grown at the permissive temperature but not for those grown at the nonpermissive temperature. Shifting of the cells from the nonpermissive to the permissive temperature caused increase in uptake rate. The change in uptake rate occurred over a period of 5 days. Miller et al. (1975) studied the rate of 2-deoxyglucose and AIB uptake into Vero cells (African green monkey cells) 8 and 100 hours following permissive infection by SV40 virus. No significant difference in the kinetic parameters of the uptake process could be determined. The authors concluded that the enhanced uptake into virally transformed cells occurs only when the host cell's genome is altered.

The study of Siddiqi and Iypes (1975) provides an interesting example of the effect of transformation by a chemical carcinogen on sugar uptake rate. Their work is especially illustrative because they used a number of transformed cell lines that were either obtained *in vivo* by feeding rats the carcinogen 4-dimethyl aminoazobenzene or *in vitro* by transforming the cells in culture with the same chemical or with aflotoxin B1 or *N*-methyl-*N*-nitrosourea. They included in their study permanent lines (HCT and B-203) in addition to the untransformed controls, which were cells of epithelial origin and were obtained by the authors from adult Vistar rat liver and newborn rat liver. The malignant cells (hepatoma cells) possessed a higher uptake rate at all stages of the culture; a decrease in uptake at confluency was observed in those lines that exhibited density inhibition of growth whether normal or malignant. A positive correlation between the increase in $V_{\max}$ for a 2-deoxyglucose uptake and the emergence of additional criteria for transformation (ability to grow in soft agar) was obtained for all lines. The comparisons between numerous lines obtained under different conditions in parallel experiments is very useful and illustrative. A very different picture emerges when the effect of transformation on the rate of nucleoside uptake is examined. With the exception of one study (Eilam and Bibi, 1977), whenever examined, the rate of uridine uptake into transformed cells was comparable, or slightly lower, than the rate of uptake into fast-growing untransformed cells. We examined the rate of uridine uptake into Nil-8 cells (Heichal *et al.*, 1979), untransformed and transformed by MSV. Both cell lines were grown for 3 days at restrictive serum concentration (0.25%). Ten percent dialyzed serum was added to half of the cultures. Whereas the uptake of uridine into untransformed cells increased $\sim$2-fold within 90 minutes after serum addition, the uptake into Nil-SV cells remained unchanged. But the uptake rate of uridine into transformed cells was more like that of the quiescent cells and significantly lower than uptake into the serum-activated cells. The measurements were obtained on a per cell basis; since the volume of transformed cells is slightly smaller than that of the untransformed cells, uptake rate measured on a volume basis may be comparable for transformed and serum-activated cells. The data accumulated in the literature lead one to conclude that no increase in uridine uptake that is specific to the transformation state can be observed in mammalian cells.

Less experimental data have been accumulated regarding the uptake of amino acids into untransformed and transformed mammalian cells. Thus it would be very hard to reach a firm conclusion regarding a transformation-specific effect. The experiments performed by Foster and Pardee (1969) with 3T3 cells and polyoma-transformed 3T3 cells (Py-3T3 cells), by Dubrow *et al.* (1978) with 3T3 cells and their SV40-transformed counterparts, and by Ronquist *et al.* (1976) with glia and glioma cells indicate that transformed or malignant cells take up AIB and cycloleucine at a faster rate than untransformed cells even at the initial phase of their growth.

One of the established "criteria" for transformation of cells in culture is their decreased adhesiveness to substrate (for a recent review, see Wright et al., 1978). It seems appropriate at this point to mention the anchorage-dependent changes in uptake reported by Otsuka and Moskowitz (1975). They found a low uptake of leucine in susupension cultures as compared to sparse (attached) monolayer cultures of 3T3 cells. The pattern of 2-deoxyglucose uptake was similar under both conditions. 3T3 cells in suspension culture responded to the addition of fresh serum by increased rate of uptake. It thus seems that the typical pattern of nutrient uptake in transformed cells is not due to loss of anchorage-dependent growth. There seems to be overwhelming evidence that uptake patterns of several nutrients are affected by the states of growth or transformation of cells in culture.

III. Are Changes in Uptake Rate Due to Changes in the Rate of Transport across the Plasma Membrane?

The problems involved in differentiating between transport across the plasma membrane and the subsequent metabolic trapping of substrates were recognized long ago (see Hatanaka, 1974; Plageman and Richey, 1974; Berlin and Oliver, 1975; Parnes and Isselbacher, 1978). But only recently, when new methods became available and new criteria were developed in various laboratories, has it become possible to attempt to answer with some degree of confidence the question at the head of this section. We will discuss four different, although related, approaches to the problem of the recognition of membrane-related changes in the sometimes complicated process of nutrient uptake.

A. Kinetic Differentiation between Transport and Metabolic Trapping

The fact that rates of transport are estimated from initial rates of uptake processes has been accepted by almost all investigators of nutrient uptake into cells in tissue culture. The difficulty lies in the question of assignment: how fast should the uptake measurement be in order to be considered a true initial rate. It turns out that no simple answer can be given, and a number of criteria have to be fulfilled before an assignment of initial rate can be made with certainty. These criteria were developed in a similar manner by a few investigators (Graff et al., 1978; Plageman et al., 1978; Heichal et al., 1979; Koren et al., 1979). The following paragraph will be devoted to the outline of the criteria needed to distinguish between transport and subsequent metabolic events. Let us first de-

TABLE I

CORRELATIONS OF UPTAKE RATES WITH GROWTH OR TRANSFORMATION STATES

Cell type	State of transformation	State of growth	Substrate	Effect	Reference
A. Sugar uptake					
1. 3T3 mouse fibroblasts	Untransformed and transformed by MSV	Growing and density inhibited	2-Deoxyglucose	The rate of uptake by untransformed cells decreased even during exponential growth and reached a minimum as the cultures approached saturation density. Cultures grown in serum-deficient medium exhibited rapid inhibition of growth and reduced uptake rates. Transformed cells failed to show either growth-dependent or density-dependent inhibition of deoxyglucose uptake	Bose and Zlotnik (1973)
2. BALB/c 3T3 mouse fibroblasts	Untransformed or transformed by SV40 and revertants	Sparse and growing, or confluent	2-Deoxyglucose	The rate of uptake by sparse untransformed cells was one-fourth and that by revertant cells was three-fourths of the uptake rate into transformed cells. The rate of sugar uptake by untransformed cells dropped to 10% at confluency; the drop was only 5-fold for the revertant cells. Subconfluent cells of all three lines had similar K_m. Confluent 3T3 and confluent revertant cells showed an increase in K_m compared to subconfluent cells	Schultz and Culp (1973)
3. a. Mouse embryo cells b. 3T3 cells	a. Untransformed b. Untransformed or transformed	Subconfluent to confluent. Density range: 2.9–16×10^4 cells/cm^2	2-Deoxyglucose	The apparent K_m for uptake was about the same for all cell types. The V_{max} for secondary cultures decreased 6-fold from sparse to density-inhibited cultures. (The same whether estimated in monolayer cultures or	Plagemann (1973)

Cell line	Cell state	Condition	Sugar	Observations	Reference
	spontaneously or by infection with murine leukemia virus or murine sarcoma virus			for cells dispersed by trypsin.) The V_{max} of transformed cells was 4 to 25 times higher than that of density-inhibited mouse embryo cells	
4. BALB/c 3T3 fibroblasts	Untransformed or transformed by SV40 and revertant cells	Growing and contact inhibited	2-Deoxyglucose	Sugar uptake diminished when untransformed cells became growth inhibited. Insulin stimulated uptake in nontransformed and density inhibited revertant cells but not in transformed cells	Bradley and Culp (1974)
5. BALB/c 3T3 fibroblasts	Untransformed and transformed by chemical carcinogens	Sparse, fast-growing, and confluent	2-Deoxyglucose	The difference between the uptake rate into exponentially growing transformed and untransformed cells was from 50 to 100%. In crowded cultures, untransformed cells ceased taking up the sugar whereas chemically transformed cells continued at the same rate. When serum concentration was reduced, untransformed cells took up sugar at reduced rates whereas transformed cells were only slightly affected. When serum concentration was increased from 2 to 10%, untransformed cells exhibited increased sugar uptake followed by cell division	Oshiro and DiPaolo (1974)
6. 3T3 cells	Untransformed	Quiescent	2-Deoxyglucose; 3-O-methyl-glucose	Stimulation of sugar uptake by calcium phosphate, insulin, and epidermal growth factor preceded the effect of the same stimulants on incorporation of thymidine into DNA	Barnes and Colowick (1977)

(continued)

137

TABLE I (continued)

Cell type	State of transformation	State of growth	Substrate	Effect	Reference
7. 3T3 cells	Untransformed and transformed by SV40; revertant lines from the transformed line	Growing and quiescent	3-O-Methylglucose	Cessation of cell growth under conditons of confluence was associated with reduced rate of uptake. Both effects on growth and uptake were eliminated by transformation. Uptake was faster in sparse, transformed cultures as compred to untransformed or revertant cultures	Dubrow et al. (1978)
8. 3T3 mouse cells. BHK (baby hamster kidney) cells	Untransformed or transformed by polyoma virus or SV40	Crowded but not confluent	2-Deoxyglucose	Uptake of the sugar after 10 minutes incubation was ~3-fold faster in transformed as compared to untransformed cells	Isselbacher (1972)
9. Swiss mouse embryo fibroblasts	Untransformed and transformed by MSV	Five days after plating	2-Deoxyglucose	Sugar uptake kinetic parameters differed in transformed and untransformed cells. K_m decreased 4-fold in infected cells and V_{max} increased 3 to 4-fold in infected cells	Hatanaka et al. (1970)
10. Nil; BHK cells	Untransformed or transformed spontaneously or by polyoma virus	Near confluence	Galactose	Enhanced (3-fold) incorporation in all transformed cells compared to untransformed cells. (Prolonged incubation with substrate: 60 minutes)	Kalckar et al. (1973)
11. Nil hamster fibroblasts	Untransformed and transformed by polyoma virus	?	Galactose	Enhanced uptake rate of transformed as compared to untransformed cells. The effect is apparently different from the one produced by hexose starvation	Kalckar and Ullney (1973)

138

12. BHK21 hamster cells	Untransformed	Growing; subconfluent and confluent	3-O-Methylglucose	Two components of transport: fast and saturable, slow and unsaturable. The first prevailed in growing cells, the second in quiescent cells	Eilam and Winkler (1976)
13. Epithelial rat liver cells in culture	Untransformed and transformed by carcinogen *in vivo*, or *in vitro*	Subconfluent and confluent	2-Deoxyglucose	The transformed cells exhibited higher uptake rates at all stages of the culture. However, a decrease in the uptake rate of the sugar at confluency was observed in those lines that exhibited density inhibition of growth	Siddiqi and Iypes (1975)
14. C-1300 mouse neuroblastoma	—	Sparse and confluent	2-Deoxyglucose	Uptake was density inhibited throughout the cell growth cycle. As the density increased from 1.5×10^4 to 13×10^4 cells/cm^2, the rate of uptake/cell decreased 4-fold	Edström and Kaine (1976a,b)
15. NRK (normal rat kidney) cells	Untransformed or transformed by Moloney sarcoma virus and a temperature-sensitive strain of the same virus	One to six days after plating	2-Deoxyglucose	Transformed cells took up 2-deoxyglucose 4–7 times faster than untransformed cells. In a line transformed by a temperature-sensitive virus, the increase was observed if the cells were grown at the permissive temperature. When the cells were shifted from the nonpermissive to the permissive temperature, uptake was increased within 5 days to the level of cells grown at the permissive temperature (correlates with morphological expression of the transformed phenotype)	May *et al.* (1973)
16. Bovine lymphocytes	—	—	3-O-Methylglucose	The uptake process is stimulated immediately upon the addition of phytohemagglutinin	Peters and Hausen (1971b)

(continued)

139

TABLE I (continued)

Cell type	State of transformation	State of growth	Substrate	Effect	Reference
17. Rat thymus lymphocytes	—	—	3-O-Methylglucose	Stimulation by concanavalin A for 30 minutes caused an increase in uptake rate from 228 to 640 nmol/10^6 cells · minute	Hume and Weidman (1978)
B. Nucleoside uptake					
1. 3T3 mouse fibroblasts	Untransformed and transformed by polyoma virus	Subconfluent and confluent	Uridine	Uridine uptake decreased more than 4-fold when untransformed cells at the confluent and nonconfluent stages were compared. In transformed cells, the decrease in uptake rate was approximately 25%. Uptake into untransformed cells increased 2 to 4-fold within 15 minutes after addition of fresh serum to confluent cultures. No such effect was seen in subconfluent or transformed cultures. Uridine uptake into transformed nonconfluent cultures was lower (~70%) than the uptake into nonconfluent untransformed cells. (Measured as uptake rate/mg cell protein)	Cunningham and Pardee (1969)
2. 3T3 mouse fibroblasts	Untransformed	Rapidly growing and density inhibited	Uridine	The uptake into acid-soluble pools decreased 3-fold as cell density increased from 0.9×10^4 cells/cm^2 to 3.4×10^4 cells/cm^2	Weber and Rubin (1973)
3. BALB/c 3T3 cells	Untransformed	Sparse; 3 days after seeding; 0.4–0.5×10^4 cells/cm^2	Uridine	Uridine uptake decreased 6-fold within 10 hours of the removal of serum from the culture. Uptake increased 4-fold 2 hours after addition of serum to serum-depleted cultures	Kram et al. (1973)

Cell	Transformation	Growth condition	Substance	Observation	Reference
4. 3T3 cells	Untransformed	Quiescent, density inhibited	Thymidine, deoxycytidine, deoxyguanosine, deoxyadenosine	Initiation of division by fresh serum brought about increased uptake of thymidine and deoxycytidine that coincided closely in time with initiation of DNA synthesis. The increase was specific, for uptake of deoxyadenosine and deoxyguanosine did not change at the time of DNA synthesis	Cunningham and Remo (1973)
5. a. Mouse embryo cells	a. Untransformed	Sparse growing (log phase)	Uridine	Uridine uptake after a 10-minute pulse was reduced 5-fold in quiescent embryo cells as compared to cells in log phase. Uptake into BALB 3T3 cells was reduced 10-fold under the same conditions. Uptake into transformed cells in log phase was comparable to that into untransformed cells at the same stage of growth	Rozengurt et al. (1978)
b. 3T3 cells	b. Untransformed or transformed by SV40 or by polyoma virus				
6. Mouse embryo cells	Untransformed	Grown for 72 hours or more in serum-free medium	Uridine (trace amounts)	Uridine uptake was enhanced (8-fold) within 24 hours and 2-fold within 30 minutes after the addition of 5% serum	Hare (1972)
7. Mouse embryo cells	Untransformed	Quiescent (low serum for 3 days)	Uridine (trace amounts)	Serum or insulin added to the cultures caused a rapid increase (2.5-fold) in uptake rate	Rozengurt and Jimenez de Asua (1973)
8. Novikoff rat hepatoma	Transformed	Varying suspension culture	Uridine	The rate of incorporation of uridine into acid-soluble pools fluctuated about 10-fold during the growth cycle of the cells. Maximum activity was observed only in the exponential phase and then decreased progressively until the stationary phase. The	Plagemann et al. (1969)

(continued)

141

TABLE I (continued)

Cell type	State of transformation	State of growth	Substrate	Effect	Reference
				rate decreased several-fold before cell division commenced following dilution of stationary phase cultures with fresh medium	
9. a. Hamster embryo cells	a. Untransformed	Growing and quiescent	Uridine	The maximum velocity or uridine uptake in growing hamster embryo cells was lower than in the methylcholantrene-transformed hamster cell line. This kinetic constant was further reduced in quiescent cells. The K_m values in growing and in quiescent hamster embryo cells, as well as in MCT cells, was of the same magnitude	Eilam and Bibi (1977)
b. MCT-established line of hamster cells	b. Transformed by methyl-cholantrene				
10. Nil-8 hamster fibroblasts	Untransformed and transformed by MSV	Quiescent (serum depleted)	Uridine	Uptake of uridine increased ~2-fold within 90 minutes of the addition of fresh serum to serum-depleted cultures. No such effect was observed with transformed cultures. After serum depletion, the rate of uridine uptake into transformed cultures was of the same order as the rate of uptake into untransformed, quiescent cultures	Heichal et al. (1979)
11. LU106 human embryonic lung cells	Untransformed	One to five days in culture	Thymidine	Uptake of thymidine into acid-soluble pools was reduced ~100-fold and incorporation into DNA some 50-fold as cell density increased by a factor of ~7 from day 1 to day 5 after plating	Kohn (1968)
12. Human lymphocytes	—	—	Uridine	Activation with phytohemagglutinin caused a 50% increase in uptake within 2 hours and	Kay and Hand-maker (1970)

(continued)

Cell type	Substance	Cell condition	Confluency	Description	Reference
				a 3.5-fold increase within 18 hours of the addition of activator	
13. Bovine lymphocytes	Uridine	—		Activation by phytohemagglutinin induced a 10-fold increase in uridine uptake within 20 hours and a 2-fold increase within 1 hour of the addition of activator. The change was in V_{max}; K_m remained constant	Peters and Hausen (1971a)
14. Pig lymphocytes	Deoxycytidine	—		Increased entry of deoxycytidine began about 12 hours after addition of phytohemagglutinin. Increased V_{max} and no change in K_m	Barlow (1976)
C. Amino acids 1. 3T3 cells	AIB (α-aminoisobutyric acid)	Untransformed and transformed by polyoma virus	Subconfluent and confluent	α-Aminoisobutyric acid and cycloleucine were accumulated about 30% less rapidly by confluent than by nonconfluent 3T3 cells. The effect was due to a change in V_{max} but not K_m. No difference between confluent Py-3T3 cells was observed. Py-3T3 cells accumulated AIB about twice as rapidly as nonconfluent 3T3 cells	Foster and Pardee (1969)
2. a. BHK hamster cells b. BALB/c 3T3 cells	AIB	a. Untransformed and transformed by polyoma virus b. Untransformed and transformed by SV40	Relatively dense, but not confluent; 3 days after seeding	2.5–3.5-fold increase in uptake of AIB and cycloleucine in transformed as compared to untransformed cultures. Effect on V_{max}. K_m remained constant	Isselbacher (1972)

TABLE I (*continued*)

Cell type	State of transformation	State of growth	Substrate	Effect	Reference
3. BALB/c 3T3 cells	Untransformed	Subconfluent and confluent	L-Alanine and L-leucine	L-Alanine was used as a test substrate for the A system and L-leucine for the L system. When cell growth was arrested by allowing cells to approach confluency or by the removal of serum. System A transport decreased and System L activity increased significantly	Oxender *et al.* (1977)
4. 3T3 cells	Untransformed	Quiescent	AIB	AIB uptake was maximally enhanced (4-fold) within 0–2 minutes of the addition of fibroblast growth factor together with insulin and dexamethasone to the cultures	Quinlan and Hochstadt (1977)
5. 3T3 cells	Untransformed and transformed by SV40 (SV-101) and revertant cells	Subconfluent and confluent	AIB	Uptake into untransformed and revertant cells decreased with cell density (assayed 1 day after seeding at different densities) and with time in culture. (Maximum effect ~4-fold decrease.) For transformed cells, the effect of a maximal increase in density was only 25% decrease in uptake rate. Uptake rate after 1 day in culture was 12 for untransformed and 20 nmol/min·mg protein for transformed cells, respectively	Dubrow *et al.* (1978)
6. W1-38 human diploid fibroblasts	Untransformed	Confluent, quiescent	Cycloleucine	Uptake of cycloleucine was increased (50%) 3 hours after these cells were stimulated to proliferate by a change to a medium + 10% serum. No such effect was observed if medium and 0.3% serum were added to logarithmically growing cultures	Costlow and Baserga (1973)

144

7. Human U-787 glia and U-251 glioma cells	Normal and neoplastic	Density: 4×10^6 glioma and 1×10^6 glia cells per 9 cm culture dish	AIB	Uptake (3-minute pulse) for neoplastic cells was 3.6 times higher than for glia cells. The difference was related to the V_{max} values for the two types of cells	Ronquist et al. (1976)
8. Mouse mammary shionogi 115 cells	Neoplastic	Grown in the absence of testosterone (density regulated); in presence of physiological concentrations of the hormone (growth regulation lost)	AIB; cycloleucine	Activity of sodium-dependent transport system was density-dependent (~3-fold decrease in rate for a ~3-fold increase in density). K_m for cycloleucine changed with density. V_{max} was the regulated parameter in AIB uptake	Jeffrey and Robinson (1976)
9. Bovine granulosa cells (primaries)	Untransformed	Subconfluent (exponentially growing) or confluent	AIB	The removal of serum from the growth medium in exponential or confluent stages of growth results in rapid (within 40 minutes) decrease in uptake rate. This decrease is rapidly and completely reversed by the addition of serum	Allen et al. (1979)
10. Rat embryo fibroblasts	Untransformed and spontaneously transformed in vitro	Serum starved	AIB	With cells incubated for 24 hours in serum-free medium, uptake occurred only by diffusion. (No finite K_m.) After stimulation by growth-stimulating serum factors, an apparent K_m could be measured after 2 hours	Frank and Ristow (1974)

fine the term "facilitated diffusion level" as that level of substrate uptake achieved when the concentration of label within the cell is equal to the concentration in the extracellular medium. In the general case, the composition of the label within the cell will be different from the composition outside since it will be distributed between the free substrate and the products of initial metabolic steps. We will assume that the uptake of the metabolite occurs by a facilitated diffusion system and obeys the kinetics of a simple carrier (Stein and Lieb, 1973; Lieb and Stein, 1974) and that this transport step is followed by metabolic conversion of the substrate to a trapped form (i.e., phosphorylation in the case of sugars or nucleosides). This conversion is assumed to obey the kinetics of a simple Michaelis–Menten equation. It was shown by Hiechal *et al.* (1979) that, when the label accumulated within the cell is plotted against uptake time, two linear regions can be expected to occur in the resulting uptake curve. When the uptake experiment commences, the internal free substrate concentration, S_2, is zero. The rate of the trapping reaction is also zero, while the rate of label accumulation within the cell is the initial value for a zero-trans experiment (Lieb and Stein, 1974) when one side of the membrane is devoid of substrate. (We cannot assume S_2 to be equal to zero in every case. This problem will be discussed in detail later). As uptake continues, S_2 rises and the rate of transport (measured as net flow of label into the cell) is reduced as substrate accumulates and begins to leave the cell. The rate of the metabolic trapping increases as S_2 increases. At some value of S_2 (defined by the kinetic parameters of transport and trapping and the extracellular concentration of substrate, S_1), the net rate of metabolic trapping will be equal to the rate of label influx, the net rate of entry of substrate by transport will just balance the consumption of substrate by the metabolic reaction, and a steady state is reached. S_2 remains constant for as long as the balance is kept. (After extended uptake times, the metabolic trapping reaction may be affected by control processes like feedback inhibition or by ATP depletion.) It is very important for the sake of this discussion to note that for such tandem processes, the uptake curve against time may be apparently linear at two different phases, i.e., apparently linear at the initial phase when the internal substrate concentration is so low that outward transport can be neglected and truly linear when the internal concentration is just sufficient for net transport to equal metabolic consumption. One must be able to distinguish between these two linear phases in order to avoid a false identification of initial rate.

The way to distinguish between the two regions is to extrapolate the uptake as a function of time back to zero time. The zero-time uptake obtained by this extrapolation should coincide with a directly determined zero-time value. In addition, this zero-time value should be that predicted from the amount of substrate trapped between the cells and substantially lower than the facilitated diffusion level. This zero-time value, which corresponds to a physical, nonspecific absorption phenomenon, should be proportional to the external substrate concentration over a wide range of concentrations. If extrapolation to zero time is attempted for "long" (minutes, in the case of uridine, for instance) uptake times, the error in the extrapolated value may itself be of the order of magnitude of the facilitated diffusion level (such extrapolation would, of course, be meaningless). Mistaken assignment of linear uptake curves was possibly very frequent in the "classic" measurements of uptake. Heichal *et al.* (1979) have shown that uptake rates obtained from both regions of the uptake curve will follow a Michaelis–Menten rate curve. The initial rate measurements will yield parameters relevant to the transport event, whereas data obtained at "long" uptake times are an estimate of the kinetic parameters for the enzyme responsible for the rate-limiting step in the metablic trapping process. The K_m obtained will be in terms of the extracellular concentration S_1, whereas S_2 is the substrate concentration for the intracellular reaction. $S_2 \leqslant S_1$; thus, the measured K_m provides an upper limit to the K_m of the metabolic reaction.

It should be noted that when the K_m values for the transport and metabolic conversion are of a different order of magnitude, an uptake measurement at a single time point may belong to a different linear region of the curve if taken at widely differing concentrations. Thus a change in the rate-limiting step may be interpreted as the existence of more than one carrier system. The criteria of initial rate assignment must be fulfilled over the whole range of concentrations. It should be noted that when the kinetic parameters for the transport and trapping reactions are very similar, the curvature in the uptake curve will not be apparent and kinetic dissection may become difficult.

B. Nonmetabolizable Substrates

One may be able to differentiate between the transport event and subsequent metabolic trapping reactions by using a substrate analog that is not metabolized by the cell. This approach has been widely used in the study of amino acid transport (AIB, cycloleucine) and in sugar transport (3-*O*-methyl-D-glucose). The following criteria should be fulfilled before a substrate analog can be considered suitable for such a role. The substrate analog must itself be a substrate for the transport system under study. In order to find out whether that condition is fulfilled, mutual inhibition studies should be performed (Stein, 1967; Christensen, 1975). The transport of substrate A should be studied in the presence of substance B and vice versa. These experiments yield both K_m values (Michaelis constants) and inhibition constants K_i for both substances in their role as substrates for the transport system or as inhibitors for the other substrate's uptake. Under conditions of competitive inhibition, if both substrates share the same transport system, the appropriate K_i and K_m values for each substrate should coincide within the error limits of the measurements. It should be noted that in order to get meaningful Michaelis and inhibition constants, the uptake for both substrates has to be assayed under conditions where transport is rate limiting.

Another accepted way of demonstrating that two substrates indeed move across the membrane using the same transport carrier is the experiment of "countertransport" or "trans acceleration." In this case, substrates A and B are at different sides of the membrane. The existence of substance A within the cell accelerates the influx of B and the outflow of A is accelerated by the presence of a high level of B in the extracellular environment. The substance used to load the cell should preferably be the unmetabolizable substance. It should be stressed that a failure to demonstrate trans acceleration between two substrates does not mean that they do not use the same transport carrier. The effectiveness of the exchange depends on the kinetic parameters of the carrier system and particularly upon the relative rates of movement of the loaded (bound to substrate) and unloaded carrier across the membrane (Stein, 1967; Stein and Lieb, 1973; Lieb and Stein, 1974; Christensen, 1975). Before a substrate can be considered as a nonmetabolite, measurements *in situ* should be applied to prove the fact. These include a chromatographic analysis of cell extracts (i.e., Graff *et al.*, 1978; Heichal *et al.*, 1979; Koren *et al.*, 1979 and estimation of intracellular water space: if indeed the substrate is not trapped within the cell by a metabolic process and is taken up by a facilitated diffusion system, the distribution of labeled substrate between the intra- and extracellular space once equilibrium has been reached should be the same no matter how much unlabeled substrate is present in the extracellular medium. (This will not be the case when a saturable reaction is involved in the trapping process; in such a case the unlabeled substrate competes with the radioactive substrate and the amount of trapped label decreases as the concentration of the "cold" substrate increases.) This criterion does not apply whenever an "uphill" or concentrative process is involved, as is the case with the transport of amino acids. As these uptake processes proceed against a concentration gradient, the distribution of label in the extra- and intracellular space does not correspond to a diffusion equilibrium.

A substrate that was not trapped within the cell by a metabolic conversion reaction can be "washed out" from the cell by an excess of buffer washing solution (Heichal *et al.*, 1979; Koren *et al.*, 1979); a failure to bring out most of the substrate after extensive washing implies the existence of a strong binding site or compartmentalization within the cell or alternatively an exchange of the label (mostly for tritiated substrates) with internal macromolecules. The kinetic data obtained under such trapping conditions are hard to interpret, and a substance that is not easily washed out from the cell should be avoided as a model substance.

The detailed molecular rate constants involving a carrier-mediated process can only be studied with a nonmetabolizable substrate analog (Stein and Lieb, 1973;

Lieb and Stein, 1974). This fact adds to the possible advantages of a substrate analog in the analysis of possible changes in a transport process.

Whenever a substrate analog is not available, or if one wishes to study the process with the natural substrate, it may be possible to inhibit the metabolic trapping rate either by eliminating its energy source (ATP depletion) or by using mutant cells that lack the enzymes necessary for the trapping process. Both approaches have been applied recently to the study of nucleoside uptake (Plagemann *et al.*, 1976, 1978; Koren *et al.*, 1978) and sugar transport (Graff *et al.*, 1978).

C. Studies in Membrane Vesicles

A third, entirely different approach to the study of membrane transport as a distinct event from subsequent metabolic trapping is the study of purified membrane vesicles. This approach has yielded some very useful results and has been discussed in detail by Parnes and Isselbacher (1978). The experimental procedure calls for a large number of cells, harvesting and nitrogen cavitation, followed by differential centrifugation to separate the subcellular fractions (Hochstadt *et al.*, 1975).

Although it seems at first that the use of isolated membrane can settle our major question of whether transport or metabolic conversion is subject to control by variations in the state of growth or transformation, it turns out that despite the effort involved in those studies their interpretation is not entirely straightforward.

The first problem one encounters is caused by the small size of the vesicles (0.1 μm) (Quinlan *et al.*, 1976). The amount of label within the vesicles at diffusion equilibrium is much smaller than that found within the original cells; this, of course, makes the measurements of initial zero-trans (Stein and Lieb, 1973; Lieb and Stein, 1974) very difficult. (A zero-trans measurement is an uptake measurement where the concentration of substrate on side 2 of the membrane is very small as compared to the dissociation constant of the substrate from the carrier molecule at side 2. Transport proceeds from the phase at side 1 to side 2). When facilitated diffusion is discussed, the concentration of label within the vesicle may be comparable to the amount of substrate bound specifically or nonspecifically to the membrane. This last problem can be overcome if one compares transport into vesicles to binding to an open membrane preparation obtained by osmotic treatment. When comparing cells from different lines (untransformed as compared to a transformed line) or even the same cell lines under different conditions (growing, as compared to quiescent, before and after serum stimulation), one assumes that the contamination of the vesicles by membranes from endoplasmic reticulum and the protein content of the membrane are the same. This may not always be the case. A more difficult question is whether the size distribution of vesicles prepared from different sources is the same. Changes in physical properties of the membrane may certainly affect the size of the vesicles and unless initial rates can be determined accurately, a different size distribution may explain differences obtained in transport measurements (Parnes and Isselbacher, 1978).

It may be the case that transport is controlled by intracellular soluble affectors that may bind at the inner face of the membrane on the transport carrier itself or at a distinct site. These affectors may not be present in the membrane preparation and the control of the transport event may thus be lost.

Despite the many experiments and theoretical difficulties inherent in the study of transport and its mode of control in membrane vesicles, it may still be the ultimate goal. Every mode of the mechanism of control will meet its final testing in a purified membrane preparation, and perhaps eventually, in a reconstituted transport system.

IV. The Mechanism of Growth-Related Control of Uptake Processes

Having outlined the kinetic and experimental criteria that may serve as useful tools in the study of an uptake process, we can now apply them to some specific systems that have been studied extensively.

A. The Nucleoside Uptake System

The mechanism of activation of nucleosides uptake, especially uridine and thymidine, upon the addition of fresh serum to quiescent cultures has been studied in great detail by a few laboratories, and its discussion may be illustrative of the kind of questions that may be answered by a detailed analysis.

The activation process of uridine uptake proceeds in two phases. We observe a rapid 2- to 4-fold increase in uptake rate, which occurs within minutes of the addition of serum, insulin, or purified growth factor to the medium (Rozengurt and Stein, 1977; Stein and Rozengurt, 1976; Heichal et al., 1979, and others). A second increase in the rate occurs when cells enter into the S phase, a few hours after the addition of serum (Hare, 1972; Rozengurt et al., 1978) or phytohemagglutinin to human lymphocytes (Kay and Handmaker, 1970). The first, rapid increase seems to be insensitive to protein synthesis inhibitors and actinomycin D (Hershko et al., 1971; Jimenez de Asua and Rozengurt, 1974) and thus seems to indicate the existence of a rapid control mechanism that does not involve the de novo synthesis of new entities involved in the uptake process. Investigators who studied the details of this activation process suggested at first that changes in the plasma membrane were involved in modulation of the transport event (Stein and Rozengurt, 1976; Rozengurt, 1976; Rozengurt and Stein,

1977). The same investigators, however, were the first to apply a severe kinetic test as to whether transport across the membrane or a subsequent metabolic event was involved in the increase of uptake rate (Rozengurt *et al.,* 1977). They used both uridine and the slowly metabolized cytosine-β-D-arabinoside. Although it has not been shown that the analog enters the cell by the same transport system as uridine, good evidence indicates that the nucleoside carrier is of a very broad specificity (Plagemann *et al.,* 1978). Under conditions where the uptake of uridine was significantly stimulated, the uptake of cytosine-β-D-arabinoside remained unchanged. In addition, the initial uptake of uridine (up to ~20 seconds) seemed to be unaffected by the addition of fresh serum to quiescent 3T3 cells, whereas the uptake at "long" (minutes) uptake times was greatly enhanced. The authors performed a chromatographic analysis of the radioactivity within the cells after short incubation with the precursors and showed that the percentage of phosphorylated uridine in the cell extract as compared to the free nucleoside was enhanced upon the addition of serum. The same chromatographic analysis was applied by Rozengurt *et al.* (1978) to mouse embryo fibroblasts and two lines of 3T3 cells (Swiss and BALB). Quiescent cultures were exposed to growth medium containing 10% fetal bovine serum for 60 minutes and then pulsed for 5 seconds with trace amounts of tritiated uridine. Cell extracts were analyzed chromatographically on Dowex columns. The relative peak values for phosphorylated and nonphosphorylated nucleoside were the following: for mouse embryo fibroblasts, 8:1.5 and 4.5:5 in the absence or presence of serum for the nonphosphorylated and phosphorylated fractions, respectively; 2.5:4.5 (no serum) and 0.5:8.5 (serum added) for Swiss 3T3 cells; and 4:1 (no serum) compared to 1.5:7.5 in the presence of serum for BALB 3T3 cells. Koren *et al.* obtained similar uridine uptake patterns for 3T3 cells (Koren *et al.,* 1978). In addition, they showed that when serum-activated cells were incubated with 10 mM 2-deoxyglucose, an uptake curve similar to that for quiescent cells was obtained (Fig. 1). Here again, it seems clear that the initial uptake rate for 5 μM uridine is identical for activated, quiescent, and 2-deoxyglucose-treated cells. The glucose analog enters the cells, is then phosphorylated without further metabolism (Renner *et al.,* 1972), and thus serves as a trap for intracellular ATP. The fact that uridine uptake curves (Fig. 1) are superimposed within the first 30 seconds indicates that serum activation or ATP depletion does not affect transport. That treatment of serum-activated cells with 2-deoxyglucose generates an uptake curve that may be superimposed on that of quiescent cells confirms the findings that serum activation of uridine uptake is concerned with intracellular metabolic events, namely, the phosphorylation of the uridine. The work described up to now deals with experimental data obtained at low concentrations of uridine and may thus be hard to interpret in terms of Michaelis–Menten kinetic parameters. Heichal *et al.* (1979) studied the effect of substrate concentration on uridine uptake by Nil-8 cells. They used quiescent cultures of untransformed

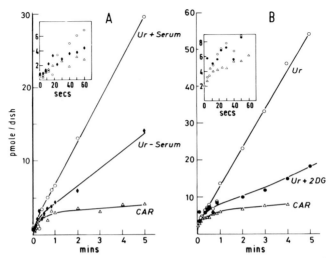

Fig. 1.   Uptake of uridine (Ur) and of its poorly metabolized analog cytosine-β-D-arabinoside (CAR) into acid-soluble pools of 3T3 cells as a function of time. (A) Uptake by activated ($\bigcirc$) and quiescent ($\blacklozenge$, $\triangle$) cells. (B) Following activation of the cells, they were further incubated at 37°C for 30 minutes with either PBS ($\bigcirc$, $\triangle$) or PBS containing 10 mM 2-deoxyglucose (2-DG; $\bullet$). Experiments A and B were performed at different times with different cell densities as reflected by the facilitated diffusion level reached by CAR. The different amount of uridine taken up by the cells in the two identical experiments (activated cells) can be accounted for by the difference in the number of cells per dish. Insets show magnification of data points for the first 60 seconds of each experiment. (From Koren et al., 1978.)

cells before and after 90 minutes stimulation by fresh serum and compared the Michaelis–Menten parameters to those for confluent cultures of transformed cells that were grown under conditions identical to those of the untransformed cells. The authors showed, using the criteria outlined earlier, that the kinetic parameters for transport and phosphorylation can be obtained separately (Figs. 2 and 3). It turns out that for Nil-8 cells the kinetics of the transport event is not affected by the addition of serum. The K_m for transport is 500 ± 200 for untransformed cells and 400 ± 60 for the transformed cells. The maximal velocity is 3600 ± 1200 pmol/min $\cdot$ 10^6 cells for untransformed cells as compared to 1500 ± 200 pmol/min $\cdot$ 10^6 cells for transformed cells. Not only is the transport event not stimulated by serum, but the rate of transport for transformed cells is lower than that for nontransformed cells, at least when estimated on the basis of the number of cells. Koren et al. (1978) have shown in a similar manner that the maximal velocity V_m of transport in SV-3T3 cells is double that observed for the untransformed cells, whether serum-stimulated or not. It seems that no consistent change in the rate of uridine transport occurs upon the transformation of mammalian cells by oncogenic virus. The nucleoside carrier is thus characterized by a much higher K_m than that previously reported for nucleoside transport. The low values were

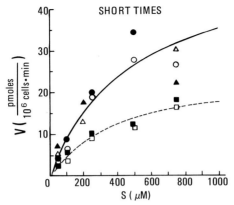

FIG. 2. Velocity of uptake of uridine at short times by normal and transformed Nil-8 cells rendered quiescent or serum-activated, determined at short times and plotted as a function of the uridine concentration. △, Normal cells − serum, Exp. 1; ○, normal cells − serum, Exp. 2; ▲, normal cells + serum, Exp. 1; ●, normal cells + serum, Exp. 2; ■, transformed cells + serum; □, transformed cells - serum. ———, Normal cells; ---, transformed cells; v, velocity; S, concentration. Velocities were derived from time courses by linear regression. The lines drawn are calculated from the Michaelis–Menten parameters obtained from a linear plot (S/v against S) of the data. (From Heichal *et al.*, 1979.)

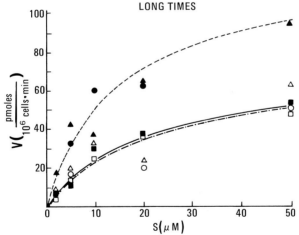

FIG. 3. Velocity of uptake of uridine, measured at long times, by normal and transformed Nil-8 cells with and without prior treatment with 10% serum after serum starvation. ▲, Normal cells + serum, Exp. 1; △, normal cells − serum, Exp. 1; ■, transformed cells + serum; □, transformed cells − serum; ●, normal cells + serum, Exp. 2; ○, normal cells − serum, Exp. 2. ---, Normal activated cells; -.-.-, normal cells (quiescent); ———, transformed cells; v, velocity; S, concentration. Velocities were derived from slopes of linear regressions of data obtained at long periods of time (more than 5 minutes) but otherwise as in Fig. 2. The lines drawn are calculated from the derived Michaelis–Menten parameters. (From Heichal *et al.*, 1979.)

estimated before rapid sampling techniques became available. The existence of a high (in the order of a few hundred micromolar) K_m for nucleosides has been demonstrated now in many cell lines using rapid sampling methodology, ATP depletion methods, and kinase-deficient cells (Wohlhueter *et al.*, 1976, 1979; Plagemann *et al.*, 1978) and seems to be a general phenomenon.

A very different pattern emerges as one examines the data under conditions where phosphorylation to the nucleotide is the rate-limiting step of uptake (long uptake time, low nucleoside concentration, see Section III, A). Figure 3 depicts the kinetic pattern of uridine phosphorylation studied *in situ* in quiescent Nil-8 cells, serum-activated cultures, and their transformed counterparts. It turns out that the kinetic parameters are significantly different for quiescent and serum-activated untransformed cells. The difference is due to a $\sim$2-fold increase of V_{max}. Values of 73 $\pm$ 23 as compared to 125 $\pm$ 13 pmol/min$\cdot 10^6$ cells for quiescent and activated cells, respectively, whereas K_m values remain unchanged 22 $\pm$ 11 and 15 $\pm$ 3 μM for quiescent and activated cells, respectively. It is interesting to note that both K_m and V_{max} for trapping are $\sim$10-fold lower than the parameters for the transport process. It is interesting to note that at very low substrate concentration ($\sim$1 μM), uptake rate is given by the quotient V_{max}/K_m. This ratio is of the same order of magnitude for both transport and metabolic trapping. This fact means that at physiologically significant concentrations (1 μM) the rates of transport and phosphorylation of uridine are comparable. As explained in Section III, the K_m obtained by kinetic analysis of metabolic trapping is an upper limit for the K_m of the enzymatic reaction. That serum stimulation affects V_{max} under conditions where trapping is rate limiting while K_m remains constant has been shown also for 3T3 cells (Goldenberg and Stein, 1978).

1. *Approaches to the Detailed Mechanism of Phosphorylation Control*

The data presented, which were accumulated in a few laboratories, provide strong evidence that the early effect of uridine uptake stimulation is indeed due to the activation of the intracellular uridine phosphorylation system. It has been repeatedly shown however, that when cells are stimulated by serum and then homogenized, the rate of uridine phosphorylation as assayed in cell extract remains unchanged when compared to extracts from nonstimulated, quiescent cells (Rozengurt *et al.*, 1977, 1978; Goldenberg and Stein, 1978). It thus seems that some intracellular control mechanism is lost upon cell disruption. This control mechanism may involve an allosteric effector of the enzyme uridine kinase that is diluted by the preparation of the extract, a subunit interaction that does not occur when the enzyme is diluted, or finally, some interaction with macromolecular structures in the intact cells. As long as the activation event cannot be demonstrated in cell extracts, its nature has to be studied by analyzing uridine uptake rates under those conditions where phosphorylation seems to be rate limiting.

The phosphorylation of uridine involves uridine as a substrate, ATP as the phosphate donor, and Mg^{2+} as the cofactor. By measuring the rate of trapping as a function of extracellular uridine concentration, a value for V_m or maximal velocity was obtained. This maximal velocity was obtained at the usual intracellular concentrations of ATP and Mg^{2+}. Goldenberg and Stein (1978) presented a method for estimating the affinity for ATP of the uridine phosphorylating system. They have used cultures of quiescent 3T3 cells and manipulated their internal ATP concentration by incubating the cells at increasing 2-deoxyglucose concentrations. This was done for serum-stimulated and nonstimulated cultures. Using a suitable extraction method, the authors have been able to measure at the same time uridine uptake and intracellular ATP concentrations. They have shown that the ATP concentration was the same in serum-stimulated and nonstimulated cells, but the sensitivity of the uptake rate to the reduction of ATP concentration was different (Fig. 4). The authors interpreted their results in the following manner. If the K_m for ATP of the uridine phosphorylating system was different for serum-activated cells as compared to quiescent cells, a certain change in ATP concentration would affect these uptake rates to different extents. This interpreta-

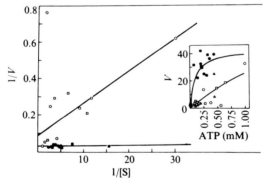

FIG. 4.   Lineweaver–Burk plot of the reciprocal velocity (pmol per 10^6 cells per minute) of uridine phosphorylation in quiescent and serum-activated 3T3 fibroblasts as a function of reciprocal intracellular ATP concentration (mM). The velocities were obtained by linear regression of uptake curves. Open symbols, quiescent cells; closed symbols, activated cells; circles and squares are separate experiments with serum as activator. The inset shows velocity of uridine phosphorylation in activated and quiescent 3T3 fibroblasts as a function of intracellular ATP concentration. The symbols are as described previously except for the triangles, which represent an experiment in which 3T3 cells were treated with insulin (500 ng/ml; 24.3 IU/mg) for 50 minutes (▲) and control quiescent cells were not exposed to insulin (△), using the same conditions of 2-deoxyglucose treatment. The lines drawn were calculated from the Michaelis–Menten equation using parameters obtained by a nonlinear least squares analysis. (From Goldenberg and Stein, 1978.)

tion enabled the authors to estimate the K_m for ATP for quiescent and activated cells and to conclude that the affinity for ATP increases by more than an order of magnitude upon the addition of serum. The intracellular concentration of ATP (~ 2 mM) is of the same order as the K_m in quiescent cells and thus may limit the rate of uptake; the same concentration is much higher than the K_m estimated for the stimulated cells, enabling the phosphorylating system of those cells to operate at its maximal rate. The model presented here provides an explanation for the effect of serum in terms of a change in affinity for ATP of the enzyme involved in the rate-limiting step of uridine phosphorylation, probably uridine kinase. This explanation cannot be considered conclusive before different methods for manipulation of ATP concentration are applied and the observation is extended to other cell lines.

We have recently (Koren and Shohami, 1980) attempted to manipulate intracellular Mg^{2+} concentrations by using the ionophore A 23187, which induces transmembrane movement of divalent metal ions. It seems that uridine uptake by serum-activated Nil-8 cells is less sensitive to Mg^{2+} depletion than uptake by quiescent cells. Taken together with the results obtained by Goldenberg and Stein, these results may mean that the affinity of the uridine phosphorylating system changes for both Mg^{2+} and ATP and that the phosphate donor may be the complex MgATP rather than free ATP.

Rozengurt and Stein (1977) tried to obtain information relevant to the mechanism of the early event of serum stimulation of uridine uptake by studying the time course of the activation event. In order to get accurate results despite the rapid response, they added radioactive substrate to quiescent 3T3 cells together with the stimulating agent (calf serum or insulin; concentration of uridine was 1 μM). Under these experimental conditions, uridine uptake as a function of uptake time proceeds for a certain lag period at a rate typical for the quiescent cells. The rate then increases until it reaches a maximum constant value. The time at which the straight line of the unstimulated system intersects with the straight line typical for the fully "switched on" system was labeled the "lag time." The lag time decreases as the concentration of the stimulating agent (insulin or serum) increases. This decrease in the lag time as a function of insulin concentration is a saturable process. But even as the maximal effect of insulin is achieved (~ 50 μg/ml), the lag time does not disappear altogether. The dependence of the activation rate of uridine on insulin concentration saturates at the same concentration range as the binding of the hormone to receptors in fibroblasts. The residual (at saturating insulin or serum concentration) lag time is sensitive to temperature, and increases upon the addition of prostaglandin El and theophylline. The authors found that the lag period, rather than the final extent of activation, is subject to control by theophylline and prostaglandin El.

The detailed analysis of the time course of the activation event reveals the fact that the stimulation of uridine uptake follows the kinetics of a cooperative phenomenon rather than that of a simple exponential law. This in turn implies the existence of a complicated control mechanism, since more than one event has to occur before stimulation of uridine uptake can be detected.

Rozengurt *et al.* (1978) extended the kinetic analysis of uridine uptake to the late events following serum addition to quiescent cells. The rate of uridine phosphorylation reached its maximal value at 6 hours after the addition of serum to quiescent mouse embryo fibroblasts or BALB 3T3 cells. This increase in phosphorylation preceded a 2-fold increase in the transport rate, which occured as cells entered into the S period some 10 hours after addition of serum to quiescent 3T3 cells.

Quinlan and Hochstadt (1974) studied the rate of uridine uptake by vesicles isolated from growing and quiescent mouse 3T3 cells. They found a 2- to 3-fold faster transport in vesicles prepared from growing cells as compared to those prepared from quiescent cells. This finding is in good agreement with the results of Rozengurt *et al.* (1978) who found a similar difference in transport rate between the quiescent and fully "switched on" mouse fibroblasts. A disturbing feature of the work of Quinlan and Hochstadt (1974) is the slow uptake rate into the vesicles (on the scale of minutes), whereas the level of facilitated diffusion in the whole cells was reached within seconds. The Michaelis–Menten constant for uptake into the vesicles is significantly lower than the K_m for transport of uridine in many cell lines. At present there is no explanation for these discrepancies.

Cunningham and Remo (1973) have studied the rate of thymidine and deoxycytidine uptake into 3T3 cells. They found an increased uptake of both nucleosides in serum-stimulated cultures as compared to quiescent cultures. This increase was not observed for either deoxyadenosine or deoxyguanosine. They found an increase in the maximum velocity of uptake and no change in K_m. Increased uptake of thymidine was prevented by cyclohexamide and further increased by actinomycin D, suggesting that the increase required newly made proteins. The uptake measurements were performed under conditions when phosphorylation is rate limiting, although this fact was not recognized at the time. An extremely interesting finding was obtained by the use of thymidine kinase-deficient (TK⁻) mutants of 3T3 cells. Confluent cultures of wild-type or TK⁻ cells were stimulated by the addition of serum and the kinetic parameters of the uptake process assayed 2 and 10 hours after the addition of serum. It turned out that, whereas the maximum velocity of uptake in 3T3 cells was stimulated by a factor of four after 10 hours, the rate was only doubled in the kinase-deficient cells. It seems reasonable to assume that the increase in uridine transport when cells enter S phase of the cell cycle can account for the observed change in thymidine transport of the mutant cells, especially since it seems that the nucleoside transport system has a broad substrate specificity (Plagemann *et al.*, 1978). Most of the change in the uptake rate of thymidine probably is due to the change in thymidine kinase activity.

In summary, the following conclusions can be drawn regarding the control of nucleoside uptake.

1. The difference in uridine and thymidine uptake between growing and quiescent cultures is mostly due to a change in the activity of the respective kinase. Uridine uptake is stimulated within minutes of the addition of serum to quiescent cells; this stimulation does not involve the synthesis of new protein and is entirely due to the increase of the rate of metabolic trapping of uridine. The increase in uridine phosphorylation rate may be concerned with a change in the affinity for ATP of the enzyme responsible for the rate-limiting step of the trapping reaction. As serum-stimulated cells enter the S phase of the cell cycle, the rate of uridine phosphorylation increases even further, the rate of thymidine phosphorylation is significantly increased, and the rate of transport for both nucleosides increases by a factor of two to three.

2. No significant effect of transformation was observed on either transport or phosphorylation of uridine.

3. Purine nucleoside uptake seems to be controlled in a manner different from that of the pyrimidine nucleoside uptake. Cunningham and Remo (1973) have shown that the uptake rate of deoxyguanosine and deoxyadenosine does not increase when quiescent cells are stimulated to enter the S phase. Rapaport and Zamecnik (1978) have shown that incorporation of adenosine into adenine nucleotide pools increases when mammalian cells are grown under conditions of serum deprivation.

B. The Sugar Uptake System

Graff *et al.* (1978) have studied the mechanism of uptake of the glucose analogs 2-deoxyglucose and 3-*O*-methylglucose into Novikoff rat hepatoma cells. They have used rapid sampling techniques, ATP-depleted cells, and chromatographic analysis. The object of their study was to obtain the kinetic parameters, both for the transport and the phosphorylation of the glucose analogs. One important result obtained by Graff *et al.* is that, in exponentially growing Novikoff cells, the rate of transport exceeds the initial rate of phosphorylation by 20 to 40% at all concentrations of deoxyglucose up to 10 mM (which is a concentration significantly higher than the K_m). This situation is indeed rather different from that reported for uridine (Heichal *et al.*, 1979) where the K_m and V_{max} for phosphorylation differ by one order of magnitude from the parameters for the transport process. That means that at high extracellular uridine concentrations ($C > 100 \, \mu M$), the rate of the phosphorylation is at its maximal value whereas the rate of transport is at less than its half maximal value. Thus the relative rates of transport and phosphorylation are greatly affected by the extracellular substrate concentration. The kinetic parameters for transport and phosphorylation in the uptake process of 2-deoxyglucose into Novikoff cells are very similar to each other. This means that there is no strong curvature in the curve of uptake against time, and thus it is not possible by straightforward kinetic criteria to decide on the conditions where metabolic trapping is rate limiting. For other cells such as those most commonly used in growth control studies, linear uptake curves for 2-deoxyglucose will not always mean that the rate-limiting step is transport across the plasma membrane.

Romano and Colbi (Romano and Colbi, 1973; Colbi and Romano, 1975; Romano 1976) tackled the problem involved in kinetic dissection of transport from phosphorylation. By chromatographic analysis of cell extracts, Colbi and Romano (1975) have shown that the phosphorylation of 2-deoxyglucose is enhanced in confluent SV40-3T3 cells as compared to confluent untransformed cultures. The cell content of the unphosphorylated sugar was not affected by transformation. The first measurement was obtained after 5 minutes of uptake, the intracellular concentration of nonphosphorylated sugar did not change as a function of time because steady state was established (see Section III); thus, no conclusion regarding the initial rate of transport can be drawn from these data. On the other hand, they measured the uptake rate of 3-O-methylglucose (which was relatively slow under their experimental conditions) and found no difference in the kinetic parameters of the transport process. Although their first measurement was obtained at 2 minutes incubation, facilitated diffusion levels were not yet reached, and although the measurement is probably not a good estimation of initial rate, the lack of any change in uptake rate upon transformation seems to suggest that in that experimental setup 3-O-methylglucose transport is not affected by transformation. It should be noted, however, that Dubrow et $al.$ (1978) found higher 3-O-methylglucose uptake rates in sparse SV40-transformed cultures of 3T3 cells than in untransformed cells. Colbi and Romano performed mutual inhibition studies of 2-deoxyglucose, 3-O-methylglucose, and the natural substrate glucose. They found that for the pair, 2-deoxyglucose and glucose, inhibition and Michaelis constants were similar, but this was not the case for the analog pair, 2-deoxyglucose and 3-O-methylglucose. They therefore challenged the use of 3-O-methylglucose as a suitable probe for the glucose transport system. This result is at variance with the observations of Graff et $al.$ (1978) in the rat hepatoma line. For this system, they have been able to show both with mutual inhibition and countertransport experiments that D-glucose, 2-deoxyglucose, and 3-O-methylglucose share the same transport system. It should not be concluded that the result obtained in the hepatoma cell line is necessarily true for other cell lines, but it is at least possible that Colbi and Romano (1975) did not reach the same conclusion because they did not measure initial uptake rates. The question remains as to whether the transport event itself is affected by transformation. It is not possible to answer that question in a conclusive manner without accumulation of a substantial amount of additional data. A qualitative answer can be given, however. Transport is always the first step in the uptake event; thus, even if phosphorylation or trapping of the sugar is greatly enhanced, the maximal experimental uptake rate will be identical or less than the rate of transport across the plasma membrane. That means that unless phosphorylation rates in untransformed cells are several-fold slower than transport rates, the significant stimulation of sugar uptake upon transformation that has been reported in many instances (Table I) could not be due to an effect on phosphorylation alone.

Lever (1976a) has studied the transport of D-glucose into vesicles prepared from 3T3 cells and their SV40-transformed counterparts. She has been able to demonstrate that D-glucose is taken up into the vesicles with no significant metabolic conversion. No effect of either transformation nor contact inhibition of growth could be detected in this transport study. These findings and similar data (Lever, 1979) may indeed mean that phosphorylation rather than transport is the step sensitive to regulation. On the other hand, as stated earlier (Section III), transport may be regulated by an intracellular substance that was lost during vesicle preparation.

In studies of glucose transport rates, one must keep in mind that the intracellular glucose concentration (in contrast to that of uridine concentration) is not insignificantly low, but may be of the same order of magnitude as the concentration of the substrate in the growth medium. If one wishes to compare transport rates under different conditions, the concentration at the inner face of the membrane must be carefully controlled. Both the K_m and V_{max} of influx could be greatly affected by the degree of occupancy of the intracellular site of the transport protein (Stein, 1967; Stein and Lieb, 1973; Lieb and Stein, 1974; Christensen, 1975).

The intracellular substrate concentration determines the degree of exchange versus zero-trans influx in an uptake experiment. Both V_{max} and K_m for an exchange phenomenon are different from the same parameters under zero-trans conditions. The effectiveness of intracellular sugar depletion by a washing procedure with a sugar-free medium is determined by the sugar concentration within the cells, by the rate of glucose transport in the particular system under study [this rate seems to be rather different for different cells; compare the results of Colbi and Romano (1975) to those of Graff et al. (1978)] and the volume ratio between the intracellular space and the washing solution. Differences in the intracellular substrate concentration may be part of the explanation of the fact that some authors report a decrease in the K_m of sugar uptake upon transformation of cells (Hatanaka et al., 1970; Schultz and Culp, 1973; Siddiqi and Iypes, 1975), whereas others do not detect the same difference (Plagemann, 1973; Bose and Zlotnik, 1973). In cases where the rate-limiting step of 2-deoxyglucose uptake changes from phosphorylation in untransformed cells to transport in the transformed cells, a change in the Michaelis constant may be detected that is not due to a qualitative change in the transport system itself.

Another effect that should be taken into account in the course of the analysis of uptake processes in untransformed as compared to transformed cells is the loss of ATP from cells deprived of glucose. Demetrakopoulos et al. (1978) have shown that untransformed cells (chick embryo fibroblasts, human skin fibroblasts, and mouse splenic lymphocytes) maintained their ATP content for 12–24 hours even in the absence of an external carbon source. This was not the case for transformed cells (Py 6, Py-Nil, Ehrlich ascites tumor cells, P388, and CHO in suspension). These cells exhibit a significant decrease in ATP level even within 1 hour of glucose deprivation. This phenomenon may become significant at long uptake times in the absence of external carbon source.

Eilam and Winkler (1976) have studied the effect of cell density and serum stimulation of 3-O-methylglucose transport in BHK hamster cells. They have applied the technique of "equilibrium exchange" to their study (Eilam and Stein, 1974). In an equilibrium exchange experiment, cells are loaded with the same concentration of substrate as that present in the extracellular medium, and one follows the equilibration of a trace amount of labeled substrate between the two sides of the membrane. The advantage of this experimental setup is that because substrate concentration does not change during the experiment, the uptake or efflux can be described by a simple exponential law and it is not necessary to obtain initial rates at very short incubation times. The authors have studied the equilibrium exchange efflux of 1 mM 3-O-methylglucose with cells that were grown for different periods after plating. Theoretically, for an efflux experiment under equilibrium exchange conditions, the logarithm of the fraction of isotope remaining in the cell should be linear with time of efflux. As can be seen in Fig. 5, this is true only for cells that were grown for 22 or 90 hours. For cells grown for 50–70 hours, curved lines were obtained indicating the presence of more than one compartment in the transporting system. In this experiment, the use of a nonmetabolized substrate is essential since the cells are loaded 1 hour prior to the experiment. We can be sure that transport is indeed the rate process under study and that the intracellular concentration is under control. The authors used the equilibrium exchange approach not only to investigate whether a change in transport occurs as a function of cell density or time in culture, but also to address the question of whether the change results from a gradual change in each individual cell's properties or whether the change in every cell is discrete. The data reported in Fig. 5 were examined in their fit to a model consisting of two exponentials in parallel, namely:

$$C_t/C_0 = F \times \exp(-k_1 t) + (1 - F) \times \exp(-k_2 t) \qquad (1)$$

C_t/C_0 is the number of counts remaining in the cells at time t divided by the number at time zero; F is the volume fraction of the fast transporting component; k_1 and k_2 are the two different rate constants for transport. The results presented in Fig. 5 are thus interpreted to mean that there is a gradual decrease in the fraction of transport that occurs at the fast rate and a parallel increase in the fraction occurring at slow rate as a function of the time cells spent in culture. If the change in each cell were due to a gradual process, linear time courses with different slopes would have been obtained as a function of the time after plating. The authors interpret their data as resulting from two types of cells, one in a ''fast'' and the other in a ''slow'' transport state. The transition of each cell from one state to the other appears to be a discrete event, reminiscent of the highly cooperative stimulation of uridine uptake discussed in the last section. Eilam and Winkler (1976) have studied the kinetic parameters for the ''slow'' and ''fast'' transport system. The fast component was characterized by a K_m of ~1 mM and a V_{max} of ~0.4 nmol min^{-1} per cell unit and an unsaturable component of ~0.1 min^{-1} per cell unit. For the slow system, no saturable component could be detected. The unsaturable component was ~0.02 min^{-1} per cell unit. Eilam and Winkler (1976) followed the equilibrium exchange efflux rate as a function of time after addition of serum to quiescent cells. They found that the fraction in the ''fast'' state increased from zero to ~87% within 2 hours after the addition of serum.

The molecular nature of the control of glucose transport by growth-related events is not known at the present. Hume and Weidmann (1978) provided evi-

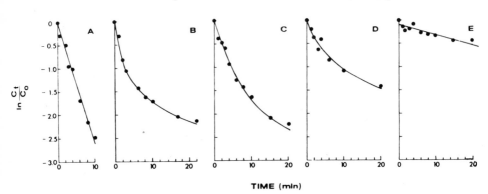

FIG. 5.   Equilibrium exchange of 1 mM 3-O-methylglucose in BHK cells at different times after plating. C_t/C_0 is the number of counts at time t divided by the number of counts at time zero. Each of the data points is the mean of a duplicate determination. The curves in B, C, and D were obtained from the computer-fitting using the model of Eq. (1). The lines in A and E were obtained by linear regression. Cells were plated at 4×10^5 cells/dish and the experiments done at (A) 22 hours; (B) 50 hours; (C) 55 hours; (D) 70 hours, and (E) 90 hours after plating. There was no further change in the slow rate of exchange (E) up to 30 minutes (not shown). (From Eilam and Winkler, 1976.)

dence for the involvement of SH groups in the stimulation of rat thymocyte 3-O-methylglucose transport by mitogenic stimuli. Detailed studies on well-defined systems in cultured cells have to be performed before a molecular model can be brought forward to account for the changes observed in glucose transport rate.

C. Uptake of Amino Acids

The uptake of amino acids differs from that of either uridine or glucose by virtue of being a concentrative process. The concentration of intracellular free substrate exceeds by several fold that of the substrate in the extracellular medium. Thus, the previously discussed criteria for differentiation between transport and metabolic trapping are not all applicable here. Oxender *et al.* (1977a) have studied the interactions among a varied group of neutral amino acids for uptake into 3T3 cells and CHO (chinese hamster ovary) cells. The results showed that here, as in the previously studied Ehrlich ascites tumor cells (Christensen, 1975), the affinities of the amino acids enable one to define two groups. One group includes the amino acids glycine, alanine, α-aminoisobutyric acid, serine, proline, and threonine: the well-known alanine-preferring system or system A. The other system, leucine-preferring or L system, includes the amino acids leucine, isoleucine, valine, phenylalanine, tyrosine, tryptophan, and histidine. This second system has a preference for the branched chain and aromatic amino acids. The two systems differ markedly in their degree of accumulation. Leucine and phenylalanine are taken up rapidly and a steady state in the uptake process is reached within 2 minutes. Glycine and alanine are concentrated to a much higher level (20-fold), and it takes from 20 to 30 minutes for these amino acids to reach a steady-state level of accumulation. These findings are similar to those reported previously for Ehrlich ascites tumor cells (Oxender and Christensen, 1963). Besides being highly concentrative, system A is sensitive to the extracellular concentration of sodium ions. System L, on the other hand, is strongly affected by exchange phenomena, namely, the intracellular concentration of amino acids that share the same transport system.

Oxender *et al.* (1977b) also studied the effect of cell density and metabolic inhibitors on the transport of neutral amino acids of the A and L systems. L-Alanine was used as a test substrate for the A system and L-leucine for the L system. The transport of the A system decreased significantly as cell growth was arrested as a result of contact inhibition or serum deprivation. The transport activity of the L system increased under the same conditions. The content of the amino acids increased 2- to 3-fold whenever cell growth was arrested. Oxender and Christensen (1963) showed that when cells were loaded with amino acids of the L system, influx of the same amino acid was enhanced as a result of exchange (countertransport). One way to explain the increase of L system transport when growth is arrested is by the increased exchange rate caused by higher intracellular levels of amino acids. The increase of the transport activity of the L system and the decrease in the A system could also be affected by the addition of cycloheximide, which caused cell growth arrest and the accumulation of free intracellular amino acids. Although the behavior of the L transport system could be explained by countertransport effects of the intracellular amino acid pool, additional factors seem to contribute to the regulation of the A system.

Robinson (1976) and Robinson and Smith (1976) obtained similar results with S 115 androgen-responsive tumor cells. The activity of the sodium-dependent transport system was found to be density dependent, whereas the activity of the sodium-independent system was not. The nonmetabolizable amino acid α-aminoisobutyric acid was used to probe for the properties of the sodium-dependent system. The transport of cycloleucine occurred by both sodium-dependent and sodium-independent pathways. These investigators found that the K_m for cycloleucine transport changed with density, and they related this finding to the facts that only the sodium-dependent system was affected by density and that the relative contribution of these two systems to the rate of transport changed with density. Only V_{max} for the transport of AIB changed with density. This finding may indicate that either the amount of functional carrier or the activity of the sodium pump caused the density-dependent changes in transport activity.

Allen *et al.* (1979) studied the effect of the removal of serum from the growth medium of ovarian granulosa cells in exponential growth phase or at confluency. They found a pronounced decrease in the rate of transport of AIB that is completely reversed by the addition of serum with no apparent lag time. Both the decrease and its reversal are insensitive to inhibitors of RNA and protein synthesis, indicating the presence of a posttranslational control mechanism. This mechanism may be specific to those highly differential cells because a purified growth hormone that is known to stimulate cell division could not stimulate serum in its uptake-stimulating activity. The same purified fibroblast growth factor did induce a rapid (within 2 minutes) increase in AIB uptake by quiescent 3T3 cells (Quinlan and Hochstadt, 1977).

Membrane vesicles were used extensively in order to elucidate the detailed mechanism of the sodium-dependent transport of the A system. Hamilton and Nilsen-Hamilton (1976) studied the uptake of AIB acid into membrane vesicles obtained from SV40-3T3 cells. They showed that uptake was time dependent and osmotically sensitive. They found a biphasic Michaelis curve and showed that only the high affinity system was stimulated by sodium. If the radioactive label was added to the vesicle system together with 50 mM sodium chloride, both the initial uptake rate and the maximal substrate level in the vesicles increased from 2- to 4-fold. The extent of stimulation by NaCl was dependent upon AIB concentration (higher stimulation at lower substrate concentration). If the substrate was added to the vesicles after equilibration of the vesicle at the same concentration of NaCl, no stimulation by NaCl was detected. In addition, the stimulation of NaCl, when added together with the substrate, decreased with uptake time. The authors explain their findings in the following manner: The stimulating effect of NaCl is probably due to the formation of a transient sodium gradient across the plasma membrane. The sodium gradient reduces the Michaelis constant of the high affinity transport system from 15 to 1 mM. V_{max}, however, is reduced ~4-fold under the same conditions. Similar conclusions have been reached by other authors (Quinlan *et al.,* 1976; Parnes *et al.,* 1976; Lever, 1976b, 1977). Amino acids of the A system seem to be cotransported with sodium ions (Christensen, 1975), and transport is driven by the energy of the electrochemical sodium gradient. This conclusion is further supported by the fact that permeable anions like SCN$^-$ or NO$_3^-$ further increase the stimulating effect of sodium by creating an internal negative membrane potential. The stimulation by sodium was blocked in the presence of nigericin, the monovalent metal inophore that causes dissipation of the sodium gradient. Lever (1976a) compared the transport of AIB into vesicles prepared from density-inhibited, proliferating or SV40-3T3 cells. She found that both the initial rate of uptake and the maximal level of accumulation of the sodium-stimulated transport system were increased by transformation and reduced by contact inhibition as compared to vesicles prepared from untransformed, growing cultures in the presence of the same initial sodium gradient. Thus differences in the extent of the sodium gradient alone do not seem to account for the transformation or growth-related regulation of the A system.

Villereal and Cook (1978) studied the effect of membrane potential on the transport of AIB into intact human diploid fibroblast cells. The Michaelis constant for growing cells was found to be ~1.2 mM and the V_{max} ~16 μmol/gm prot·minute as compared to ~2 mM and ~9 μmol/gm prot·minute for quiescent cells. When quiescent cells were incubated for 1 hour with fresh serum, the K_m was reduced to the value found for growing cells, but V_{max} was not affected. The cells were exposed to varying concentrations of extracellular potassium at a constant concentration of sodium (50 mM). (Choline was used to keep the osmotic balance.) In the presence of valinomycin (the monovalent metal ionophore that induces potassium movement), the membrane behaved as a potassium electrode and membrane potential was determined by the ratio of intracellular and extracellular potassium concentration. Villereal and Cook (1978) were able to reproduce the effect of serum stimulation (1 hour) by causing hyperpolarization of the membrane of the quiescent cells. The rate of uptake of quiescent cells can be set equal to that of stimulated cells when membrane potential decreases from -22 to -49 mV as calculated from the potassium concentration ratio. Changes in membrane potential can cause the reduction of the K_m of the transport system from ~2 to ~1.2 mM. But the difference in V_{max} between growing and quiescent cells, as shown by Lever (1976), is not explained by changes in sodium gradient or changes in membrane potential.

V. Transport of Ions

Although ions cannot be considered metabolites, we have already seen in the case of amino acids that their transport may be closely related to that of other molecules. It seems appropriate to devote a short section to some studies regarding growth and transformation-related changes in ion transport.

Phosphate transport into near-confluent or confluent cultures is higher in transformed 3T3 cells than in the untransformed cultures (Cunningham and Pardee, 1969; Harel et al., 1975). Phosphate uptake rates are greatly reduced as untransformed cells approach confluency (Cunningham and Pardee, 1969; Weber and Eldin, 1971). Weber and Eldin (1971) studied the kinetics of phosphate uptake into 3T3 cells as a function of time after plating and cell density. They have measured the rate of transport into the inorganic phosphate pool, the rate of labeling of the nucleotide pool, and the rate of labeling of RNA. They have found that the rate of phosphate transport drops 5-fold as cells become density inhibited. On the other hand, the level of intracellular inorganic phosphate pools remains constant at around 21,000 pmol/10^6 cells, and those of organic acid-soluble phosphate pools drop slightly from $\sim53,000$ to $\sim52,000$ pmol/10^6 cells.

Jimenez de Asua et al. (1974) studied the kinetics of the early stimulation of phosphate uptake by serum. They have shown that the rate of phosphate transport increased rapidly and linearly after the addition of serum. No apparent lag time could be detected. The rate reached its maximal value within 10 minutes of the addition of serum. Intracellular levels of cAMP were estimated at the same time and their reduction coincided in time with the enhancement of phosphate transport rate. The activation of phosphate transport by serum is due to increase of the maximum velocity of the transport process with no apparent change in the K_m value. The elevation of phosphate transport rate precedes that of uridine uptake by some 10 minutes (see Section IV,A). Jimenez de Asua et al. (1974) showed that the extent of serum stimulation of uridine uptake was greatly affected by the concentration of phosphate in the extracellular medium. Harel and Jullien (1976) used their own data and the data obtained by others to define the term "proximity inhibition." They suggest the use of the term $1/\sqrt{c}$ (where c is cell density) as a measure of the mean intercellular distance. They obtained two parallel straight lines when the rate of phosphate uptake into 3T3 cells was plotted as a function of the "mean intercellular distance," in the presence or absence of serum. On the other hand, the slopes of the lines that describe the rate of phosphate uptake in 3T3 and SV-3T3 cells are significantly different.

The role of monovalent cation transport in the control of proliferation of mammalian cells was recently reviewed by Kaplan (1978). No attempt will be made to present a comprehensive survey of the field. We will mention some representative results and some recent developments that may be relevant to other topics discussed in this chapter.

Some investigators who studied the process of Rb^+ uptake into untransformed and transformed cells reported elevated ouabain-sensitive uptake in the transformed cells (Banerjee and Bosmann, 1976; Kimbelberg and Mayhew, 1975, 1976). These studies were criticized by Parnes and Isselbacher (1978) for their failure to measure either initial rates or to take into account density-related changes in transport rates. Brown and Lamb (1978) studied the uptake of $^{86}Rb^+$ (which was used as a K^+ tracer) into sparse 3T3, SV-3T3, and Py-3T3 cells. The time course of Rb^+ uptake into the cells was identical when sparse cultures were used for the experiment. The first time point in their measurement was at ~ 10 minutes; thus, initial rates of uptake were not estimated, although if judged by the work of other investigators (Rozengurt and Heppel, 1975), uptake is still linear with time after 10 minutes. On the other hand, Spaggiare et al. (1976) find reduced potassium uptake rate into SV-3T3 cells as compared to the untransformed cultures when assayed at low cell density. Rozengurt and Heppel (1975) studied the stimulation of $^{86}Rb^+$ influx rates upon the addition of fresh serum to quiescent 3T3 cells. Uptake rates were enhanced some 4-fold by the addition of serum. With growing cells, the stimulation by serum is 20–40%. Rozengurt and Stein (1977) showed, by exposing the cells to the stimulating serum and the radioactive tracer at the same time, that the activation of Rb^+ uptake was complete within 2 minutes. The increase of Rb^+ uptake rate was reversible within 15 minutes on removal of serum (Rozengurt and Heppel, 1975). These investigators showed that the rate of influx was subject to stimulation by serum, whereas the rate of efflux of the isotope remained constant. Both the stimulated and the basal rates were strongly inhibited by ouabain, indicating that the stimulation occurs via the Na,K-ATPase pump. The enhanced rate of Rb^+ uptake is due to a change in V_{max}, whereas the Michaelis constant is not sensitive to the addition of serum. The stimulation was not affected by the addition of 15 $\mu g/ml$ of cycloheximide. Effectors that were expected to elevate intracellular cAMP did not cause any inhibition in the stimulation of Rb^+ uptake by the addition of serum to the quiescent cultures.

Recent work by Smith and Rozengurt (1978a, b) provides some information regarding the mechanism of activation of Rb^+ uptake via the Na,K-ATPase by serum. They showed that serum stimulated 2- to 3-fold the rate of Li^+ entry into quiescent 3T3 cells or mouse embryo cells. This stimulation was very rapid (2–3 minutes). This effect on Li^+ was not inhibited by inhibitors of macromolecular synthesis. Forty percent of the Li^+ uptake was sensitive to ouabain and seemed to be mediated by Na,K-ATPase. The residual uptake was inhibited by a high concentration of Na^+ and the inhibitor of sodium transport, amiloride, and thus indicated the presence of a specific "sodium porter." Both components of Li^+ uptake were stimulated upon the addition of serum or other growth factors.

Smith and Rozengurt (1978b) have shown as well, in a complementary work, that sodium influx can induce activation of the Na^+,K^+ pump in quiescent fibroblasts. For this purpose they used Monensin, a carboxylic acid ionophore that forms an uncharged complex with Na^+ and proton and catalyzes the electroneutral exchange of the two ions across membranes. In the presence of 0.3 μg Monensin, the Na^+,K^+ pump was stimulated to the same extent as the stimulation caused by the addition of serum. From these data, the authors concluded that there is an Na^+ channel in the plasma membrane of fibroblasts and that the rate of Na^+ influx via that channel is enhanced in the presence of serum. If intracellular Na^+ concentration is limiting the activity of the pump, this additional influx can cause activation of the pump, enhance the rate of K^+ influx, and possibly cause hyperpolarization of the membrane. This model for regulation seems self-consistent and may provide a useful working hypothesis for further research in the area.

VI. Is Nutrient Uptake Directly Involved in the Control of Cell Growth?

Results obtained in recent investigations indicate that the answer to the question presented in the heading of this section is negative, at least for hexose, phosphate, and nucleoside uptake. The argument for nucleoside uptake is indeed trivial; the concentration of uridine in tissue culture media is negligible, and it seems clear that the growth of cells is not affected by this fact. Arguments for phosphate and glucose are more complicated. We have already mentioned the work of Dubrow *et al.* (1978) who demonstrated the existence of a revertant cell line from transformed 3T3 cells that regained both density and serum regulation of growth, but growth arrest did not occur by entry into the G_1/G_0 state that is the mechanism typical to the untransformed parent cell line. These revertant cells did not exhibit density or serum regulation of 3-O-methylglucose or AIB transport.

These results mean that cell growth can be arrested with no change in nutrient uptake pattern. Similar results were obtained by Bush and Shudell (1978). These investigators were able to slow down the rate of growth of SV40-3T3 cells by serum depletion. Under conditions in which the cells were viable but their growth rate decreased several fold, no parallel change in the uptake rates of glucose and amino acid analogs could be detected.

Brownstein *et al.* (1975) used cytochalasin B as a tool to study the interrelationship between DNA synthesis and uptake rates. While DNA synthesis was not inhibited by concentrations of cytochalasin B up to 10 μg/ml, this concentration was high enough to block both glucose uptake and the stimulation of thymidine uptake that occurs as serum-stimulated quiescent cells entered the S phase. Of special interest is the series of studies performed by Cunningham and co-workers (Greenberg *et al.*, 1977; Barsh *et al.*, 1977; Naiditch *et al.*, 1977). The studies were performed with 3T3 cells, HF cells, and polyoma-transformed 3T3 cells. The growth of the cells was arrested when the cells were grown at very low phosphate levels (0.005 mM) or in the absence of glucose. On the other hand, the rate of phosphate or glucose uptake into serum-stimulated or growing cells could be set below the uptake rate into quiescent cells by reducing phosphate and glucose below the levels in culture media (0.1 mM phosphate, 0.7 mM glucose). When nonconfluent cultures were switched to low glucose or low phosphate media, the rates of DNA synthesis or cell division were not affected over a several days period.

In another set of experiments, quiescent cells were switched to media containing high serum concentrations, but low phosphate and glucose concentrations. Nutrient uptake rates in these media remained lower than uptake rates into quiescent cells, but the subsequent onset of DNA synthesis and increase in cell number was the same as that observed in media containing optimal glucose and phosphate levels. These results show that growth-related changes of glucose or phosphate uptake do not control cell proliferation rates under usual culture conditions.

VII. Concluding Remarks

From the data presented in this article, it is possible to conclude that a strong correlation exists between the rate of nutrient uptake and the patterns of cell growth. The rate of sugars, amino acid, and nucleosides are decreased as cell growth is arrested. Sugar and amino acid uptake rates seem to increase when cells in culture are transformed to the malignant state. These increased uptake rates are due to changes in metabolic trapping reactions and, in some cases, to changes in the transport process itself. Future work will probably concentrate upon the elucidation of the relevant control mechanisms on the molecular level.

ACKNOWLEDGMENT

Special thanks are due to Professor W. D. Stein who introduced me to this field. His encouragement, helpful discussions, and critical evaluations of my work during the past years, and particularly during the preparation of this manuscript, have been invaluable to me.

REFERENCES

Abercrombie, M. (1962). *Cold Spring Harbor Symp. Quant. Biol.* **27,** 427–431.

Allen, W. R., Nielsen-Hamilton, M., Hamilton, R. T., and Gospodarowicz, D. (1979). *J. Cell. Physiol.* **98,** 491–502.

Banerjee, S. P., and Bosmann, H. B. (1976). *Exp. Cell Res.* **100,** 153–158.

Barlow, S. D. (1976). *Biochem. J.* **154,** 395–403.

Barnes, D. D., and Colowick, S. (1977). *Proc. Natl. Acad. Sci. U.S.A.* **74,** 5593–5597.

Barsh, G. S., Greenberg, D. B., and Cunningham, D. (1977). *J. Cell. Physiol.* **92,** 115–128.

Berlin, R. D., and Oliver, J. M. (1975). *In* "International Review of Cytology" (G. H. Bourne and J. F. Danielli, eds.), Vol. 42, pp. 287–334. Academic Press, New York.

Boone, C. W. (1975). *Science* **188,** 68–70.

Boone, C. W., Takeichi, N., Paranjpe, M., and Gilden, R. (1976). *Cancer Res.* **36,** 1626–1633.

Bose, S. K., and Zlotnik, B. J. (1973). *Proc. Natl. Acad. Sci. U.S.A.* **70,** 2374–2378.

Bradley, W. E. C., and Culp, L. A. (1974). *Exp. Cell. Res.* **84,** 335–350.

Brown, K. D., and Lamb, J. F. (1978). *Biochim. Biophys. Acta* **510,** 292–297.

Brownstein, B. L., Rozengurt, E., Jimenez de Asua, L., and Stoker, H. (1975). *J. Cell. Physiol.* **85,** 579–586.

Burns, F. J., and Tannock, J. F. (1970). *Cell Tissue Kinet* **3,** 321–334.

Bush, H., and Shodell, M. (1978). *Exp. Cell Res.* **114,** 27–30.

Christensen, H. N. (1975). "Biological Transport." Benjamic, Reading, Massachusetts.

Colby, C., and Romano, A. H. (1975). *J. Cell. Physiol.* **85,** 15–24.

Costlow, M., and Baserga, R. (1973). *J. Cell. Physiol.* **82,** 411–420.

Cunningham, D. D., and Pardee, A. B. (1969). *Proc. Natl. Acad. Sci. U.S.A.* **64,** 1049–1056.

Cunningham, D. D., and Remo, R. A. (1973). *J. Biol. Chem.* **248,** 6282–6288.

Demetrakopoulos, G. E. V., Linn, B., and Amos, H. (1978). *Biochem. Biophys. Res. Commun.* **82,** 787–794.

Dubrow, R., Pardee, A. B., and Pollack, R. (1978). *J. Cell. Physiol.* **95,** 203–212.

Edström, A., and Kaine, M. (1976a). *Exp. Cell Res.* **97,** 6–14.

Edström, A., and Kaine, M. (1976b). *Exp. Cell Res.* **97,** 15–22.

Eilam, Y., and Bibi, O. (1977). *Biochim. Biophys. Acta* **467,** 51–64.

Eilam, Y., and Stein, W. D. (1974). *In* "Methods in Membrane Biology" (E. D. Korn, ed.), Vol. 2, pp. 283–354. Plenum, New York.

Eilam, Y., and Winkler, C. (1976). *Biochim. Biophys. Acta* **433,** 393–403.

Foster, D. O., and Pardee, A. B. (1969). *J. Biol. Chem.* **244,** 2675–2681.

Frank, W., and Ristow, H. J. (1974). *Eur. J. Biochem.* **49,** 325–332.

Goldenberg, G. J., and Stein, W. D. (1978). *Nature (London)* **274,** 475–477.

Graff, J. C., Wohlhueter, R. M., and Plagemann, P. G. W. (1978). *J. Cell. Physiol.* **96,** 171–188.

Greenberg, D. B., Barsh, G. S., Ho, T. S., and Cunningham, D. (1977). *J. Cell. Physiol.* **90,** 193–210.

Hamilton, R. T., and Nielsen-Hamilton, M. (1976). *Proc. Natl. Acad. Sci. U.S.A.* **73,** 1907–1911.

Hare, J. D. (1972). *Biochim. Biophys. Acta* **255,** 905–916.

Harel, L., and Jullien, M. (1976). *J. Microsc. Biol. Cell.* **26,** 75–78.

Harel, L., Jullien, M., and Blat, C. (1975). *Exp. Cell Res.* **90,** 201–210.

Hatanaka, M. (1974). *Biochim. Biophys. Acta* **355,** 77–104.

Hatanaka, M., Auge, C., and Gilden, V. (1970). *J. Biol. Chem.* **245,** 714–717.

Heichal, O., Ish-Shalom, D., Koren, R., and Stein, W. D. (1979). *Biochim. Biophys. Acta* **551,** 169–186.

Hershko, A., Mamont, P., Shields, R., and Tomkins, G. M. (1971). *Nature (London) New Biol.* **232,** 206–211.

Hochstadt, J., Quinlan, D. C., Rader, R., Li, C. C., and Dowd, D. (1975). *Methods Membr. Biol.* **5**, 117–162.

Holley, R. W. (1972). *Proc. Natl. Acad. Sci. U.S.A.* **69**, 2840–2841.

Holley, R. W. (1975). *Nature (London)* **258**, 487–490.

Hume, D. A., and Weidman, M. J. (1978). *J. Cell. Physiol.* **96**, 303–308.

Isselbacher, K. J. (1972). *Proc. Natl. Acad. Sci. U.S.A.* **69**, 585–589.

Jeffrey, H., and Robinson, H. (1976). *J. Cell. Physiol.* **89**, 101–110.

Jimenez de Asua, L., and Rozengurt, E. (1974). *Nature (London)* **251**, 624–626.

Jimenez de Asua, L., Rozengurt, E., and Dulbecco, R. (1974). *Proc. Natl. Acad. Sci. U.S.A.* **71**, 96–98.

Kalckar, H. M., and Ullrey, D. (1973). *Proc. Natl. Acad. Sci. U.S.A.* **70**, 2502–2504.

Kalckar, H. M., Ullrey, D., Kijamoto, S., and Hakomori, S. (1973). *Proc. Natl. Acad. Sci. U.S.A.* **70**, 839–843.

Kaplan, J. G. (1978). *Annu. Rev. Physiol.* **40**, 19–41.

Kay, J. E., and Handmaker, S. D. (1970). *Exp. Cell Res.* **63**, 411–421.

Kimbelberg, H. K., and Mayhew, E. (1975). *J. Biol. Chem.* **250**, 100–104.

Kimbelberg, H. K. and Mayhew, E. (1976). *Biochim. Biophys. Acta* **455**, 865–875.

Kohn, A. (1968). *Exp. Cell Res.* **52**, 161–172.

Koren, R., and Shohami, E. (1980). *Exp. Cell Res.* **127**, 55–61.

Koren, R., Shohami, E., Bibi, O., and Stein, W. D. (1978). *FEBS Lett.r* **86**, 71–75.

Koren, R., Shohami, E., and Yeroushalmi, S. (1979). *Eur. J. Biochem.* **95**, 333–339.

Kram, R., Mamont, P., and Tomkins, G. M. (1973). *Proc. Natl. Acad. Sci. U.S.A.* **70**, 1432–1436.

Lever, J. E. (1976a). *J. Cell. Physiol.* **89**, 779–787.

Lever, J. E. (1976b). *Proc. Natl. Acad. Sci. U.S.A.* **73**, 2614–2618.

Lever, J. E. (1977). *J. Biol. Chem.* **252**, 1990–1997.

Lever, J. E. (1979). *J. Biol. Chem.* **254**, 2961–2967.

Lieb, W. R., and Stein, W. D. (1974). *Biochim. Biophys. Acta* **373**, 178–196.

May, J. T., Somers, K. D., and Kit, S. (1973). *Int. J. Cancer* **11**, 377–384.

Miller, M. S., Kwock, L., and Wallach, F. H. (1975). *Cancer Res.* **35**, 1826–1829.

Moolten, F. L., Moolten, D. N., and Caparell, N. J. (1977). *J. Cell. Physiol.* **93**, 147–152.

Naiditch, W. P., and Cunningham, D. D. (1977). *J. Cell. Physiol.* **92**, 319–332.

Oshiro, Y., and DiPaolo, J. A. (1974). *J. Cell. Physiol.* **83**, 193–202.

Otsuka, H., and Moskowitz, M. (1975). *J. Cell. Physiol.* **86**, 379–388.

Oxender, D. L., and Christensen, H. N. (1963). *J. Biol. Chem.* **238**, 3686–3699.

Oxender, D. L., Lee, M., Moore, P. A., and Cecchini, G. (1977a). *J. Biol. Chem.* **252**, 2675–2679.

Oxender, D. L., Lee, M., and Cecchini, G. (1977b). *J. Biol. Chem.* **252**, 2680–2683.

Pardee, A. B. (1974). *Proc. Natl. Acad. Sci. U.S.A.* **71**, 96–98.

Pardee, A. B. (1975). *Biochim. Biophys. Acta* **417**, 153–172.

Parnes, J. R., and Isselbacher, K. J. (1978) *In* "Progress in Experimental Tumor Research" (D. F. H. Wallach and F. Homburger, eds.), Vol. 22, pp. 79–122. Karger, Basel.

Parnes, J. R., Garvey, T. Q., III, and Isselbacher, K. J. (1976). *J. Cell. Physiol.* **89**, 789–794.

Peters, J. H., and Hausen, P. (1971a). *Eur. J. Biochem.* **19**, 502–508.

Peters, J. H., and Hausen, P. (1971b). *Eur. J. Biochem.* **19**, 509–513.

Plagemann, P. G. W. (1973). *J. Cell. Physiol.* **82**, 421–434.

Plagemann, P. G. W., and Richey, D. P. (1974). *Biochim. Biophys. Acta* **344**, 263–305.

Plagmann, P. G. W., Ward, G. A., Mahy, B. W. J., and Korbecki, M. (1969). *J. Cell. Physiol.* **73**, 233–250.

Plagemann, P. G. W., Marz, R., and Erbe, J. (1976). *J. Cell Physiol.* **89**, 1–18.

Plagemann, P. G. W., Marz, R., and Wohlhueter, R. M. (1978). *J. Cell. Physiol.* **97**, 49–72.

Quinlan, D. C., and Hochstadt, J. (1974). *Proc. Natl. Acad. Sci. U.S.A.* **71**, 5000–5003.

Quinlan, D. C., and Hochstadt, J. (1977). *J. Cell. Physiol.* **93**, 237–246.

Quinlan, D. C., Parnes, J. R., Shalom, R., Garvey, T. Q., III, Isselbacher, K. J., and Hochstadt, J. (1976). *Proc. Natl. Acad. Sci. U.S.A.* **73**, 1631–1635.

Rapaport, E., and Zamecnik, P. (1978). *Proc. Natl. Acad. Sci. U.S.A.* **75**, 1145.-1147.

Renner, E. D., Plagemann, P. G. W., and Bernlohr, R. W. (1972). *J. Biol. Chem.* **247**, 5765–5776.

Robinson, J. H. (1976). *J. Cell. Physiol.* **89**, 101–110.

Robinson, J. H., and Smith, J. (1976). *J. Cell. Physiol.* **89**, 111–122.

Romano, A. H. (1976). *J. Cell. Physiol.* **89**, 737–744.

Romano, A. H., and Colbi, C. (1973). *Science* **179**, 1238–1240.

Ronquist, G., Åren, G., Ponten, J., and Westermark, B. (1976). *J. Cell. Physiol.* **89**, 433–440.

Rozengurt, E. (1976). *J. Cell. Physiol.* **89**, 627–632.

Rozengurt, E., and Heppel, L. A. (1975). *Proc. Natl. Acad. Sci. U.S.A.* **72**, 4492–4495.

Rozengurt, E., and Jimenez de Asua, L. (1973). *Proc. Natl. Acad. Sci. U.S.A.* **70**, 3609–3612.

Rozengurt, E., and Stein, W. D. (1977). *Biochim. Biophys. Acta* **464**, 417–432.

Rozengurt, E., Stein, W. D., and Wigglesworth, N. (1977). *Nature (London)* **267**, 442–444.

Rozengurt, E., Mierzejewski, K., and Wigglesworth, N. (1978). *J. Cell. Physiol.* **97**, 241–252.

Rudland, P. S., and Jimenez de Asua, L. (1979). *Biochim. Biophys. Acta* **560**, 91–133.

Sander, G., and Pardee, A. B. (1972). *J. Cell. Physiol.* **80**, 267–272.

Schultz, A. R., and Culp, L. A. (1973). *Exp. Cell. Res.* **81**, 95–103.

Siddiqi, M., and Iypes, P. T. (1975). *Int. J. Cancer* **15**, 773–780.

Smith, J. A., and Martin, L. (1973). *Proc. Natl. Acad. Sci. U.S.A.* **70**, 1263–1267

Smith, J. B., and Rozengurt, E. (1978a). *J. Cell. Physiol.* **97**, 441–450.

Smith, J. B., and Rozengurt, E. (1978b). *Proc. Natl. Acad. Sci. U.S.A.* **75**, 5560–5564.

Spaggiare, S., Wallach, M. J., and Tupper, J. T. (1976). *J. Cell. Physiol.* **89**, 403–416.

Stein, W. D. (1967). "The Movement of Molecules Across Cell Membranes." Academic Press, New York.

Stein, W. D., and Lieb, W. R. (1973). *Isr. J. Chem.* **11**, 325–339.

Stein, W. D., and Rozengurt, E. (1976). *Biochim. Biophys. Acta.* **419**, 112–118.

Stocker, M. G. P., and Rubin, H. (1967). *Nature (London)* **215**, 171–172.

Todaro, G. J., Matsuya, Y., Bloom, S., Robbins, A., and Green, H. (1967). *In* "Growth Regulating Substances for Animal Cells in Culture" (V. Defendi and M. Stocker, eds.), pp. 87–101. Wistar Institute Symposium Monograph 7. Wistar Institute Press, Philadelphia, Pennsylvania.

Villereal, M. L., and Cook, J. S. (1978). *J. Biol. Chem.* **253**, 8257–8262.

Wassenaar, W., Tator, C. M., and Batty, H. P. (1975). *Cancer Res.* **35**, 785–790.

Weber, M. J., and Eldin, G. (1971). *J. Biol. Chem.* **246**, 1828–1833.

Weber- M. J., and Rubin, H. (1973). *J. Cell. Physiol.* **77**, 157–168.

Weber, M. J., Hale, A. H., You, T. M., Buckman, T., Johnson Terrance, M., Brady, M., and Larosa, D. D. (1976). *J. Cell. Physiol.* **89**, 711–722.

Wolhueter, R. M., Marz, R., Graff, J. G., and Plagemann, P. G. W. (1976). *J. Cell. Physiol.* **89**, 605–612.

Wohlhueter, R. M., Marz, R., and Plagemann, P. G. W. (1979). *Biochim. Biophys. Acta* **553**, 262–283.

Wright, T. C., Ukena, T. E., and Karnovsky, M. J. (1978). *In* "Prog. Exp. Tumor Res." (F. Homburger and D. F. H. Wallach, eds.), Vol. 22, pp. 1–27.

Intracellular Source of Bioluminescence

BEATRICE M. SWEENEY

Department of Biological Sciences,
University of California,
Santa Barbara, California

I. Introduction

Bioluminescence, the emission of light by living organisms, is a phenomenon so beautiful and so surprising that it has attracted the attention of biologists from the beginning of this science (Harvey, 1952). Now we have a fairly complete knowledge of what organisms have this capability, what cells of the organism are responsible, and a great many details of the biochemistry of light emission. There is, however, a gap in our understanding. Bioluminescent organelles have been seen in a few organisms [*Noctiluca,* for example (Eckert and Reynolds, 1967)], but the structure, even the identification of these intracellular sources at the electron microscopic level, is almost completely unknown (Buck, 1978). In all organisms except perhaps the bacteria and fungi (Airth *et al.,* 1966; Berliner and Hovnanian, 1963), light clearly does not originate from the cytoplasm as a

173

whole. Discrete subcellular flashes were seen in the light microscope in *Noctiluca* as early as Quatrefages (1850).

It is clear that the biochemistry of light emission is not the same in all organisms. The molecules from which light is emitted, the luciferins, are quite different chemically in bacteria, earthworms, and fireflies. Although the structure of the luciferin in dinoflagellates is still unknown, the absence of cross reactions between the dinoflagellate luciferin and luciferase and those from other organisms argues for a different biochemistry in this group as well. Different luciferins imply different luciferases. The biochemical diversity among bioluminescent organisms may not be as great as was once thought, since coelenterate-type luciferin ["coelenterazine" (Shimomura and Johnson, 1979)], which is a rather complex molecule containing five rings, is found not only among the Cnidaria (including *Aequorea, Renilla,* and *Cavernularia*), but also in the Ctenophores and even in more distantly related organisms such as the squid *Watasenia* (Cormier, 1978). The diversity in the biochemistry of bioluminescence implies a similar variety in subcellular structures emitting light. Even among organisms sharing the same luciferin, different modes of light production can be found, some lighting internally (*Porichthys*), whereas others secrete luciferin and luciferase separately to react only once outside the animal (*Cypridina*).

In 1952, Harvey summarized what was known at that time concerning the bioluminescence of all organisms. In rereading this monumental work, I found an overwhelming number of references to the cells that were thought from light microscopy to emit light. Two common features of these suspected photocytes that were observed by many early workers in this field were their granular appearance and their yellow or yellow–green color. In *Cypridina,* which secretes both luciferin and luciferase, two types of glandular cells were described by Harvey; one contained rather large, yellow granules, whereas the granules of the other cell were smaller and apparently not colored. The presence of yellow granules was also noted in Hydromedusae, *Renilla,* the Ctenophores, several Annelids including *Tomopteris* and *Odontosyllis,* the snail *Dyakia,* copepods, and the tunicate *Pyrosoma.* That bioluminescent structures should be yellow in color is perhaps not too surprising since the emission of most of these organisms is blue or blue–green. The luciferin would be expected to absorb in the blue region of the spectrum and hence appear yellow. However, in none of these organisms has luminescence been proven to come from the yellow granules seen in the light microscope.

We should expect to find subcellular structures to be the sites of bioluminescence rather than the whole cytoplasm, both on the basis of these numerous observations and from an empirical point of view as well. Only bacteria and fungi emit light continuously. In all others, the discontinuous bioluminescence implies some kind of control mechanism and at least the separation of luciferin and luciferase.

Our knowledge concerning the subcellular sites of bioluminescence has not advanced greatly from the "yellow granule" stage of 1952. Admittedly, it is difficult to identify these sources with certainty, since they can probably only really be seen in fixed material examined in the electron microscope. Image intensification combined with the highest magnification possible in the light microscope can be helpful but requires some means to correlate the light-emitting structures visualized with features seen in electron micrographs. So far the necessary correlation has only been possible in a very few organisms, for example, the polynoid worm *Acholoë,* the firefly, and the dinoflagellate *Noctiluca.* We report here preliminary results of our investigation of another organism, the dinoflagellate *Pyrocystis fusiformis,* which is especially favorable material with regard to the possibility of correlating image intensification with electron microscopy (see Section II,C). It is my purpose here to review what little is known or suggested concerning intracellular sources of bioluminescence. Perhaps a review of the state of the art with regard to the cytology of bioluminescence will stimulate investigation in this fertile field.

Omitted from consideration are all cases where the bioluminescence is known to arise from symbiotic bacteria. Such cases are common, particularly among fish. The bacteria fulfill the role of photocyte in these associations. Nothing is known about the subcellular structure of bacterial light emission. Only those bioluminescent organisms in which something is known of the intracellular source of bioluminescence will be discussed.

II. Dinoflagellates

Among the dinoflagellates, there are many species that are capable of light emission on mechanical stimulation (Sweeney, 1979a). The striking displays of bioluminescence in the sea are most often due to these organisms, which are common members of the plankton; sometimes dinoflagellates are the only component, as in "red tides" and the "bays of fire" found in the Caribbean islands.

The dinoflagellates are unicellular; thus, there is no question of distinguishing a photocyte. None the less, the identification of an intracellular source of light emission is not simple. Studies with homogenized cells have further complicated the problem rather than helped it. Light can be obtained from both a soluble and a particulate fraction, which separate when extracts are centrifuged. The luciferin and the luciferase are the same in both fractions, which have the same emission spectrum. The particulate material, once its luminescence has been spent, can be recharged in the presence of added luciferase and luciferin free from its binding protein (Fogel and Hastings, 1972; Fuller *et al.,* 1972). The particles responsible for the bioluminescence in the sediment have not yet been identified although it has been suggested that they are probably membranous (Hastings, 1978). There is still the possibility that they are vesicularized membrane fragments that have trapped luciferin and luciferase during homogenization. This is rendered unlikely because light is emitted under different conditions from the particulate and soluble fractions. The identification of particles as chevron-shaped crystals of guanine-like material named "scintillons" (DeSa *et al.*, 1963) has now been rescinded (Fogel *et al.*, 1972). However, the unknown particles are still sometimes referred to as "scintillons," a usage that causes some confusion.

A. *Gonyaulax polyedra*

Image intensification studies (Reynolds, 1972) gave the impression that *Gonyaulax* emitted light from regularly arranged discrete areas. However, the theca of this dinoflagellate is not homogeneous but is indented with circular pits in a regular pattern that might simulate bright spots. It does not seem likely, though, that the entire cytoplasm of *Gonyaulax* emits light. The cell is densely packed with chloroplasts, mitochondria, and trichocysts (Herman and Sweeney, 1975), each of which can be eliminated as a source of bioluminescence because many nonluminescent dinoflagellates contain these structures (Bouck and Sweeney, 1966), and dinoflagellates that are not photosynthetic may luminesce. The suggestion that "polyvesicular bodies" present in the peripheral cytoplasm of *Gonyaulax* (Schmitter, 1971) are the source of bioluminescence can probably be eliminated on the grounds that this structure is absent in some luminescent dinoflagellates. Freeze–fracture images of the polyvesicular body vesicles show their membranes to be free of protein particles, which suggests membrane recycling rather than bioluminescence as their function (Sweeney, unpublished observations). The luminescence of *Gonyaulax polyedra* is very much brighter at night under the control of a circadian oscillator (Sweeney, 1969). On the basis of studies of freeze–fractured cells in which a large particle appeared during the night phase, I suggested (Sweeney, 1976) that the luciferase might be contained in the membrane of the large peripheral vesicle in *Gonyaulax*. The evidence for this site is purely circumstantial at present.

B. *Noctiluca miliaris*

The identification of "microsources" of bioluminescence may prove much simpler in other dinoflagellates than *Gonyaulax*, particularly those which have a less compact internal structure. In the large, spherical *Noctiluca miliaris*, Eckert and Reynolds (1967) were able to photograph localized intracellular flashes and correlate these with particles that fluoresced on excitation with light of 365 and 405 nm (wavelength of emission not specified). These microsources could be seen under phase microscopy in the peripheral cytoplasm of *Noctiluca*, which forms a thin layer against the cell wall outside the large vacuole. Microsources, identified by their fluorescence, comprised about one-third of the phase-retarding bodies found there. The size of the microsources was difficult to determine by light microscopy, but they were estimated to be less than 1.5 μm in diameter. No electron microscopy was done in connection with Eckert's studies. However, more recently, a bioluminescent *Noctiluca miliaris* from East Asia, which contains green flagellate symbionts, was studied by electron microscopy (Sweeney, 1978). In the peripheral cytoplasm of this cell, membrane-bound bodies were seen with about the correct position and dimensions to be the microsources of Eckert. However, they could also be nonluminescent phase-retarding bodies that share the dimensions and location of microsources. In appearance, these bodies are spherical or slightly oval and contain electron-dense material (cells fixed with OsO_4).

C. *Pyrocystis fusiformis*

Recently another large vacuolated dinoflagellate *Pyrocystis fusiformis,* isolated in Indonesia, has been studied in my laboratory both from a physiological and an ultrastructural point of view (Sweeney, 1979b). My studies of this dinoflagellate have been complimented by an electrophysiological investigation by Edith Widder and James Case in this department.

Pyrocystis fusiformis is of particular interest because, like *Gonyaulax,* its luminescence is under the control of a circadian oscillator, so that, in a light–dark cycling environment, luminescence is very bright at night; and this behavior continues for at least several cycles in constant light (Sweeney, 1979c). As in a number of other algae [including *Pyrocystis lunula* (Swift and Taylor, 1967)], during the light period, the chloroplasts are dispersed about the cytoplasm near the cell periphery or in transvacuolar strands, but at the beginning of the night, all migrate to the center of the cell around the nucleus. This movement too is under circadian control (Sweeney, 1979b). For our purpose here, the most interesting feature of *Pyrocystis* is the reverse migration of the bioluminescence from the cell center just exterior to the nucleus during the day to the peripheral cytoplasm at night (Sweeney, 1979b). This change in the distribution of bioluminescence can be clearly seen in the light microscope in darkness because of a fortuitous difference between *Gonyaulax* and *Pyrocystis*. Whereas the flash in *Gonyaulax* is a single short burst of light that lasts about 0.1 second, in *Pyrocystis,* when stimulated by the addition of acid, not only a short flash but also a long-continued and visible glow is emitted. During the day, this glow, which is correlated with blue–green fluorescence on excitation with ultraviolet light, arises only in the cell center in a distinct spherical zone just external to the nucleus. At night, separate microsources scattered through the whole cell continue to emit small visible flashes for a long time after the addition of acid. This migration has been followed by image intensifier by E. Widder who is presently continuing this investigation.

The localization and migration of microsources in *Pyrocystis fusiformis* offers the opportunity to identify the structure responsible for light emission in electron microscopy. The subcellular source must be a structure or structures that are found in the cell center just external to the nucleus during the day and dispersed throughout the cytoplasm at night. We have tentatively identified a structure with these characteristics. The spherical structure that is found just external to the nucleus during the day but not at night (Fig. 1a) consists of a spherical group of a large number of tightly packed membrane-bound particles about 0.25–0.5 μm. The contents of these vesicles are electron dense under the fixation conditions used. They are short, rounded rods approaching spheres surrounded by a single unit membrane (Fig. 1b). Sometimes they are divided by longitudinally or transversely oriented clear lines. Their size, shape, and occasional division into halves make it possible to identify these particles in the peripheral cytoplasm in cells fixed at night. It seems highly likely that they are the intracellular source of bioluminescence in *Pyrocystis*. Experiments on homogenates are in progress in my laboratory. Such homogenates are highly luminescent but have no particulate fraction with the properties of that in *Gonyaulax*. Nothing exactly similar to the vesicles of *Pyrocystis* has been located in electron micrographs of *Gonyaulax* and *Noctiluca*. However, in both these species there are darkly staining bodies bounded by a single membrane, and those in *Noctiluca* are of the same size and location as the microbodies described for this species by Eckert and Reynolds (1967) (Section II,B). In addition to lacking a particulate bioluminescent fraction, *Pyrocystis (P. noctiluca* and *P. lunula)* also lacks luciferin-binding protein (Schmitter *et al.*, 1976). The addition of hydrogen ions does strongly stimulate bioluminescence, however. The mechanism for the control of bioluminescence suggested by Hastings (1978) cannot be completely correct for *Pyrocystis* in the absence of luciferin-binding protein. Until details of the luciferin and luciferase content of the vesicles of *Pyrocystis* are known, speculation regarding the mechanism of luminescence control are premature.

III. Coelenterates: Cnidaria

A large number of coelenterates, including hydrozoa such as *Aequorea* and *Obelia* and anthozoa such as *Renilla* and the sea pens, are capable of bioluminescence. The biochemistry of light emission is now fairly well known (Cormier, 1978). All luminescent members of this group contain the same luciferin, and all

---

FIG. 1.   Vesicles of the unicellular bioluminescent dinoflagellate *Pyrocystis fusiformis,* possibly the intracellular source of light. (a) Vesicles (V) in a sphere in the center of a cell fixed at the end of a 12-hour light period. Note that the collection of vesicles is not held together by any visible structure. ×7200. (b) Enlarged view of the vesicles in a spherical structure like that in (a); fixed at the end of a 12-hour light period. ×60,000. (Electron micrographs by the author.)

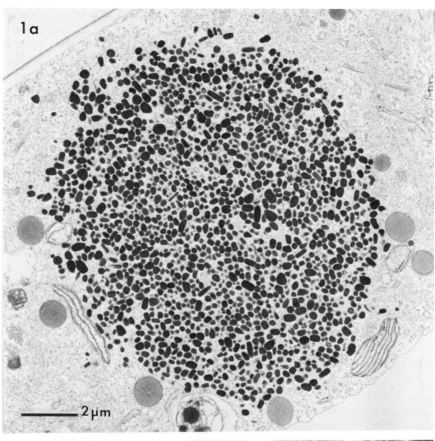

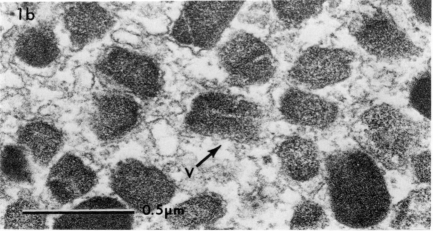

respond to the addition of Ca^{2+} with a flash, but the mechanism by which calcium activates bioluminescence is somewhat different in hydrozoans and anthozoans. In the former, there is a precharged photoprotein, aequorin, which is a complex containing luciferin stabilized in an excited state. Light is emitted upon the addition of calcium ion (Morin and Hastings, 1971a). Two similar photoproteins, mnemiopsin and beroin, have been isolated from Ctenophores. The light emitted is blue. In the anthozoans, luciferin is bound to a luciferin-binding protein reminiscent of that in dinoflagellates and is inactive until released by the addition of calcium (Anderson et al., 1974). In many Anthozoans, there is an additional component of bioluminescence, the green fluorescent protein (Morin and Hastings, 1971b; Charbonneau and Cormier, 1979). In these species, the excitation energy is transferred to this protein from luciferin (Hart et al., 1979) and the emitted light is green, not blue. In all coelenterates, then, it is of the utmost importance for bioluminescence that calcium be rigorously excluded from contact with photoproteins or luciferin–luciferin-binding protein complexes until light is to be emitted. We would expect that any subcellular structure containing the bioluminescence system would incorporate this feature.

In extracts of *Renilla* (Morin and Hastings, 1971a; Anderson and Cormier, 1973, 1976a), luminescence is associated with a sedimenting particle that emits light on the addition of water, presumably because this treatment renders vesicles permeable to calcium. Vesicular structures, which are $0.2~\mu$m in diameter with a single membrane, can be seen in electron micrographs of extracts of ten different coelenetrates and have been called "lumisomes" (Anderson and Cormier, 1973). When vesicles are present, the light emitted by *Renilla* extracts is green as *in vivo*, whereas the soluble luminescent system emits in the blue, suggesting that the green fluorescent protein is only associated with the vesicles. Lumisomes can also be shown to contain luciferin-binding protein and presumably also the other components of the light-emitting system since the vesicle preparation is luminescent on lysing of the vesicles. The establishment of a sodium gradient across the lumisome membrane (high sodium inside) also makes the vesicle permeable to calcium, as shown by the production of a flash milliseconds after calcium addition (Anderson and Cormier, 1976b).

While the evidence for the presence of lumisomes in extracts of biolumines-cent coelenterates is convincing, no such structures have been identified with certainty in sections of photogenic tissue. There is always the possibility that vesicles are formed during homogenization, a well-known phenomenon when membranes are broken. Collections of large numbers of vesicles, of about the same dimensions as lumisomes and surrounded as a group by a second mem-brane, have been described from electron micrographs of the photogenic tissue of *Renilla* (Spurlock and Cormier, 1975). They attributed the bioluminescence to this structure, which they called the "luminelle." However, identical structures are regularly found in nonluminescent Anthozoans, an observation that casts doubt on their interpretation of the function of the luminelle. Furthermore, neither lumisomes nor luminelles could be found in electron micrographs of the photogenic tissue of the siphonophore *Hippopodius hippopus* (Bassot *et al.*, 1978). Luminescence was not correlated with the presence of any other subcellular structure in this animal either.

IV. Insects: Fireflies

A. LARVAE

Although it is not the most common of bioluminescent organisms, for many people the firefly is the most familiar because it is terrestrial rather than marine. There are many genera and species of this esthetically pleasing beetle, which inhabit the tropics of the world and extend their range into eastern and central North America. The larva, which lives in moist soil, is less often noticed than the adult, but it too is bioluminescent. The light is emitted as a glow from a pair of light organs in the posterior segment of the abdomen of the larva. In *Photuris pennsylvanicus* from the eastern United States, a small pair of light organs, 0.5 mm in diameter (Oertel *et al.*, 1975), is found in the eighth abdominal segment. The organ is quite simple and is composed of two cell layers: a dorsal layer of cells containing crystals and probably serving as a reflector and a photocyte layer beneath. We are concerned with these photocytes, which are the source of bioluminescence.

The photocytes are 12–20 μm in diameter and are transparent *in vivo*. Their cytoplasm is filled with vesicles or "photocyte granules" (Oertel *et al.*, 1975). Some of these vesicles appear in electron micrographs to be filled with opaque material (Fig. 2), whereas others look partially or completely empty. The photocytes are very irregular in shape and interdigitate with neighboring photocytes, small nerve endings, and tracheoles (Fig. 2a). Mitochondria are also plentiful, which is not surprising, since ATP is required in the firefly light reaction.

The photocyte vesicles are a feature that the photocytes of all species of larval fireflies have in common (Peterson, 1970). In *Photuris*, the vesicles are about 0.6 μm in diameter and are bounded by a unit membrane (Oertel *et al.*, 1975). The contents of these vesicles are homogeneous except for a small dense region to one side and sometimes small groups of 2–5 microtubules are also present (Fig. 2b). Microtubules may be associated with invagination of the vesicle membrane. Photocyte vesicles have been shown to contain catalase activity (Hanna *et al.*, 1976) and so may be considered as a type of peroxisome or microbody.

It is difficult to prove that the photocyte vesicles of firefly larvae play a role in bioluminescence. However, their number, position in the photocyte, vesicular nature, and the presence of adjacent nerve endings all suggest that they are the

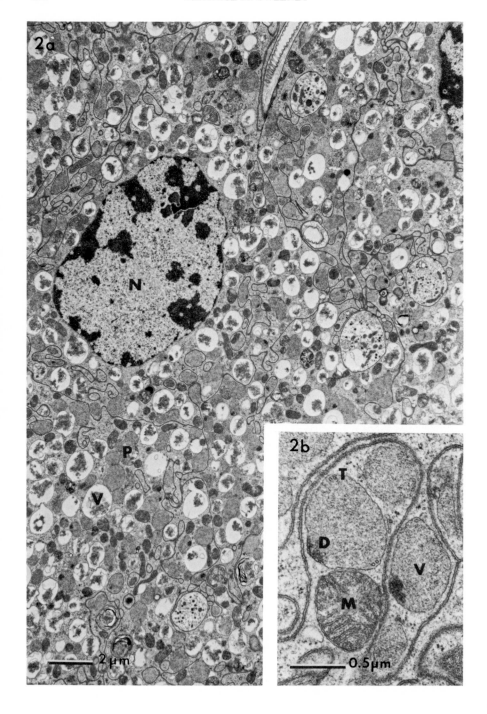

subcellular bioluminescent organelle in the firefly. It is not possible, however, to alter their appearance in electron micrographs by extensively stimulating the larvae to luminesce before fixation (Oertel *et al.*, 1975).

The larval light organs may sometimes persist through pupation and continue to emit light, although now without innervation (Strause *et al.*, 1979). Thus they can overlap in time with the development of the adult light organ, to which they appear not to contribute either structurally or biochemically.

The photocytes of the light organ of the glow worm *Lampyris noctiluca* resemble quite closely those of the larval firefly (Barber and Dilly, 1966). Each photocyte contains some 3000 vesicles, which are 0.6 μm in diameter and are bounded by a single membrane. As in the firefly, profiles of the vesicles in electron micrographs show dark regions and invaginations of the membrane.

B. Adults

In contrast to the simple structure of the larval light organ in the firefly, the adult organ, which also occupies the ventral part of the posterior segments of the abdomen, is far from simple. Photocytes are paired cells, are triangular in cross section, and form six-pointed, fluted columns around central tracheoles and nerves (see diagrams in Smith, 1963, and Hansen *et al.*, 1969). Photocytes are considerably larger than in the larva [24 × 8–10 μm (Hansen *et al.*, 1969)] and are in contact with tracheoles but not with nerve endings (Ghiradella, 1977).

The photocytes themselves are differentiated into regions (Fig. 3a). The central part of each photocyte contains closely packed vesicles: 1.5 × 2 μm in *Photuris* and 3 × 7 μm in *Pteroptyx* (Hopkins and Hanna, 1972; Case and Lindberg, unpublished). Like the vesicles of the larval photocyte, these are membrane-bound and contain more dense regions and microtubules in contact with the vesicle membrane (Fig. 3b). Crystals are also sometimes present. The lateral parts of the photocytes are free from vesicles and contain mitochondria and small unidentified granules. Studies utilizing an image intensifier clearly show that only the central parts of the photocytes emit light [this is one bit of evidence for a function in bioluminescence for the photocyte vesicles (Reynolds, 1972)]. Smalley *et al.* (1980) have observed that fluorescence, which can be correlated with bioluminescence, is also limited to the central portion of each photocyte. If the light organs are stimulated to luminesce with synephrine, the luminescence is bright enough to identify vesicles as its source.

Fig. 2.   Photocytes from the light organ of the larva of the firefly *Photuris pennsylvanica*. (a) Several interdigitating photocytes (P) containing photocyte vesicles (V) and a nucleus (N). ×6250. (b) Enlarged view of photocyte vesicles (V) with dense inclusions (D) and tubules (T) within the vesicles and a mitochondrion (M). ×30,000. (Electron micrographs courtesy of K. A. Linberg and J. F. Case.)

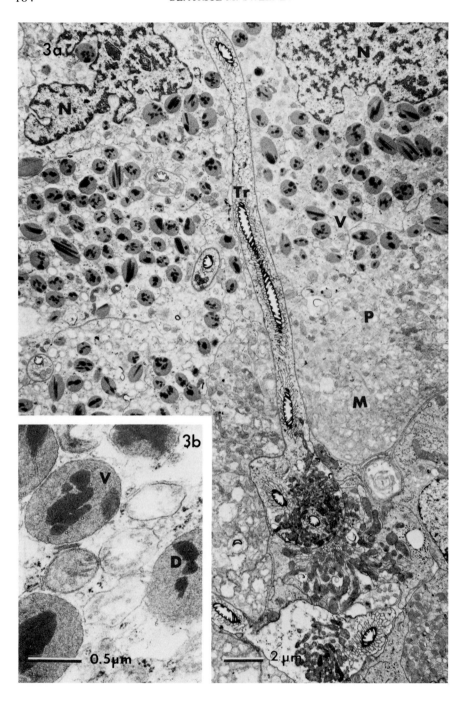

The case for the photocyte vesicles as the intracellular source of bioluminescence is quite strong. We do not know, however, how vesicles may be stimulated to emit light without direct contact with nerve endings. Even nerve terminals to the photocytes are absent in the adult. There are also questions concerning how ATP, presumably synthesized in the mitochondria, enters the photocytes vesicles to adenylate the luciferin or how the reactions leading to light production are controlled.

V. Annelids

A. POLYCHAETS

1. Polynoid Worms: Acholoë

Among the annelids, there are a number of bioluminescent species. The light-producing structures are quite different in different groups; in some, luminescence is intracellular, whereas in others, a luminous slime is secreted.

One of the best known from the ultrastructural point of view is the scale worm Acholoë. The body of this creature is covered by a double row of thin, oval scales or elytra that luminesce brightly on stimulation (Nicol, 1953). The fine structure of the elytra of Acholoë astericola has been thoroughly investigated by Bassot and his collaborators. Each elytron is composed of a dorsal and a ventral cell layer covered on the surfaces by cuticle. Between these cell layers is an open network of fibers through which runs the elytral nerve, which controls the flashing of the scale (Nicol, 1953; de Ceccatty et al., 1977). The cells of the dorsal layer contain melanin and probably act as reflectors, whereas the photocytes make up the central layer that is rich in cysteine (Bassot, 1966c). Within each photocyte are 20–30 paracrystalline arrays, which are clearly resolved in electron micrographs (Bassot, 1966a; Bassot and Nicolas, 1978) (Fig. 4a and d). These arrays are most unusual in structure. They appear to be continuous with the endoplasmic reticulum. These tubular membranes are folded into figure eights in the manner of Christmas ribbon candy and are aligned in close packing more than one layer deep to give the appearance of a crystal. Each loop is about 80 nm across and 86 nm high (Bassot, 1966a). The register of these loops in layers at different depths is not perfect, so that, in thick section in some planes, the arrays appear to consist of large tubules. The paracrystalline structures develop last during the differentia-

---

FIG. 3. Photocytes of the adult firefly Photinus greeni. (a) Several photocytes (P) separated by a tracheole (Tr) and containing photocyte vesicles (V), a nucleus (N), and mitochondria (M). ×5200. (b) Enlarged view of photocyte vesicles with dense inclusions (D). ×30,000. (Electron micrographs courtesy of K. A. Linberg and J. F. Case.)

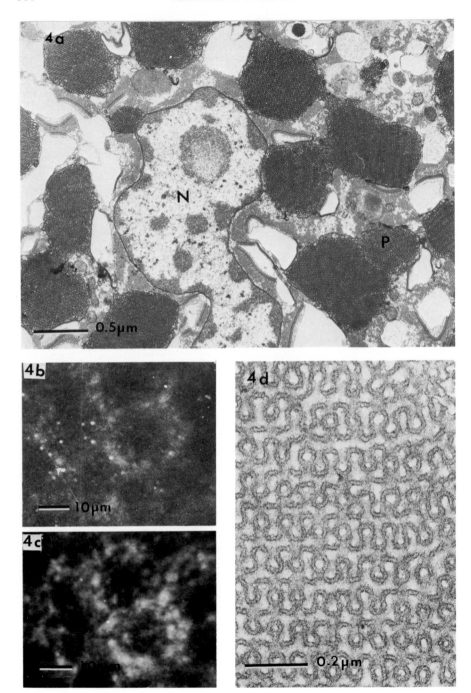

tion of the photocyte and their presence coincides with the appearance of bioluminescence (Nicolas, 1977).

By combining observations of fluorescence and bioluminescence observed by image intensifier (Fig. 4b and c), Bassot and Bilbaut (1977b) were able to correlate the location of light emission with the paracrystalline arrays, which they named "photosomes" in consequence. In some unknown manner, perhaps through excitation of the photosome plasma membrane coupled to nerve stimulation, light emission is stimulated and spreads in a wave through the photocytes and thus over the elytron. The migration of bioluminescence can be seen at low magnification and can also be followed by the spread of the green fluorescence of the photosomes, the maximum emission of which is at 525 nm (Bassot and Bilbaut, 1977a; Bassot and Nicolas, 1978). The structure of the photosomes permits a remarkable proliferation of membrane. Since the membranes of the photosomes are double and enclose a lumen between, two volumes (inner and outer separated by membrane) are present. The division of the photosomes into two compartments may well be important for the function of the luminescent biochemistry. Bassot (1966a) has calculated that the volume within the two membranes of the endoplasmic reticulum of the photosome is about 1/43 of the volume between the folds of the paracrystalline array. Stimulation of bioluminescence may be accompanied by a change in the permeability of the photosome membranes. Recently, Bassot and Nicolas (1978) have found similar photosomes in six other luminous scale worms, *Harmothoë impar, H. longisetis, H. lunulata, Gattyana cirrosa, Lagisca extenuata,* and *Polynoë scolopendrina.* Interestingly, the lens of the eye of both luminous and nonluminous scale worms (Bassot and Nicolas, 1978) contains a similar array of folded membranes, as does the eye of the mollusc *Limax* (Eakin and Brandenburger, 1975). However, these neither fluoresce nor bioluminesce (Bassot and Nicolas, 1978).

Elytra of *Acholoë astericola* and *Lagisca extenuata* have been homogenized and the extract separated into pellet and supernatant fractions by centrifugation (Lecuyer and Arrio, 1975). Only the pellet showed the fluorescence characteristic of photosomes, but these were not purified further. Bioluminescence of the pellet could be stimulated by the addition of $FeSO_4$ ($3 \times 10^{-4}M$). The reduction in the osmolarity of the extract on the addition of the $FeSO_4$ solution may have caused lysis of the vesicles and hence a nonspecific stimulation of bioluminescence.

FIG. 4. Photocytes of the polynoid worms, *Acholoë astericola* and *Harmothoë lunulata,* showing the paracrystalline photosomes within. (a) Photocyte of *Harmothoë* with a nucleus (N) and many photosomes (P). ×27,000. (b) Luminescence of a photocyte of *Acholoë* with image intensification. ×750. (c) Fluorescence of the same region as in b. Note the subcellular origin in discrete photosomes of both the bioluminescence and the fluorescence. ×750. (d) Enlarged view of the folded membranes of the photosome of *Acholoë.* ×80,000. (Electron micrographs courtesy of J.-M. Bassot.)

Photosomes from the elytra of *Harmothoë lunulata* have been purified by sucrose density gradient centrifugation (Nicolas, 1979). Light emission, stimulated by the addition of dithionite, is well correlated with the presence of these photosomes in the gradient, as shown by electron microscopy of the fractions. Care was taken to exclude Ca^{2+} during the extraction procedures, but none of the isolated fractions containing photosomes were stimulated to produce light on the addition of calcium.

The identification of the photosomes of *Achloë* and other scale worms as intracellular sources of bioluminescence by a combination of electron microscopy, image intensification, fluorescence, and extraction make the scale worms one of the few organisms in which the light-emitting organelle is known.

2. *Chaetopterus*

Chaetopterus is a polychaet that can discharge a luminous secretion upon electrical or mechanical stimulation. This worm lives within an opaque tube, a circumstance that makes it difficult to understand the function of bioluminescence in this animal. The luminous secretion orignates from the epithelium of the anterior feelers, the aliform notopedia of the middle region, and the posterior notopedia. It appears to come from glandlike cells, the content of which is dense and finely granular when viewed with the light microscope (Nicol, 1952a) since the number and the distribution of these cells corresponds with that of bioluminescence (Nicol, 1952b). The granules can be distinguished as points of light in the secreted material, which seem to disappear as light emission progresses.

Anctil (1979a) has recently published an electron microscopic study of the epithelial luminescent gland of the aliform notopedia of *Chaetopterus variopedatus*. From the coincidence of bioluminescence, fluorescence, and light microscopic images, he identified the orthochromatic goblet cells as the cells responsible for the luminous secretion. He considers that the membrane-bound goblets of these cells are responsible for the synthesis of the photoprotein. The only contradictory evidence is (a) the lack of a color match between the fluorescence of goblet cells, which is yellow–green, and the blue fluorescence of the purified photoprotein and the blue *in vivo* bioluminescence; and (b) the common occurrence of goblet cells identical in appearance in nonluminous annelids. Although luminescence is under nervous control in *Chaetopterus,* Anctil did not observe any direct contact between nerve cells and secretory cells. Supportive cells between the goblet cells are innervated and may contract to expel luminous material from the secretory cells on stimulation.

3. *Odontosyllids*

The Ondontosyllids or Fire Worms, like *Chaetopterus,* secrete luminous material. The cells from which the secretion arises are thought to be glandular, with ducts to the exterior. These cells are filled with granules that are light yellow (Harvey, 1952). There is no difinitive recent study, however, to demonstrate whether these granules are active in light production.

B. Oligochaets

A number of luminous earthworms are known. The light-emitting cells come from the coelomic fluid, by which they are excreted to the exterior through the mouth of other orifices (Harvey, 1952). The photocytes contain granules that are refractile and greenish or yellowish in color. The coelomic fluid fluoresces greenish-yellow at a wavelength not very different from the bioluminescence (Johnson *et al.*, 1966).

Rudie and Wampler (1978) describe subcellular loci of bioluminescence within the photogenic cells of the coelomic fluid of the earthworm *Diplocardia longa* observed with the aid of image intensification. These cells contain several sizes of granules, some of which are greenish. Homogenized coelomic cells emit light on the addition of H_2O_2, but no distinct separation of bioluminescent granules could be achieved by sucrose density centrifugation. The large, prolate granules, one of several particles found in the coelomic cells, were ruled out as the source of bioluminescence because their numbers were not correlated with bioluminescence in fractions from the sucrose gradients. Much of the luminescence remained at the top of the gradients and so could possibly be associated with membraneous material. Obviously an investigation of the luminous coelomic cells by electron microscopy is in order.

VI. Molluscs

A. Cephalopods: Squid

Squid show great variety in the manner in which they emit light and in the number and placement of the photophores. Some harbor luminous bacteria, others secrete luminous material into the ink sac, whereas still other species are decorated with many small photophores in intricate patterns (Herring, 1977; Herring and Morin, 1978). *Watasenia scintillans* is one of the more available squid in which bioluminescence is not bacterial in origin. It possesses three luminous organs at the tip of each of the pair of ventral arms. All the photocytes of this organ are filled with rods, which are square in cross section (Okada, 1966), 2.5–5 μm long and 1–3 μm wide as seen in electron micrographs. They do not resemble bacteria. The crystal-like rods are surrounded by membrane lamellae, but the fixation is not sufficiently good in these early micrographs to make sure of the relationship between these core cubes and the lamellae and cell membranes. Interestingly, the luciferin in *Watasenia* is identical to that of coelenterates (Cormier, 1978).

Two other squid, *Bathothauma lyromma* (Dilly and Herring, 1974) and *Heteroteuthis dispar* (Dilly and Herring, 1978), luminesce without the aid of symbiotic bacteria. The photophore of the former covers the ventral surface of the eye, whereas that of the latter is inside the mantle cavity on the ink sac. In the photocytes of *Bauthothauma* are strange paracrystalline lattices which are in the form of rods 0.2–7 μm long, and are reminiscent of those in *Watasenia* photocytes. Whether they have any function in bioluminescence is conjectural. However, similar structures have been found in the photophores of the crustacean *Euphasia* (Herring, 1978).

In *Heterteuthis,* the photophore is a sac that has a much-folded epithelial wall and is surrounded by reflective tissue. The cells of the epithelium have dense brush borders and contain large vesicles with either pale or dense interiors. Similar vesicles, 0.3–10 μm in length, are also seen free in the lumen of the photophore. They cannot be confused with bacteria because each vesicle contains a portion of a flagellum. Whether or not these vesicles are the source of bioluminescence is not known.

B. OTHER BIOLUMINESCENT MOLLUSCS

Both the clam *Pholas* and the limpet *Latia* secrete luminous material. In *Pholas,* the luminous organs contain only two types of glandular cells (Bassot, 1966b). The secretions of both of these cell types may play a role in light production, but this has not been unequivocally demonstrated. The luminous organs of *Latia neritoides* also contain two types of glandular cells (Bassot, 1966b) with different staining properties.

VII. Vertebrates: Fish

A. *Porichthys*

Fish are the only vertebrates known to emit light. Among the marine fishes, particularly the deep-dwelling ones, bioluminescence is very common. The light-emitting structures are extremely varied, and some are very complex, with reflecting layers, lenses, and light guides, in addition to the photocytes themselves (Bassot, 1966b; Herring and Morin, 1978). Light may originate from symbiotic luminous bacteria, from glandular secretions, or from within photocytes. Bacteria have been cultured from the light organs of both shallow-water and bathypelagic fish, including Leiognathids, Monocentrids, and the esci of angler fishes. Bacteria-like bodies occupy the light organ of the Anomalopidae but have never been cultured from this source. There are many fishes in which light emission is not of bacterial origin, including the coastal *Porichthys* and the bathypelagic sharks, Myctophids, hatchet fish, and the barbels of angler fish.

The availability and ease of maintenance of *Porichthys* have made this fish a favorite object of study. Among the fishes, it is thus the best known from a morphological and physiological point of view. The epidermal photophores are 1–2 mm in diameter (Strum, 1969a), occur in lines on the ventral side of the head and body, and include a lens and a reflective layer. The photocytes are not in direct contact with nerve endings but can be excited to luminesce by the addition of adrenalin (Strum, 1969b) or noradrenalin (Anctil, 1979b).

In *Porichthys,* the photocytes are bordered by microvillae and contain many cytoplasmic vesicles (Anctil, 1979b). On exhaustive stimulation, these vesicles coalesce. In the nonluminous *Porichthys* which are from the northwest coast of the United States and are deficient in the luciferin that the fish obtain from their diet of *Cypridina* (Tsugi *et al.,* 1972), the photocyte vesicles are present but contain less flocculent material (Case and Strause, 1978). These authors thus conclude that the photocyte is adapted for taking up and storing luciferin from the blood stream. Such a function of the vesicles is consistent with the observation (Anctil, 1979b) that, with exhaustive chemical stimulation, the flocculent contents disappear.

During the development of larval *Porichthys,* the ability to emit light is strikingly correlated with the earliest evidence of vesicles in the photocytes (Anctil, 1977) although luciferin is known to be present even in the egg of luminous *Porichthys* (Tsugi *et al.,* 1972). Anctil (1979c) considers that the membrane of the photocyte vesicles, which are constantly being replenished from the endoplasmic reticulum, are the sites of synthesis of the luciferase (which is not the same as that of *Cypridina*) by analogy with the situation in *Apogon,* which is another fish that obtains its luciferin from *Cypridina* (Tsugi and Hanada, 1966).

B. BATHYPELAGIC FISHES

The bioluminescent midwater fishes are much more difficult to study than *Porichthys.* While dead or dying specimens [brought up in thawls or netted at the surface at night or under unusual circumstances during violent upwelling that sometimes occurs in the Straits of Messina (Christophe *et al.,* 1979)] can be fixed for electron microscopy, the fish or their isolated light organs rarely survive long enough to make possible the correlation of structure with light emission. Photophores of *Chauliodus* contain photocytes that flash more rapidly than those of *Porichthys* and appear to be secretory (Baquet, 1975). Searsiids have been seen to emit a jet of blue particulate luminous material (Herring and Morin, 1978). Basson (1966b) described a secretory type of cell, the "A" cell, from the photophores of Gonastomatid fishes, which contained highly organized endoplasmic reticulum (ER). In *Maurolicus,* this rough ER is very extensive, filling the entire cell with parallel bands. Similar proliferations of ER are present in the photocytes of other luminous Gonastomatids. However, only their prominence in the photocytes suggests a function in bioluminescence for this membranous system.

VII. Concluding Remarks

It is clear from the preceding discussion that there are only a few organisms in which anything is known concerning bioluminescent organelles or intracellular sources of luminescence. The identification of subcellular light-emitting sources was not possible until photogenic tissue could be examined with the electron microscope, since at best only point sources can otherwise be seen. The photocyte must of course first be identified among the other cells of the photophore. Even this step has been difficult to take in some luminous creatures. A knowledge of the structure of the photocyte alone is not sufficient for the identification of a subcellular site. At the least, correlation with light emission, as seen most easily with the aid of an image intensifier operating through a high power lens of a light microscope, is needed as confirming evidence. Fluorescence can sometimes be established as an indicator of bioluminescence. This simplifies the problem, since fluorescence is relatively long-lasting, whereas the luminescent flash is often very brief. The separation of subcellular components on sucrose density gradients or by other means may or may not be helpful. If the light-emitting fractions of the gradient are not homogeneous, an erroneous identification may be made, as for example in the case of the dinoflagellate ''scintillon,'' or no identification may be possible.

With these considerations in mind, I think that the most complete identification of a subcellular source of bioluminescence is probably that in the scale worm *Acholoë*. Here the folded membrane structure of the paracrystalline arrays is very easily recognized as distinctive. The photosomes are large enough to be recognized in the light-producing fractions of extracts separated by sucrose density gradient centrifugation. The photocyte granules of fireflies are somewhat more problematic, although they are consistently present in both larval and adult photocytes, which arise independently during development.

The particular structures from *Pyrocystis fusiformis,* described here for the first time, are found in the same intracellular location as the bioluminescence, both when the luminescence is confined to the center of the cell during the day and when widely scattered points of light are observed at night. Fluorescence can be correlated with bioluminescence when the particles are concentrated in a central sphere during the day. Image intensification confirms the change in the position of bioluminescent sources. In other luminous organisms, the identification of the intracellular sources of bioluminescence is more doubtful.

All three examples above, and many of the postulated but less secure sources of bioluminescence as well, have in common the prominence of single-membrane-bound internal spaces, which may be either the lumen of endoplasmic reticulum, as in *Acholoë,* or more conventional vesicles. In many electron micrographs, the contents of these structures are not homogeneous and may change their appearance during long-continued stimulation of bioluminescence. That bioluminescence requires a vesiculate membranous structure is certainly in keeping with our developing understanding of membrane biochemistry and membrane integral proteins. The production of hydrogen gradients requires such an arrangement. The separation of luciferin and luciferase or Ca^{2+} and photoprotein until the moment of light emission also presupposes separate subcellular compartments.

Many variants on the membrane vesicle theme are seen among the photocytes that have been examined. Secreting vesicles of ER origin are well known in many organisms. The essential feature appears to consist of two compartments separated by a membrane and, as long as this prerequisite is met, many different arrangements may be equally effective. We should perhaps not expect to find uniformity of substructure among photocytes of different species in which the capacity to bioluminesce has perhaps evolved independently. Certainly no such uniformity of bioluminescent subcellular sources is evident.

ACKNOWLEDGMENT

The author wishes to acknowledge with thanks the support of the National Science Foundation Grant No. PCM77-07709.

REFERENCES

Airth, R. L., Foerster, G. E., and Behrens, P. Q. (1966). *In* "Bioluminescence in Progress" (F. H. Johnson and Y. Hanada, eds.), pp. 203–223. Princeton Univ. Press, Princeton, New Jersey.

Anctil, M. (1977). *J. Morphol.* **151**, 363–396.

Anctil, M. (1979a). *Can. J. Zool.* **57**, 1290–1310.

Anctil, M. (1979b). *Rev. Can. Biol.* **38**, 67–80.

Anctil, M. (1979c). *Rev. Can. Biol.* **38**, 81–96.

Anderson, J. H., and Cormier, M. J. (1973). *J. Biol. Chem.* **248**, 2937–2943.

Anderson, J. M., and Cormier, M. J. (1976a). *Biochem. Biophys. Res. Commun.* **68**, 1234–1241.

Anderson, J. M., and Cormier, M. J. (1976b). *Fed. Proc.* **35**, 1584.

Anderson, J. M., Charbonneau, H., and Cormier, M. J. (1974). *Biochemistry* **13**, 1195–1201.

Baquet, F. (1975). *Prog. Neurobiol.* **5**, 97–125.

Barber, V. C., and Dilly, P. N. (1966). *Z. Zellforsch.* **73**, 286–302.

Bassot, J. M. (1966a). *J. Cell Biol.* **31**, 135–158.

Bassot, J. M. (1966b). *In* "Bioluminescence in Progress" (F. H. Johnson and Y. Hanada, eds.), pp. 557–610. Princeton Univ. Press, Princeton, New Jersey.

Bassot, J. M. (1966c). *Cah. Biol. Mar.* **7**, 39–52.

Bassot, J. M., and Bilbaut, A. (1977a). *Biol. Cell.* **28**, 155–162.

Bassot, J. M., and Bilbaut, A. (1977b). *Biol. Cell.* **28**, 163–168.

Bassot, J. M., and Nicolas, M. T. (1978). *Experimentia* **34**, 726–728.

Bassot, J. M., Bilbaut, A., Mackie, G. O., Passano, L. M., and de Ceccatty, M. P. (1978). *Biol. Bull.* **155**, 473–479.

Berliner, M. D., and Hovanian, H. P. (1963). *J. Bacteriol.* **86**, 339–341.

Bouck, G. B., and Sweeney, B. M. (1966). *Protoplasma* **61**, 205–223.

Buck, J. B. (1978). *In* "Bioluminescence in Action" (P. J. Herring, ed.), pp. 419–460. Academic Press, New York.

Case, J. F., and Strause, L. G. (1978). *In* "Bioluminescence in Action" (P. J. Herring, ed.), pp. 331–336. Academic Press, New York.

Ceccatty, M. P., de, Bassot, J. M., Bilbaut, A., and M. T. Nicolas (1977). *Biol. Cell.* **28**, 57–64.

Charbonneau, H., and Cormier, M. J. (1979). *J. Biol. Chem.* **254**, 769–780.

Christophe, B., Baquet, F., and Marechal, G. (1979). *Comp. Biochem. Physiol. A* **64**, 367–372.

Cormier, M. J. (1978). *In* "Bioluminescence in Action" (P. J. Herring, ed.), pp. 75–108. Academic Press, New York.

DeSa, R., Hastings, J. W., and Vatter, A. E. (1963). *Science* **141**, 1259–1270.

Dilly, P. N., and Herring, P. J. (1974). *J. Zool.* **172**, 81–100.

Dilly, P. N., and Herring, P. J. (1978). *J. Zool.* **186**, 47–59.

Eakin, R. M., and Brandenburger, J. L. (1975). *J. Ultrastruct Res.* *53*, 382–394.

Eckert, R., and Reynolds, G. T. (1967). *J. Gen. Physiol.* **50**, 1429–1458.

Fogel, M., and Hastings, J. W. (1972). *Proc. Natl. Acad. Sci. U.S.A.* **69**, 690–693.

Fogel, M., Schmitter, R. E., and Hastings, J. W. (1972). *J. Cell Sci.* **11**, 305–317.

Fuller, C. W., Kreiss, P., and Seliger, H. H. (1972). *Science* **177**, 884–885.

Ghiradella, H. (1977). *J. Morphol.* **153**, 187–204.

Hanna, C. H., Hopkins, T. A., and Buck, J. (1976). *J. Ultrastruct. Res.* *57*, 150–162.

Hansen, F. E., Miller, J., and Reynolds, G. T. (1969). *Biol. Bull.* **137**, 447–464.

Hart, R. C., Matthews, J. C., Hori, K., and M. J. Cormier (1979). *Biochemistry* **18**, 2204–2210.

Harvey, E. N. (1952). "Bioluminescence." Academic Press, New York.

Hastings, J. W. (1978). *In* "Bioluminescence in Action" (P. J. Herring, ed.), pp. 139–170. Academic Press, New York.

Herman, E. M., and Sweeney, B. M. (1975). *J. Ultrastruct. Res.* **50**, 347–354.

Herring, P. J. (1977). *Symp. Zool. Soc. London* **38**, 127–159.

Herring, P. J. (1978). *In* "Bioluminescence in Action" (P. J. Herring, ed.), pp. 199–240. Academic Press, New York.

Herring, P. J., and Morin, J. G. (1978). *In* "Bioluminescence in Action" (P. J. Herring, ed.), pp. 273–329. Academic Press, New York.

Hopkins, T. A., and Hanna, C. H. (1972). *Physiologist* **15**, 171.

Johnson, F. H., Shimomura, O., and Haneda, Y. (1966). *In* "Bioluminescence in Progress" (F. H. Johnson and Y. Haneda, eds.), pp. 385–389. Princeton Univ. Press, Princeton, New Jersey.

Lecuyer, B., and Arrio, B. (1975). *Photochem. Photobiol.* **22**, 213–215.

Morin, J. G., and Hastings, J. W. (1971a). *J. Cell. Physiol.* **77**, 305–312.

Morin, J. G., and Hastings, J. W. (1971b). *J. Cell. Physiol.* **77**, 313–318.

Nicol. J. A. C. (1952a). *J. Mar. Biol. Assn. U.K.* **30**, 433–452.

Nicol, J. A. C. (1952b). *J. Mar. Biol. Assn. U.K.* **31,** 113–144.

Nicol. J. A. C. (1953). *J. Mar. Biol. Assn. U.K.* **32,** 65–84.

Nicolas, M. T. (1977). *Arch. Zool. Exp. Gen.* **118,** 103–120.

Nicolas, M. T. (1979). *C. R. Acad. Sci. D* **289,** 177–180.

Oertel, D., Lindberg, K. A., and Case, J. F. (1975). *Cell Tissue Res.* **164,** 27–44.

Okada, Y. K. (1966). *In* "Bioluminescence in Progress" (F. H. Johnson and Y. Haneda, eds.), pp. 611–625. Princeton Univ. Press, Princeton, New Jersey.

Peterson, M. K. (1970). *J. Morphol.* **131,** 103–116.

Quartrefages, A., de (1850). *Ann. Sci. Nat. Zool. 3* **14,** 236–281.

Reynolds, G. T. (1972). *Q. Rev. Biophys.* **5,** 295–347.

Rudie, N. G., and Wampler, J. E. (1978). *Comp. Biochem. Physiol. A* **59,** 1–8.

Schmitter, R. E. (1971). *J. Cell Sci.* **9,** 147–173.

Schmitter, R. E., Njus, D., Sulzman, F. M., Gooch, V. D., and Hastings, J. W. (1976). *J. Cell. Physiol.* **87,** 123–134.

Schimomura, O., and Johnson, F. H. (1979). *Comp. Biochem. Physiol. B* **64,** 105–108.

Smalley, K. N., Tarwater, D. E., and Davidson, T. (1980). *J. Histochem. Cytochem.* **28,** 323–329.

Smith, D. S. (1963). *J. Cell Biol.* **16,** 323–359.

Spurlock, B. O., and Cormier, M. J. (1975). *J. Cell Biol.* **64,** 15–28.

Strause, L. G., DeLuca, M., and Case, J. F. (1979). *J. Insect Physiol.* **25,** 329–347.

Strum, J. M. (1969a). *Anat. Rec.* **164,** 433–462.

Strum, J. M. (1969b). *Anat. Rec.* **164,** 463–478.

Sweeney, B. M. (1969). "Rhythmic Phenomena in Plants." Academic Press, New York.

Sweeney, B. M. (1976). *J. Cell Biol.* **68,** 451–461.

Sweeney, B. M. (1978). *J. Phycol.* **14,** 116–120.

Sweeney, B. M. (1979a). *In* "Biochemistry and Physiology of Protozoa" (M. Levandowsky and S. H. Hutner, eds.), 2nd ed., pp. 288–306. Academic Press, New York.

Sweeney, B. M. (1979b). *J. Phycol.* **15** (Suppl.), 23.

Sweeney, B. M. (1979c). *In* "Encyclopedia of Plant Physiology" (W. Haupt and M. E. Feinleib, eds.), New Series 7, pp. 71–93. Springer-Verlag, Berlin and New York.

Swift, E., and Taylor, W. R. (1967). *J. Phycol.* **3,** 77–81.

Tsugi, F. I., and Haneda, Y. (1966). *In* "Bioluminescence in Progress" (F. H. Johnson and Y. Haneda, eds.), pp. 137–149. Princeton Univ. Press, Princeton, New Jersey.

Tsugi, F. I., Barnes, A. T., and Case, J. F. (1972). *Nature (London)* **237,** 515–516.

INTERNATIONAL REVIEW OF CYTOLOGY, VOL. 68

Differentiation of MSH-, ACTH-, Endorphin-, and LPH-Containing Cells in the Hypophysis during Embryonic and Fetal Development

JEAN-PAUL DUPOUY

Laboratory of Animal Physiology,
Faculty of Sciences,
University of Picardie,
Amiens, France

I. Introduction

The ontogenesis of the hypophysis has been extensively studied at the light microscopic level (review in Hanström, 1966; Wingstrand, 1966) and more recently at the electron microscopic level for various species.

There are difficulties in the accurate identification of the pituitary cells en-gaged in the differentiation process and in estimation of onset of hormone syn-thesis and secretion. Electron microscopic demonstration of secretory granules has been considered as the first and most reliable expression of cellular dif-ferentiation. However, the lack of morphological signs of cytodifferentiation does not preclude functional differentiation of hormone synthesis and secretion prior to secretory granule formation and storage.

Originally investigators postulated total similarity of adult and fetal or em-bryonic pituitary cells; they based cell identification on the morphological criteria that were applied to the adult hypophysis, e.g., shape and size of the cells, granule dimensions and density, size and development of the Golgi complex, and amount of rough endoplasmic reticulum. Immunoenzyme techniques, first de-scribed by Nakane and Pierce (1966), have been used extensively to localize pituitary hormones at the light and electron microscopic level (review in Moriarty, 1973; Girod, 1977). When applied to embryonic material, these cytochemical techniques demonstrated that the identification of hormone-containing cells in the developing pituitary cannot be exclusively based on morphological characteristics that are continuously changing during ontogenesis. However, it appeared that immunocytological techniques applied at light and electron microscopic levels could provide accurate information for the analysis of morphological and functional cytodifferentiation in the developing pituitary gland.

We can ask whether, during pituitary morphogenesis one cell begins to pro-duce several hormones and subsequently differentiates to elaborate only one kind of hormone. Moreover, does pituitary development result from proliferation of undifferentiated and/or differentiated cells? Is cytodifferentiation a self-determining event, and when does it occur in the developing pituitary gland? In order to answer the last question, the chronology of fetal or embryonic develop-ment has to be known with the greatest precision.

Jost and Picon (1970) emphasized the great importance of exact reckoning of fetal age in laboratory animals such as rats, mice, and rabbits, in which day-by-day physiologic changes may be occurring. They discussed different ways to determine the onset of gestation in mammals.

In animals such as rabbits, in which ovulation is provoked by copulation, the time of mating has been considered as time 0 and the fetal age can be expressed with reliability as the number of days or weeks postcoitus (Schechter, 1970, 1971; Chatterjee, 1974, 1975). In spontaneously ovulating animals such as rats and mice, the exact time of fertilization is known, with some imprecision. Usually males and females were caged overnight, fertilization was assumed to occur in the middle of the period when males and females were together (Jost and Picon, 1970).

Important discrepancies in reckoning fetal age or pregnancy stage can be observed in the literature. The day on which sperm was found in smears was considered as day 0 of gestation (Stoeckel *et al.*, 1973a; Watanabe *et al.*, 1973; Osamura, 1977; Chatelain *et al.*, 1979; Dupouy *et al.*, 1979; Watanabe and Daikoku, 1979, in rat; Stoeckel *et al.*, 1979, in mouse; Thompson and Trimble 1976, in hamster), whereas other investigators erroneously considered it as day 1 (1-day-old embryos) (Nemeskery *et al.*, 1976; Setalo and Nakane, 1976, in rat; Sano and Sasaki, 1969; Eurenius and Jarskar, 1975; Dearden and Holmes, 1976, in mouse). Such differences in the dating methods do not facilitate the comparisons of data relative to the earlier investigations on fetal development. Moreover, spontaneous variations in the time of fertilization and implantation of ova (e.g., litter size) can produce age variation within and between litters. These differences in time probably have great influence in the earliest stages of embryonic development. In order to minimize but not extinguish them, it appears necessary to study several animals from various litters.

In species with a long pregnancy (farm animals, monkeys, humans), the exact fetal age determination is less critical, except for investigations during early stages of development. In man, gestational age was usually estimated by menstrual history and/or crown–rump (CR) length of the fetus. For nonmammalian vertebrates, the stages of embryonic development were based on the number of somites (birds) or the size and morphological features of the tadpoles (amphibians).

Every study on cytodifferentiation of the pituitary gland performed in early stages of fetal or embryonic development needs a knowledge of precise age; this determination is of critical importance.

During the last 10 years, prodigious improvements in the understanding of pituitary hormones resulted from the development of various sophisticated investigation methods. Structural relationships between several hormones have been demonstrated and specified for ACTH, MSH, and LPH; new peptides such as endorphins have been discovered.

The aim of this chapter is to review the recent progress on the differentiation of ACTH-, MSH-, endorphin-, and LPH-containing cells in the hypophysis during embryonic and fetal development and to recapitulate the main relationships between these hormones.

II. First Signs of Cytological Differentiation during Embryonic and Fetal Life

A. Structural Signs of Cytological Differentiation

It is well established that the anterior lobe of the fetal pituitary gland arises from the Rathke's pouch, which develops as an invagination of the oral ectoderm, while an infundibular evagination differentiates into the posterior lobe. Afterward, anterior pituitary cells differentiate from the epithelial cells of the Rathke's pouch; these cells appear initially as undifferentiated (i.e., lacking secretory granules) chromophobe cells with large nuclear:cytoplasmic ratios and few organelles.

1. *Human Fetus*

Several authors studied the development of the human fetal adenohypophysis; however, the use of different methods to determine the fetal age makes it difficult to compare reported data.

According to Satow *et al.* (1971) and Fukuda (1973), Rathke's pouch appears by the end of the first month of pregnancy. Conklin (1968) studied its growth in human embryos and fetuses whose ages, based on crown–rump (CR) length, ranged from 11.5 (5–6 weeks) to 365 mm (term). In the youngest specimen, Rathke's pouch was lined by a layer of stratified cuboidal epithelial cells that exhibited many mitotic figures; in older fetuses, cell cords growing into the richly vascularized mesenchyme were produced by intensive cellular proliferation. Conklin (1968) reported that the first chromophil cells, which were identified by their characteristic staining with PAS and orange G, appeared in the region of the epithelial–mesenchymal junction of one 20-mm CR fetus (7-week-old); later carminophilic-type cells and Alcian blue-PAS-positive-type cells were first observed in glands from 65 (11-week-old) and 78 mm CR fetuses (12-week-old), respectively. According to Falin's investigations (1961) with the light microscope, the differentiation of the cells in the anterior lobe of the human hypophysis starts also as early as the seventh to eighth week. Using histochemical methods, Anderson *et al.* (1970) observed the first alcianophilic cells in a 28-mm CR fetus; they were forming small groups at the epithelial–mesenchymal junction and showed a pronounced maltase-resistant PAS positivity. By application of the PAS method, Andersen *et al.* (1970) showed glycogen in all parenchymal cells of the early anlage of the adenohypophysis. Dubois (1971), using ultrastructural methods, reported the presence of α- or β-glycogen particles in all the undifferentiated cells of the human fetal hypophysis; the beginning of cell differentiation, which is characterized by Golgi apparatus and rough-surfaced endoplasmic reticulum development, was associated with a sharp decrease in glycogen content. In one human fetus studied by Fukuda (1973), 41 days after ovulation no granulated cells were yet observed in the primitive adenohypophysis, although marginal cells and other primordial cells exhibited more active cytological features; the mitochondria were more numerous, the free ribosomes and particularly the polysomes were increasing their size and number, the Golgi apparatus became more prominent and showed flattened cisternae and minute vesicles, the rough-surfaced endoplasmic reticulum became elongated, and glycogen granules aggregated in certain areas of the cytoplasm.

In the pituitary gland of a 25-mm CR fetus (eighth week of pregnancy), Dubois (1968) observed only a few differentiating cells with poorly granulated cytoplasm.

All these reported structural changes in the anterior pituitary primordium appear as primary evidence of cytodifferentiation. Although the presence of granules in the cytoplasm of chromophils is usually taken as an expression of secretory activity of the adenohypophyseal cells, it cannot be excluded that hormone production and even release may occur without the formation of granules (Grieshaber and Hymer, 1968). Which hormone is produced by the first granulated cells remains to be determined.

2. Sheep Fetus

Alexander et al. (1973a) studied the adenohypophysis of fetal sheep by light and electron microscopy. Whereas the adenohypophysis of the younger animals (54 days) contained predominantly chromophobic cells when examined by light microscopy, the cells showed abundant rough-surfaced endoplasmic reticulum, polyribosomes, large Golgi areas, and secretory granules. They tried to identify the cells according to granule sizes and first saw corticotrophs in 95-day-old fetuses.

3. Pig Fetus

In the pig embryo, the formation of all morphological elements of the pituitary gland took place from the nineteenth to the forty-sixth day of fetal life (Bielanska-Osuchowska and Liwska, 1975; Liwska, 1975). The growth of the gland and the functional differentiation of the pars distalis cells were beginning approximately on the forty-sixth day of gestation (Liwska, 1978). In the youngest embryos observed (46–51 days of pregnancy), several types of differentiated cells were present in the pars distalis (Liwska, 1978); therefore, the process of cytodifferentiation must start earlier.

4. Rabbit Fetus

In the fetal rabbit, the first appearance of the Rathke's pouch was reported to occur on the tenth day postcoitus (Schechter, 1970); the cytoplasm of its cells was very rich in free ribosomes and polysomes until day 11. Secretory granules were first observed in some cells of the pars distalis anlage on day 16 (Schechter, 1970, 1971). The first differentiating cells contained secretory granules of about 90–200 nm in diameter, closely resembling those found in basophils of the adult rabbit pars distalis (Schechter, 1970, 1971); thus, these cells were considered as thyrotroph and gonadotroph precursors (Schechter, 1971). The beginning of pars distalis cytodifferentiation was contemporaneous with the mesenchymal infiltration of this pituitary lobe and with the establishment of vascular channels (Schechter 1970, 1971). Chatterjee (1974, 1975) extensively studied the development and the cytodifferentiation of the rabbit pars intermedia. As early as day 14 postcoitus primitive blood vessels were beginning to develop at the junctional zone of the presumptive pars intermedia and pars nervosa; at 15 days, secretory granules averaging 70 nm in diameter, appeared for the first time in some cells of the developing pars intermedia, at the side nearest to the pars nervosa. Chatterjee reported that the cytodifferentiation of this pituitary lobe was starting earlier than the establishment of either a nerve or vascular supply, which occurred 1 day and 1 week later, respectively.

5. *Rat Fetus*

In the rat, the Rathke's pouch appeared as a V-shaped invagination of the oral ectoderm as early as the 19-somite stage (Schwind, 1928), on 8–9 days (Satow *et al.*, 1971) or 10.5 days of gestation (Negm 1970, 1972). The presumptive adenohypophyseal tissue changed from flat to cuboidal epithelium during embryonic days 11–12 (Gash *et al.*, 1975) and, near the days 14–15, the rudimentary pituitary was transformed into a closed vesicle without direct communication with the epipharynx (Negm, 1970, 1972; Stoeckel *et al.*, 1973a; Svalander, 1974). An infundibulum evagination, which was differentiating into the posterior lobe of the pituitary, appeared around day 11 (Satow *et al.*, 1971). The first indications of cytodifferentiation of the anterior pituitary anlage have been found between the fourteenth and fifteenth day, with the appearance of faintly PAS-positive granules in the cells around the cavity of the Rathke's pouch and the cells of the outgrowths arising from the anterior wall of the pouch (Negm, 1970)

Svalander (1974) extensively studied the ultrastructure of the fetal rat adenohypophysis during embryonic development. After the fourteenth gestational day, he reported that the cells lining the lumen of the growing pituitary vesicle (so-called primordial cells) accumulated glycogen and presented a large Golgi system and lipid droplets in their cytoplasm but no secretory granules; from the sixteenth day, he observed the formation of follicles in the rudimentary pars distalis and the arrangement of granule-containing cells at their periphery. During the period 16–17 days, the earliest granulated cells with small electron-dense spherical granules, 120 nm in diameter, were found exclusively at the periphery of the primordial cell cords, rostrally in the gland, and at the epithelial–mesenchymal border close to blood vessels (Svalander, 1974). Similar cells, although containing smaller granules (50–100 nm of diameter), have been observed by Yoshimura *et al.* (1970) at the corresponding stage of fetal development, and Yoshida (1966) reported cells with small granules similar to those of "basophils" on the sixteenth day of gestation and cells with larger granules as those of "acidophils" on the seventeenth day. Other investigators also demonstrated secretory granules by the seventeenth day of gestation in the developing rat pituitary (Fink and Smith, 1971; Satow *et al.*, 1971). From day 18, Maillard (1963) found two granulated cell types she called type A (granules

320–350 nm in diameter) and type B (granules 160 nm in diameter), whereas Svalander (1974) observed three different types of granulated and highly differentiated cells with structural characteristics similar to those of the hormone-producing cells of the adult.

The presumptive pars tuberalis initially appeared in 14-day-old rat fetuses (Stoeckel et al., 1973a,b) or on the sixteenth day (Svalander, 1974) as an anterior process of primordial cells. The pars tuberalis cells contained lysosome-like particles and abundant glycogen and accumulated numerous secretory granules from the sixteenth to seventeenth day until term (Stoeckel et al., 1973a,b; Svalander, 1974).

The ontogenesis of the pars intermedia has been followed extensively by Svalander (1974). The hypophyseal-developing vesicle and the rudimentary neural lobe established first contact by the thirteenth day of gestation; on the seventeenth day, an intimate contact was rostrally observed between pars intermedia and pars nervosa while vascularized mesenchyme separated the two processes dorsocaudally. The glycogen content of the intermediate cells decreased until the seventeenth day in the very actively proliferating pars intermedia. At this last stage, Svalander (1974) noted many cells with a well-developed Golgi apparatus and small secretory granules. Late in gestation, numerous granules with a mean maximum diameter of 165 nm were demonstrated in the cells of the pars intermedia.

According to ultrastructural signs, the pars intermedia cells seemed to differentiate later than the pars distalis and the pars tuberalis cells; however, conventional electron microscopic methods were unsuitable for identification of the hormone-producing cells in the differentiating pituitary gland.

6. Hamster Fetus

The development of the golden hamster pituitary gland has been described by Thompson and Trimble (1976) at the ultrastructural level. At 8.5 days of gestation, Rathke's pouch was formed but still open to the stomodeum; at 10.5 days, when pars nervosa appeared as a cellular evagination of the diencephalon, pars tuberalis anlage had begun to form. At 11.5 days, Rathke's pouch was closed but not yet vascularized. At this gestational stage, which preceded the formation of granules, more Golgi complexes, more rough-surfaced endoplasmic reticulum, many ribosomes, and a considerable amount of glycogen were observed in the differentiating cells. Thompson and Trimble noted the first secretory granules in cells of the pars distalis at 12.5 days of gestation when blood vessels penetrated the parenchyma of the gland.

7. Mouse Fetus

In the mouse embryo, the formation of the Rathke's pouch started as early as day 10 (Stoeckel et al., 1979) or day 12 (Sano and Sasaki, 1969) by upward evagination of the epithelium bordering the ceiling of the primary oral cavity. Rathke's pouch was closed by day 11 (Stoeckel *et al.*, 1979) or day 13 (Sano and Sasaki, 1969). Yet on day 12, Stoeckel *et al.* (1979) were able to distinguish the anlage of all the hypophyseal lobes (neurohypophysis, pars intermedia, pars distalis, and pars tuberalis); most of the cells of the pars tuberalis anlage were already characterized by numerous glycogen particles and PAS-stainable material. In contrast, rarely a few pars distalis cells contained glycogen and were PAS positive. Twenty-four hours later, on day 13, the fetal pars tuberalis cells appeared as differentiated as adult cells on the basis of the development of the rough-surfaced endoplasmic reticulum and the Golgi system, the appearance of dense secretory vesicles (200 nm in diameter), and the glycogen content (Stoeckel *et al.*, 1979). At this gestational stage, the first signs of secretory differentiation were observed by Stoeckel *et al.* (1979) in the ventral part of the pars distalis. Although only occasional granulated cells were found in the adenohypophysis of the 15-day-old fetus (Dearden and Holmes, 1976), the granulation of secretory cells was only conspicuous from the sixteenth day of gestation (Sano and Sasaki, 1969; Dearden and Holmes, 1976) and predominantly in the upper and anterior adenohypophysis (Dearden and Holmes, 1976). A transitory glycogen storage was observed in undifferentiated cells (chromophobes) and was sparse or absent in cells showing secretory activity (granule-containing cells) (Sano and Sasaki, 1969; Stoeckel *et al.*, 1979).

The first appearance of granulated cells was observed before (Dearden and Holmes, 1976) or contemporarily with the development of internal vascularization of the parenchyme.

In the pars intermedia anlage, the first differentiated cells were observed in 17-day-old embryos (Eurenius and Jarskär, 1975); they were characterized by small and highly osmiophilic granules (mean diameter, 110 nm) adjacent to the Golgi complex or aligned along the cell membrane. Later in gestation, in 18- to 19-day-old fetuses, a sharp increase in the number and granulation of the pars intermedia cells was observed (Eurenius and Jarskär, 1975).

From these structural observations, it appears that in the mouse pituitary anlage, as in the rat, pars tuberalis is the first to differentiate, whereas pars intermedia is the last.

8. *Lower Nonmammalian Vertebrates*

In the literature, few investigations were reported on the ultrastructural development of the pituitary in lower nonmammalian vertebrates, including amphibia and fish.

In *Triturus viridescens,* the first secretory granules appeared prior to hatching in the presumptive pars intermedia; in contrast, the cells of the pars distalis were agranular up to hatching, and only secretory granules were seen in some cells of recently hatched larvae (Dent and Gupta, 1967). In the hypophyseal primordium of the *Xenopus laevis* tadpoles, the appearance of the earliest secretory granules was observed in stage 33/34 (Nyholm, 1972). In *Rana temporaria*, the presumptive pituitary gland was formed of undifferentiated embryonic cells in 7-mm tadpoles; the first granulated cells appeared in the pars intermedia anlage in stage 10 mm and the differentiation of the pars distalis was evident near premetamorphosis (Doerr-Schott, 1968). During the larval growth of *Bufo bufo* (toad), pars distalis showed five types of granulated cells (Mira-Moser, 1972).

In fish, cellular differentiation and histogenesis of the adenohypophysis were investigated in the Chondrichthyen *Scyllium canicula* by Alluchon-Gérard (1970, 1971). She recognized Rathke's pouch at stage 10 mm and saw the first secretory granules simultaneously in the four hypophyseal lobes (rostral, ventral, medial and neurointermediate) at state 35 mm; the end of embryonic development was characterized by a great increase in the granular content of the adenohypophyseal cells.

B. SUBCELLULAR LOCALIZATION OF POLYPEPTIDIC HORMONES

The presence of granules in the cytoplasm of pituitary cells has been commonly considered as an accurate expression of cellular differentiation. Biochemical studies and, more recently, cytochemical investigations provided evidence that pituitary hormones were mainly localized in the cytoplasmic granules. ACTH activity was associated with the fraction containing secretory granules isolated by differential centrifugation and filtration from rat pituitary glands (Hymer and McShan, 1963; Perdue and McShan, 1966; Costoff and McShan, 1969; Ohtsuka *et al.*, 1972). LPH and endorphin activities were also associated with cytoplasmic granules obtained from homogenates of bovine pituitary (Queen *et al.*, 1976; Labella *et al.*, 1977); moreover, the authors reported that nearly all of the activity was contained within the secretory granules. Similar investigations have been performed for other pituitary hormones (GH, prolactin, TSH, LH, FSH).

In recent years, immunocytochemical techniques have been used for the ultrastructural localization of several hormones in the pituitary cells (see review of Moriarty, 1973). Most of the investigations were performed in adult animals. ACTH was localized in the secretory granules of the corticotrophs in rat (Moriarty and Halmi, 1972a,b; Moriarty, 1973; Moriarty and Moriarty, 1973; Moriarty *et al.*, 1973; Moriarty and Garner, 1977a; Pelletier *et al.*, 1977; Weber *et al.*, 1978a; Martin *et al.*, 1979), in monkey, ox, sheep, and pig (Pelletier *et al.*, 1977), and in man (Moriarty and Moriarty, 1973; Pelletier *et al.*, 1977, 1978). MSH immunoreactivity was also associated with cytoplasmic granules in rat (Moriarty and Halmi, 1972b; Moriarty, 1973), man (Moriarty, 1973), and amphibians (Doerr-Schott, 1974). Recent studies performed in several species localized β-endorphin (Martin *et al.*, 1979, in rat pituitary) and LPH (Pelletier *et al.,* 1977, 1978, in man; Pelletier *et al.,* 1977, in monkey, ox, sheep, pig, and rat pituitaries) in secretory granules.

In human fetal adenohypophysis, cells immunoreactive with anti-α-(17–39)-ACTH, β-(1–24)-ACTH, β-MSH (Li *et al.,* 1976), anti-β-LPH, and α- and β-endorphin (Li *et al.,* 1979) were identified at the ultrastructural level using the peroxidase–antiperoxidase complex method; hormones and peptides were detected in the secretory granules (Fig. 1).

Morphologically and biochemically, the presence of cytoplasmic granules appeared as evidence of hormonal activity of the pituitary cells, but it cannot be excluded that some hormone production, storage, and release may occur without the formation of granules. Moreover, in addition to immunostainable granules, the cytoplasm can contain nonreactive granules as was reported for corticotrophs of the rat anterior pituitary (Weber *et al.,* 1978a) and some reactive material not associated with secretory granules (Pelletier *et al.,* 1977).

C. IDENTIFICATION OF THE VARIOUS HORMONE-PRODUCING CELLS ACCORDING TO CONVENTIONAL ULTRASTRUCTURAL FEATURES

Until quite recently, the identification of pituitary cell types was based on ultrastructural criteria such as cell shape, granule size and distribution, and organelle development. Now, in the light of the immunocytochemical observations, it appears that this old system of cell identification does not provide rigorous information concerning the hormone contained in the pituitary cells of the adult and, of course, of the fetus or the embryo in which cells are continuously changing during histo- and cytogenetic processes.

Granule dimensions and abundance are subject to large variations that are dependent on physiological state, as reported for adult rat corticotrophs, which were immunologically identified by Moriarty and Halmi (1972a) and Moriarty (1973). Ohtsuka *et al.* (1972) also reported that there was no direct correlation between the size of free granules and their hormonal activity. The size of secretory granules was observed to increase with the progression of gestational or embryonic age in many species (Sano and Sasaki, 1969, in mouse; Alluchon-Gerard 1971, in fish embryo) or to vary with the fixation method employed for electron microscopic study (Bugnon *et al.,* 1975, in human fetus). Finally, one and the same cell may elaborate and store in cytoplasmic granules several kinds of hormones, as will be specified in the following sections.

These observations can explain why pituitary cells as corticotrophs have

---

FIG. 1.   Ultrastructural localization of immunoreactive α-endorphin and corticotropin in cells of the human fetal anterior pituitary. (a) Fourteen-week-old fetus; anti-α-endorphin serum; lead citrate counterstain. (b) Fifteen-week-old fetus; anti-β-(1–24)-ACTH serum; lead citrate counterstain. Reaction product is mainly over the secretory granules. ×43,000. (From J. Y. Li *et al.,* 1979.)

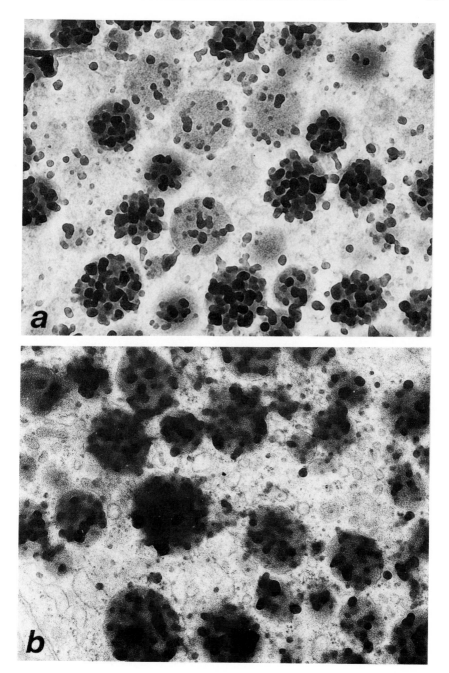

been assigned to the classic groups, chromophobes, acidophils, or basophils by light and electron microscopists before the use of immunocytological methods. The identification of pituitary cells using conventional ultrastructural features applied to the adult hypophysis appears very hazardous in embryonic and fetal material.

III. Time of Appearance of Cells Containing Polypeptides (ACTH, MSH, LPH, Endorphins) during Normal Embryonic and Fetal Development

Immunocytological methods have been extensively used to reveal hormone-containing cells in the embryo and fetal adenohypophysis during its development (Table I).

A. IN MAMMALIAN VERTEBRATES

1. *In Man*

a. *ACTH-Containing Cells.* The use of antibodies to α-(17–39)-ACTH has enabled the detection of immunoreactive cells in the pituitary gland of 5- (Osamura, 1977) or 7-week-old fetuses (eighth week of gestation) (Dubois *et al.*, 1973; Baker and Jaffe, 1975; Bugnon *et al.*, 1976a,b; Bégeot *et al.*, 1977, 1978a,b,c). Corticotrophs were also revealed as early as the eighth week with anti-β-(1–24)-ACTH sera (Dubois *et al.*, 1973; Bugnon *et al.*, 1976a,b). The immunoreactive cells were located in both anterior and posterior lamina of the Rathke's pouch (Dubois *et al.*, 1973; Bugnon *et al.*, 1974, 1976a,b; Bégeot *et al.*, 1977, 1978a,c; Osamura, 1977) and confined chiefly to the borders of well vascularized regions of the mesenchyma (Bugnon *et al.*, 1974, 1976a,b; Baker and Jaffe, 1975). The fluorescent cells of the anterior pituitary of the 7-week-old human fetus were always chromophobes (Dubois *et al.*, 1973). However, Bugnon *et al.* (1976a,b) and Bégeot *et al.* (1977) reported that corticotrophs showed an affinity for PAS.

A great increase in the number and the cytoplasmic area of the ACTH-containing cells was noticeable from 9 to 23 weeks in the pars distalis (Bégeot *et al.*, 1977, 1978a; Osamura, 1977). The human fetus develops a rudimentary pars inter media, which involutes shortly after birth (Wingstrand, 1966). Numerous cells of the intermediate lobe were strongly immunoreactive after application of anti-α-(17–39)-ACTH serum (Bégeot *et al.*, 1978c).

A close relationship between the level of morphological differentiation of the adenohypophysis and the bioactive ACTH content was reported by Pavlova *et al.* (1968); whereas no hormone was observed in 8-week-old embryos, the first signs of adrenocorticotropic activity appeared in the glands of 9- to 10-week-old fetuses, and subsequently, the amount of ACTH in the hypophysis increased gradually. Likewise, Skebelskaya (1965) [cited by Levina *et al.* (1968)] reported corticotropin activity in extracts of hypophysis in a 9-week-old fetus and Kastin *et al.* (1968) reported activity in an 11-week-old human fetus. Winters *et al.* (1974) reported relatively high levels of immunoreactive ACTH in the plasma of the umbilical cord of 12- to 19-week-old human fetuses.

b. *β-MSH-Containing Cells.* Using anti-β-MSH serum, Bugnon *et al.* (1976a,b) were able to reveal melanotrophs as early as the eighth week of gestation. According to Dubois *et al.* (1973), the β-MSH activity appeared some hours later than the α-(17–39)-ACTH activity in the human fetal hypophysis. Baker and Jaffe (1975) and Osamura (1977) showed β-MSH-containing cells only from 13 to 14 weeks. The first melanotrophs were revealed in the regions near the nervous lobe close to the hypophyseal cleft or its vestiges (Dubois *et al.*, 1973), in the median zone and the lateral areas of the pars distalis (Baker and Jaffe, 1975) and also in the pars intermedia (Baker and Jaffe, 1975). The immunoreactive cells were always near capillaries (Dubois *et al.*, 1973; Baker and Jaffe, 1975). Levina (1968) and Levina *et al.* (1968) showed first appearance of bioactive MSH in whole extracts of fetal anterior hypophysis at the tenth week, and Kastin *et al.* (1968) reported melanocyte-stimulating activity in pituitary glands of 11-week-old human fetuses. With the use of contiguous thin pituitary sections treated with various antisera [α-(17–39)-ACTH, β-(1–24)-ACTH, β-MSH] or immunocytochemical double staining with antisera to two hormones (ACTH, β-MSH), it was shown that numerous cells contained both ACTH and β-MSH (Dubois *et al.*, 1973; Baker and Jaffe, 1975; Bugnon *et al.*, 1975, 1976a,b).

c. *α-MSH-Containing Cells.* While anti-β-(1–24)-ACTH, anti-α-(17–39)-ACTH, and anti-β-MSH sera revealed the same cells in the fetal hypophysis, Dubois *et al.* (1973) and Bugnon *et al.* (1976a) were unable to reveal melanotrophs with anti-α-MSH serum regardless of the age of the fetus and the hypophyseal region investigated.

In contrast, using fractionation of a pituitary extract from a 24-week-old female fetus and a 31–week-old male infant, Silman *et al.* (1976) presented for the first time evidence that the human fetal pituitary, for most of intrauterine life, contained only small amounts of intact ACTH but relatively larger quantities of peptides closely resembling α-MSH and CLIP (corticotrophin-like intermediate lobe peptide).

d. *LPH- and Endorphin-Containing Cells.* With anti-β-LPH, anti-γ-(1–58) LPH, and anti-γ-(1–46)-LPH sera, Dubois *et al.* (1973) were unable to reveal fluorescent cells in the anterior hypophysis whatever the age of the human fetus, whereas other investigators (Bugnon *et al.*, 1974, 1976b; Bégeot *et al.*, 1978a,b,c) showed immunoreactive cells as early as the eighth or ninth week. According to Celio (1979), the time of appearance of β-endorphin immunoreac-

TABLE I

IMMUNOCYTOLOGICAL STUDY OF THE α-MSH-, β-MSH-, ACTH, α-ENDORPHIN-, β-ENDORPHIN-,
AND β-LPH-CONTAINING CELLS IN THE HYPOPHYSIS OF EMBRYOS AND FETUSES
OF DIFFERENT SPECIES

Species	Methods[a]	Antisera used[b]	References
Mammals			
Human	IF	Anti-α-MSH	Dubois (1973)
		Anti-β_b-MSH	
		Anti-β-(1–24)-ACTH	
		Anti-α-(17–39)-ACTH	
		Anti-β_p-LPH	
		Anti-γ_p-(1–58)-LPH	
		Anti-γ_p-(1–46)-LPH	
	IF; IP	Anti-β-(1–24)-ACTH	Bugnon *et al.*
		Anti-β_p-LPH	(1974)
	IP	Anti-β_h-MSH	Baker and Jaffe
		Anti-β-(1–24)-ACTH	(1975)
		Anti-β_p-(17–39)-ACTH	
	IF; IP	Anti-β_b-MSH	Bugnon *et al.*
		Anti-β-(1–24)-ACTH	(1975)
		Anti-α-(17–39)-ACTH	
		Anti-β_p-LPH	
	IF; IP	Anti-α-MSH	Bugnon *et al.*
		Anti-β-MSH	(1976a)
		Anti-β-(1–24)-ACTH	
		Anti-α-(17–39)-ACTH	
	IP	Anti-β-MSH	Bugnon *et al.*
		Anti-β-(1–24)-ACTH	(1976b)
		Anti-α-(17–39)-ACTH	
		Anti-β-LPH	
	IF; IP	Anti-$_p$-ACTH	Begeot *et al.*
		Anti-α-(17–39)-ACTH	(1977)
	IP	Anti-β-MSH	Osamura (1977)
		Anti-(1–39)-ACTH	
		Anti-α-(17–39)-ACTH	
	IF; IP	Anti-α-(17–39)-ACTH	Begeot *et al.*
	— —	Anti-β_p-LPH	(1978a,b,c)
	IF	Anti-α-endorphin	
		Anti-β-endorphin	
	IP	Anti-β-endorphin	Celio (1979)
	IP	Anti-β-(1–24)-ACTH	Li *et al.* (1979)
		Anti-α-(17–39)-ACTH	
		Anti-β-LPH	
		Anti-α-endorphin	
		Anti-β-endorphin	

TABLE I (*continued*)

Species	Methods[a]	Antisera used[b]	References
Pig	IF	Anti-α-MSH Anti-β-MSH Anti-(1–24)-ACTH Anti-$_p$-(25–39)-ACTH Anti-$_p$-(17–39)-ACTH Anti-β_p-LPH Anti-γ_p-LPH Anti-α-endorphin Anti-β-endorphin	Dubois (1977)
Rat	IP	Anti-ACTH	Setalo and Nakane (1972)
	IP	Anti-ACTH	Nakane *et al.* (1973)
	IF	Anti-α-MSH Anti-β-MSH Anti-β-(1–24)-ACTH Anti-α-(17–39)-ACTH	Begeot *et al.* (1975)
	IF	Anti-α-MSH Anti-β-MSH Anti-β-(1–24)-ACTH Anti-α-(17–39)-ACTH	Dupouy and Dubois (1975)
	IP (organ culture)	Anti-$_p$-ACTH Anti-(17–34)-ACTH	Nemeskery *et al.* (1976)
	IP	Anti-$_p$-ACTH Anti-(17–34)-ACTH	Setalo and Nakane (1976)
	IP (organ culture)	Anti-$_p$-(1–39)-ACTH Anti-$_p$-(17–39)-ACTH	Watanabe and Daikoku (1976)
	IP	Anti-ACTH	Osamura (1977)
	IP	Anti-β-(1–24)-ACTH Anti-α-(17–39)-ACTH Anti-β-LPH Anti-α-endorphin Anti-β-endorphin	Begeot *et al.* (1979)
	IP	Anti-α_b-MSH Anti-β_b-MSH Anti-β-(1–24)-ACTH Anti-α_p-(17–39)-ACTH Anti-β_p-LPH Anti-γ_p-(1–58)-LPH Anti-α-endorphin Anti-β-endorphin	Chatelain *et al.* (1979) Dupouy *et al.* (1979)
	IP	Anti-$_p$-(1–39)-ACTH	Watanabe and Daikoku (1979)

(*continued*)

TABLE I (*continued*)

Species	Methods[a]	Antisera used[b]	References
	IP (organ culture)		Begeot (personal communication)
Birds			
Chick embryo	IP	Anti-$_p$-(1–39)-ACTH	Ferrand *et al.* (1974)
	IF; IP	Anti-β_b-MSH Anti-β-(1–24)-ACTH Anti-α-(17–39)-ACTH	Bloch *et al.* (1975) Fellman *et al.* (1975)
Quail embryo	IF	Anti-$_p$-(1–39)-ACTH	Ferrand *et al.* (1975)
Amphibia			
Rana temporaria tadpole	IF	Anti-α-MSH Anti-β_b-MSH Anti-β-(1–24)-ACTH Anti-α-(17–39)-ACTH	Doerr-Schott and Dubois (1973)
Alytes obstetricans tadpole	IF	Anti-α_b-MSH Anti-β_b-MSH Anti-β-(1–24)-ACTH Anti-α_p-(17–39)-ACTH	Remy and Dubois (1974)
Xenopus laevis tadpole	IF	Anti-α-MSH Anti-β_b-MSH Anti-β-(1–24)-ACTH Anti-α-(17–39)-ACTH	Nyholm and Doerr-Schott (1977)

[a] IF, Immunofluorescence; IP, immunoperoxidase.

[b] b, Bovine; h, human; p, porcine.

tive cells was between the sixth and the eighth week, in both the anterior and posterior lamina of Rathke's pouch. Moreover, as the same cells reacted with the different antibodies used in these studies [anti-β-(1–24)-ACTH, anti-(17–39)-ACTH, anti-β-LPH, and anti-β-MSH (Bugnon *et al.*, 1974, 1976b); anti-α-(17–39)-ACTH, anti-β-LPH, anti-α-endorphin, and anti-β-endorphin (Bégeot *et al.*, 1978a,b,c)] it was demonstrated that α- and β-endorphins, β-lipotropin, β-MSH, and ACTH were localized in the same cells (Fig. 2).

Such immunocytochemical observations were in agreement with biochemical evidence that ACTH and β-LPH were arising from a much larger precursor protein (Mains *et al.*, 1977; Roberts and Herbert, 1977a) and that the β-LPH (a 91-amino acid-residue protein) contained the entire sequences of α-endorphin (61–76 β-LPH) and β-endorphin (61–91 β-LPH) as the total hormonal peptide β-MSH (41–58 β-LPH) (Fig. 3).

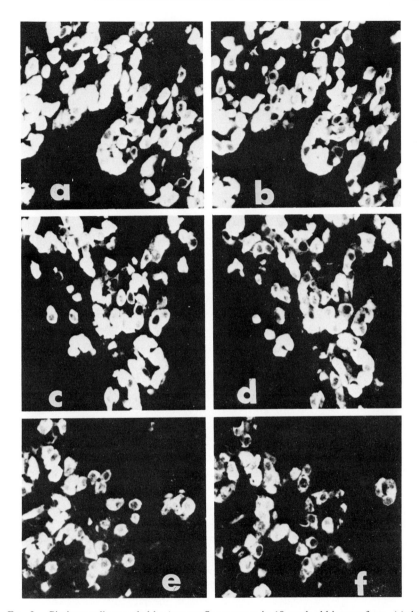

FIG. 2.   Pituitary cells revealed by immunofluorescence in 18-week-old human fetus. (a) Anti-α-(17–39)-ACTH serum. (b) Anti-β-LPH serum. Area corresponding to that of (a); the same cells were stained after application of the two antisera. (c) Anti-β-LPH serum. (d) Anti-β-endorphin serum. Area corresponding to that of (c); the same cells were stained after application of the two antisera. (e) Anti-β-endorphin serum. (f) Anti-α-endorphin serum. Area corresponding to that in (e); the same cells were stained after application of the two antisera. ×220. (From Bégeot *et al.*, 1978c.)

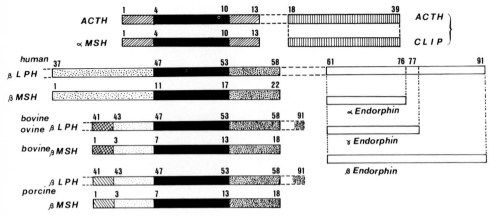

Fɪɢ. 3. Schematic representation of the structural relations between ACTH, MSH, LPH, endorphins, and CLIP. The common sequences of amino acids between the polypeptides have been indicated.

2. *In Sheep and Pig*

In fetal sheep, plasma ACTH has been detected from at least day 54 of gestation and the pituitary gland showed ACTH concentration comparable with adult on day 80 (Alexander *et al.*, 1973b). During the last 2 months of gestation, fetal hypophysis was able to release corticotrophin in response to hemorrhage (Alexander *et al.*, 1971) or hypoxia (Boddy *et al.*, 1974). In extracts of fetal sheep pituitaries from 120 days to term, β-melanocyte-stimulating hormone, β-LPH, γ-LPH, β-endorphin, and ACTH were identified by chromatography and radioimmunoassay, whereas little or no α-MSH and corticotrophin-like intermediate lobe peptide (CLIP) were detected (Silman *et al.*, 1979). ACTH and LPH in high-molecular-weight forms were present in the fetal pituitary (Silman, 1979) and big ACTH was present in the fetal circulation (Jones and Roebuck, 1979).

In the pig fetus, corticomelanotropic cells were revealed by immunofluorescence using polypeptidic hormones (ACTH, MSH, LPH, endorphins) antisera as early as the thirtieth day of pregnancy (Dubois, 1977).

3. *In Rat*

a. *ACTH-Containing Cells.*   Using the enzyme antibody method, Sétalo and Nakane (1972, 1976) localized ACTH cells in the pars distalis of the fetal rat hypophysis starting on day 16. With the immunofluorescence method performed on the hypophysis of fetuses collected in the morning hours from day 15 to day 21 of gestation, we observed the first appearance of ACTH-containing cells on day 17 in the pars distalis and on day 18 in the pars intermedia (Bégeot *et al.*, 1975; Dupouy and Dubois, 1975). In more recent studies, the ontogenesis of corticotrophs was investigated by immunohistochemistry using the peroxidase-antiperoxidase complex. The first cells stainable with anti-α-(17-39)-ACTH and anti-β-(1-24)-ACTH sera appeared in the morning of day 16 in the pars distalis and only in the afternoon of day 17 in the pars intermedia (Chatelain *et al.*, 1979; Dupouy *et al.*, 1979; Bégeot *et al.*, 1979). With an antiporcine-(1-39)-ACTH, Watanabe and Daikoku (1979) were able to detect earlier ACTH cells in the fetal adenohypophysis; they appeared on day 15 of gestation in the pars distalis, although they were not completely absent on day 14 and on day 16 toward the posterior end of the pars intermedia. The immunoreactive cells appeared initially at the periphery of the cord (Setalo and Nakane, 1972, 1976), at the opposite side of the pituitary cleft, before being distributed throughout the pars distalis (Chatelain *et al.*, 1979; Dupouy *et al.*, 1979; Watanabe and Daikoku, 1979); they were in contact with the ventral mesenchymal tissue and some cells were seen abutting on the blood capillary and/or connective tissue that penetrated more deeply into the gland (Watanabe and Daikoku, 1979). There was no chronological difference between both sexes regarding the onset of hormone synthesis (Setalo and Nakane, 1972, 1976; Watanabe and Daikoku, 1979).

In the pars intermedia, corticotrophs were always clustered (Dupouy and Dubois, 1975; Chatelain *et al.*, 1979; Dupouy *et al.*, 1979). The increase in the number of corticotrophs can be correlated with the rise in bioactive ACTH levels in the fetal pituitary gland from day 17 of gestation (Dupouy, 1976).

b. *β-MSH-Containing Cells.* β-MSH cells were first observed in the basal part of the pars distalis on day 16 with the immunofluorescence method (Bégeot *et al.*, 1975; Dupouy and Dubois, 1975) or on day 15 in the afternoon with the peroxidase-antiperoxidase complex (Chatelain *et al.*, 1979; Dupouy *et al.*, 1979). They were revealed in the morning of day 17 in the pars intermedia with both methods (Dupouy and Dubois, 1975; Chatelain *et al.*, 1979; Dupouy *et al.*, 1979).

c. *α-MSH-Containing Cells.* The first cells revealed with an anti-α-MSH serum and peroxidase-antiperoxidase complex were observed at 16 days 8 hours in the pars distalis of some hypophyses, near the sella turcica, and on day 17 in the pars intermedia (Chatelain *et al.*, 1979; Dupouy *et al.*, 1979). In this latter lobe, α-MSH-containing cells were demonstrated 24 hours earlier than with the immunofluorescence method (Bégeot *et al.*, 1975; Dupouy and Dubois, 1975).

d. *LPH- and Endorphin-Containing Cells.* Anti-α-endorphin or anti-β-endorphin and anti-p-(1-58)-γ-LPH or anti-p-(1-91)-β-LPH sera were used to reveal lipotropic cells; these cells were previously observed as early as day 15 in the basal part of the pars distalis and in the morning of day 17 in the pars intermedia (Chatelain *et al.*, 1979; Dupouy *et al.*, 1979). Bégeot *et al.* (1979) also observed lipotropic cells on day 16 of gestation in the pars distalis but not in younger rat fetuses.

From all these histological observations, it appeared that the cytodifferentiation of corticotrophs, melanotrophs, and lipotropic cells occurred earlier in the pars distalis than in the pars intermedia (Dupouy and Dubois, 1975; Bégeot *et al.*, 1979; Chatelain *et al.*, 1979; Dupouy *et al.*, 1979). Based on the treatment of serial sections with various antisera, it was clearly shown that MSH, ACTH, and LPH occur in the same cells located in the pars distalis as well as in the pars intermedia (Bégeot *et al.*, 1979; Chatelain *et al.*, 1979; Dupouy *et al.*, 1979) (Figs. 4 and 5).

4. *In Mouse*

The immunocytological study of the mouse pituitary gland during gestation is in progress (J. Doerr-Schott and M. E. Stoeckel, personal communication). Using anti-(1–24)-ACTH, anti-(17–39)-ACTH, and anti-α- and β-endorphin sera, they were able to reveal the first immunoreactive cells, in the pars distalis only, on day 14 (day 0 is defined as the day when a vaginal plug was observed). The same cells were revealed with all the antisera; they were located at the periphery of the gland near the Atwell-recess, at the posterior and basal edge of the pars distalis. The differentiation of the pars intermedia cells was observed to begin 2 days later than that of the pars distalis cells. At term (days 18–19 of gestation), the pars intermedia still showed a great number of nonreactive cells. α-MSH-containing cells were very numerous in the pars distalis of the mouse fetus.

The presence of melanocyte-expanding activity in the pituitary gland of the fetal mouse was investigated by Enemar (1963) using transplantation of pars distalis or pars intermedia anlage to the skin of frog tadpoles. Biological activity was demonstrated in the pars distalis as early as the fifteenth day of gestation and in the pars intermedia from the sixteenth day.

B. In Nonmammalian Vertebrates

1. *In Birds*

In the *chick* embryo, ACTH-containing cells were observed immunocytologically as early as the eighth (Bloch *et al.*, 1975; Fellmann *et al.*, 1975) or the ninth (Ferrand *et al.*, 1974) day of incubation. These cells remain located in the anterior part of the adenohypophysis at the periphery of cellular cords and near capillary vessels (Fellmann *et al.*, 1975). They were not revealed by anti-β-MSH or anti-β-LPH sera (Bloch *et al.*, 1975; Fellman *et al.*, 1975).

Pedernera (1972) used the *in vitro* secretion of embryonic adrenals as a reliable test to determine the corticotropic activity of the chick embryo pituitary gland at different stages of development (from 6 to 16 days); he was able to demonstrate ACTH activity from day 8 of development. These data were in agreement with those of Székely *et al.* (1958) who employed the ascorbate

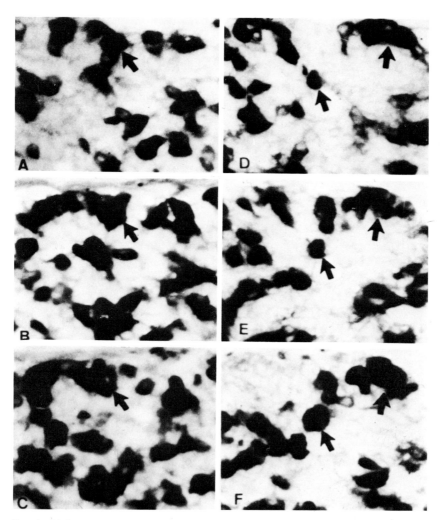

FIG. 4.   Adjacent serial sections of the pituitary from a rat fetus at 17 days 15 hours treated with various antisera. (A) Anti-α-MSH serum; (B) anti-β-MSH serum; (C) anti-(17–39)-ACTH serum. As shown by the arrows, melanotropic and corticotropic activities are found in adjacent sections containing portions of one and the same cell. (D) Anti-(17–39)-ACTH serum; (E) anti-α-endorphin serum; (F) anti-β-endorphin serum. As shown by the arrows, corticotropic and endorphin activities are found in the same cells in adjacent sections. ×1000. (From Chatelain et al., 1979.)

depletion of the rat adrenals to measure the ACTH content of chick embryo pituitaries.

By immunofluorescence, Ferrand et al. (1975) revealed corticotrophs in the adenohypophysis of 14-day-old *quail* embryos; however, since investigations

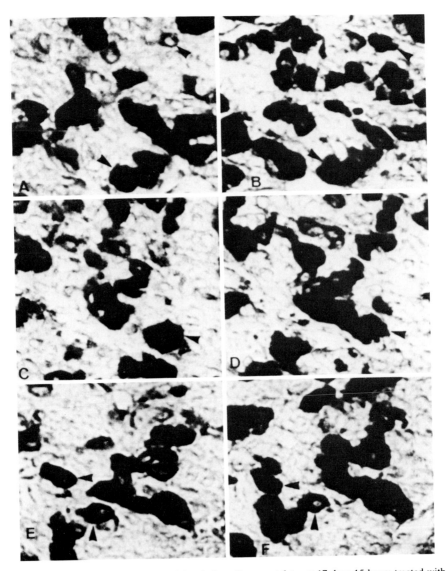

FIG. 5. Adjacent serial sections of the pituitary from a rat fetus at 17 days 15 hours treated with various antisera. (A) Anti-β-MSH serum; (B, D, and F) anti-β-LPH serum; (C) anti-α-endorphin serum; (E) anti-(17–39)-ACTH serum. It is apparent that β-MSH and β-LPH (A and B), α-endorphin and β-LPH (C and D), ACTH and β-LPH (E and F) are present in the same cells found in adjacent sections (arrowheads). ×1000. (From Chatelain *et al.*, 1979.)

were not performed in the early stages of development, the time of differentiation of ACTH-containing cells is still unsettled.

In chick as in quail embryo adenohypophysis, APUD (amine precursor uptake and decarboxylation) cells, which are situated largely in the cephalic lobe, were shown by immunofluorescence to be corticotrophs (Ferrand *et al.*, 1975). Such an observation raises the question of the possible neural crest origin of the ACTH-containing cells.

2. *In Amphibia*

Doerr-Schott and Dubois (1973) investigated by immunofluorescence the cytodifferentiation of corticotrophs and melanotrophs in *Rana temporaria L.* tadpole pituitary. The first ACTH-containing cells were revealed with anti-β-(1–24)-ACTH and anti-α-(17–39)-ACTH sera, as early as the embryonic 10 mm stage and earlier than the differentiation of the pituitary primordia into pars distalis and pars intermedia, which was assumed to occur between stages 10 and 12 mm. In the pars intermedia, the first immunoreactive corticotrophs appeared only at stage 12 mm. From this latter stage, all the intermediate cells reacted with both anti-α-MSH and anti-β-MSH sera; moreover, they showed (like corticotrophs) PAS-positive and Alcian blue-negative reactions. The reacted also with paraldehyde fuchsine without oxidation. Melanotrophs were lacking in the pars distalis. During the premetamorphic stage (from 13 to 25 mm) and the prometamorphic stage (from 25 to 32 mm), the number of corticotrophs was increasing in the pars distalis and in the rostral half of the lobe, near the surrounding capillaries.

The use of anti-β-(1–24)-ACTH, anti-α-MSH, and anti-β-MSH sera allowed Rémy and Dubois (1974) to detect by immunofluorescence in *Alytes obstetricans* tadpole adenohypophysis, ACTH- and MSH-containing cells as early as the time of hatch. Corticotrophs appeared in the pars intermedia and in the rostral part of the pars distalis in the neighborhood of the pituitary portal vessels. While the α-MSH and β-MSH-containing cells occupied the whole pars intermedia, the same cells of the rostral part of the pars distalis reacted with both anti-β-MSH and anti-β-(1–24)-ACTH sera.

In the adenohypophysial primordium of *Xenopus laevis* tadpoles, β-MSH-producing cells were revealed by immunofluorescence from stage 37/38 onward and α-MSH-producing cells from stage 39; moreover, α- and β-MSH occurred in the same cells (Nyholm and Doerr-Schott, 1977). In contrast, these authors reported failure to demonstrate corticotrophs with anti-α-(17–39)- or anti-β-(1–24)-ACTH sera. Elsewhere, it was experimentally determined (Etkin, 1941; Nyholm, 1969) that the hypophyseal primordium of *Xenopus* tadpoles was secreting a melanophore-active substance earlier than the demonstration, by immunohistological techniques, of MSH-containing cells (Nyholm and Doerr-Schott, 1977). A similar discrepancy exists between the onset of MSH secretion

(Pehlemann, 1965) and the time of identification of melanotrophs (Doerr-Schott and Dubois, 1973) in *Rana temporaria* tadpoles. In Anuran tadpole (*Pelobates fuscus, Rana esculenta, Rana temporaria, Xenopus laevis*) adenohypophysis, the earliest demonstrable function was the melanophore-stimulating function of the pars intermedia (Pehlemann, 1965).

3. *In Reptiles and Fish*

Whereas immunohistochemical methods have been used extensively for the cytological analysis of the adult adenohypophysis in reptiles and fish (for extensive bibliography, see the review of Girod, 1977), unfortunately no investigation was performed in the developing pituitary gland of these nonmammalian vertebrates. Therefore, we can't report when and how cytodifferentiation of the hormone-containing cells occurs in those species.

IV. Multiple Hormonal Production by a Single Pituitary Cell

A. IMMUNOCYTOLOGICAL PROOFS

There is a large amount of immunocytological, biochemical, and physiological evidence that a single hypophyseal cell type can elaborate several kinds of hormones. Most of the cytological investigations have been performed on adult pituitary gland of several species (see review in Girod, 1977). In contrast, few observations were made on the developing embryonic or fetal pituitary. Corticomelanotrophic cells, revealed by anti-ACTH, anti-α-MSH, and anti-β-MSH sera, were observed in the pars intermedia of rat fetuses (Dupouy and Dubois, 1975; Chatelain *et al.*, 1979; Dupouy *et al.*, 1979) and amphibian tadpoles (Doerr-Schott and Dubois, 1973; Rémy and Dubois, 1974). In the human fetus early in pituitary organogenesis, the pars intermedia can be identified as the caudal wall of the Rathke's pouch; some corticotrophic cells of this lobe were also stained with anti-β-MSH (Baker and Jaffe, 1975; Bugnon *et al.*, 1976a). In the developing pars distalis, corticomelanotrophs can be revealed with both anti-ACTH and anti-β-MSH sera, as was reported for man (Dubois *et al.*, 1973; Moriarty, 1973; Moriarty and Moriarty, 1973; Baker and Jaffe, 1975; Bugnon *et al.*, 1975, 1976a,b), rat (Chatelain *et al.*, 1979; Dupouy *et al.*, 1979) (Fig. 4), and amphibian tadpoles (Rémy and Dubois, 1974). Corticotrophs of both pars distalis and pars intermedia, also contained β-LPH and endorphins [Bugnon *et al.*, 1974, 1975, 1976b; Bégeot *et al.*, 1978a,b,c,; Celio, 1979 (in man) Bégeot *et al.*, 1979; Chatelain *et al.*, 1979; Dupouy *et al.*, 1979 (in rat)] (Figs. 4 and 5).

The presence of several hormones in the same cell of the developing and mature hypophysis would explain why the plasma concentrations of two hormones tend to parallel each other under many conditions as, for example, it was reported in adults for ACTH and β-MSH (Abe *et al.*, 1967, 1969; Orth *et al.*, 1973; Hirata *et al.*, 1975), ACTH and β-LPH (Gilkes *et al.*, 1975; Hirata *et al.*, 1976; Tanaka *et al.*, 1976; Jeffcoate *et al.*, 1978), ACTH and β-endorphin (Guillemin *et al.*, 1977), and β-endorphin and β-LPH (Scherrer *et al.*, 1978). Evidence has been presented that β-MSH immunoreactivity in human pituitary and plasma might be an artifact formed, during extraction procedures, by enzymic degradation of β-lipotropin, which contains the entire amino acid sequence of β-MSH (Scott and Lowry, 1974; Bloomfield and Scott, 1974; Bloomfield *et al.*, 1974; Bachelot *et al.*, 1976, 1977; Tanaka *et al.*, 1976) (Fig. 3).

B. Structural Relations between ACTH, MSH, LPH, Endorphins, and CLIP

Adrenocorticotrophins (ACTH) and melanocyte-stimulating hormones (MSH) belong to two classes of structurally related peptides; one class includes ACTH, α-MSH, and CLIP; the other class includes β-MSH, β-LPH, γ-LPH, and endorphins (α, β, γ) (Fig. 3). All of these peptides except CLIP and endorphins share a common heptapeptide core (Met-Glu-His-Phe-Arg-Trp-Gly). Nevertheless, "it is generally accepted that the two groups are distinct and have probably been formed by the duplication of a single gene with subsequent progressive substitutions leading to their divergence" (Lowry and Scott, 1975).

The peptide sequence of mammalian α-MSH is identical to the first 13 amino acid residues of ACTH with acetylation of the N-terminal serine and amidation of the C-terminal valine (Harris and Lerner, 1957).

The 24 N-terminal amino acids of ACTH are identical in all the studied species and possess full biological activity, whereas the 15 C-terminal amino acids (25–39 sequence) vary from one species to another.

The CLIP is identical to the fragment (18–39) of ACTH (Scott *et al.*, 1973).

β-MSH is a polypeptide containing 22 amino acids in man (Harris, 1959); all other species have only 18 (Fig. 3). In all the investigated species, the entire sequence of β-MSH is found within β-LPH, a 91-amino acid protein showing lipotropic and melanocyte-stimulating activity (Fig. 3). The sequence of 56 C-terminal residues of β-LPH is surprisingly homologous in various species, whereas the N-terminal (1–35) amino acid sequence exhibits great variability (Li and Chung, 1976) (Fig. 3).

In addition to β-MSh, β-LPH contains the complete sequences of γ-LPH (fragment 1–58) and endorphins (a generic term for opiate-like peptides).

β-Endorphin or C-fragment is a 31-amino acid polypeptide formed of residues (61–91) of β-LPH; α- and γ-endorphins correspond, respectively, to the residues (61–76) and (61–77) of the β-LPH (see review of Chrétien *et al.*, 1973; Jacob, 1977; Bertagna and Girard, 1978) (Fig. 3).

Such structural relations between β-LPH, β-MSH, and endorphins suggest that β-LPH could be the precursor for these peptides and could explain why immunocytology reveals them in the same pituitary cells.

C. Evidence for Prohormones of Corticotrophins, Melanotrophins, Lipotrophins, and Endorphins in the Pituitary Gland

1. Isolation of Common Precursors from Pituitaries and Tumors

High-molecular-weight glycoprotein present in the mouse pituitary tumor cell line (AtT-20/D-16v) contained amino acid sequences similar to both (1–39)-ACTH and β-LPH (Eipper et al., 1976; Eipper and Mains, 1977; Mains et al., 1977; Allen et al., 1978; Roberts et al., 1978). It appeared as the common precursor for various ACTH and endorphin-containing peptides characterized by their apparent molecular weight. Structural analysis of this common precursor, referred to as "pro-ACTH/endorphin" or proopiocortin (Rubinstein et al., 1978), indicated that the corticotrophin-like segment was located in the middle of the molecule, whereas the β-lipotrophin-like segment was located in the COOH terminal of the precursor (Eipper and Mains, 1978a).

Immunological evidence for common precursors to corticotrophins, lipotrophins, and endorphins has been obtained for pituitaries of rat (Crine et al., 1978; Eipper and Mains, 1978b; Yoshimi et al., 1978), beef (Nakanishi et al., 1977), frog (Loh and Gainer, 1978; Pezalla et al., 1978; Loh, 1979), man (Yoshimi et al., 1978) and human nonpituitary tumor (Bertagna et al., 1978; Orth et al., 1978).

From fetal sheep pituitary extracts subjected to chromatography on Sephadex G-100 superfine, Silman (1979) and Silman et al. (1979) isolated three peaks of large-molecular-weight material; the immunoassay profiles of two peaks between the three peaks mentioned suggested that the former peaks could be stem hormone precursors for both adrenocorticotrophin- and lipotrophin-related peptides.

As yet, no information was available on the fetal pituitary gland of other species except for rat fetus in which immunoreactive ACTH was associated with high-molecular-weight material in pituitary extract (Chatelain and Dupouy, 1980).

2. Biosynthesis of Large-Molecular-Weight Precursors by Cell-Free System and Cloned cDNA

When messenger RNA was isolated from cultures of mouse pituitary tumor cell line (AtT-20/D-16v) and translated in a reticulocyte cell-free system, the product synthesized was shown to be a common precursor of corticotrophin (Roberts and Herbert, 1977a) and β-lipotrophin (Roberts and Herbert, 1977b). Tryptic peptide analysis showed that the β-LPH peptides were located carboxy-terminally to the corticotrophin peptides (Roberts and Herbert, 1977b). Recently, Nakanishi *et al.* (1979) reported the nucleotide sequence of cloned cDNA coding for bovine ACTH and β-LPH precursor; moreover, they reported evidence that a new peptide, like α-MSH, or β-MSH, and so-called γ-MSH, could be formed by proteolytic processing of the precursor.

V. Determination of Cytological Differentiation

A. INVESTIGATION METHODS

1. *In Vivo Observations and Experiments*

a. *Normal Development.*   It appears very difficult to determine what presumptive factors are influencing cytodifferentiation by single observations of the normal *in vivo* development of the pituitary anlage. However, by light and electron microscopy, with or without the assistance of immunocytological methods, it is possible to identify where the first differentiated cells appear and what tissues are surrounding them. It is generally observed that cytodifferentiation of the hormone-containing cells takes place near the vascularized mesenchyme surrounding pituitary anlage.

b. *Congenital Anencephaly and Experimental Encephalectomy.*   Congenital anencephaly results from a defect of neural tube formation in early development. Infants with anencephaly showing complete absence of hypothalamus (review in Nakano, 1973) offer an opportunity to investigate pituitary differentiation and function in the absence of an intact hypothalamic–pituitary axis.

Experimental anencephaly has been performed in fetuses of laboratory animals by surgical extirpation of whole brain including the hypothalamus while the pituitary was left *in situ* (technique described in Jost *et al.*, 1966); these operated fetuses were referred to as encephalectomized fetuses (Fig. 6).

Other anomalies in brain development have been induced experimentally in animals by hypervitaminosis A (Giroud and Martinet, 1957; Fujita *et al.*, 1970; Eguchi *et al.*, 1973), neutron irradiation (Satow *et al.*, 1971), or drug treatment (Osamura, 1977) in early stages of gestation.

c. *Implant of Pituitary Anlage.*   A method for examining *in vivo* the determination of pituitary cytodifferentiation has been to directly implant into the hypothalamus of an adult either pituitary anlage (Gash *et al.*, 1975) or clonal cells derived from the Rathke's pouch epithelium (Shiino *et al.*, 1977) removed from young rat embryos. Coelomic and/or chorioallantoidic implants also have been used with mouse (Ferrand and Nanot, 1968) or chiefly with bird pituitary anlage (Le Douarin *et al.*, 1967a,b; LeDouarin and Ferrand, 1968; Ferrand and Le Douarin, 1968; Ferrand, 1969a,b, 1970a,b, 1971a,b, 1972).

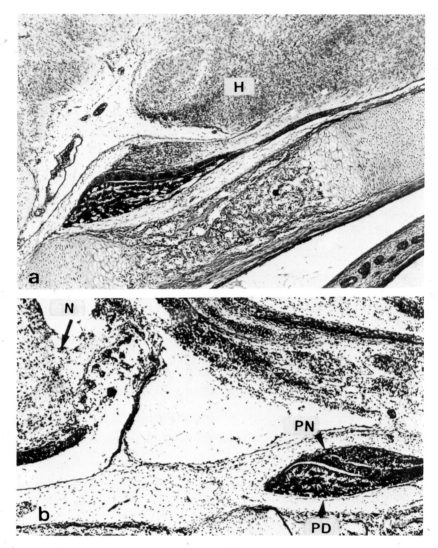

FIG. 6. Longitudinal sections through the heads of an intact (a) and a littermate encephalectomized (b) rat fetuses. Surgery was done on day 16, sacrifice on day 21. The hypothalamus (H) and
other nervous structures normally covering the pituitary gland (PD, pars distalis; PN, pars nervosa)
have been removed. The interrupted nervous stem (N) is seen on the left, in the posterior part of the
head. ×50.

2. In Vitro Experiments

a. *Adenohypophysial Primordium in Organ Culture.*   To see whether pituitary
gland has the capacity of self-differentiation and to test whether or not hypothala-

mus is involved in the cytodifferentiation of the adenohypophysis, Rathke's pouches were isolated from young rat embryos and cultivated *in vitro* for several days (Watanabe *et al.*, 1973; Nemeskery *et al.*, 1976; Watanabe and Daikoku, 1976; Bégeot *et al.*, 1979). In all the experiments performed by Le Douarin and Ferrand, the implantation of bird pituitary anlages into the host tissues was generally preceded by a short stay in culture *in vitro*. Dubois *et al.* (1976) also put fragments of human fetal adenohypophysis in tissue culture to study cyto-differentiation.

b. *Isolated Pituitary Cells in Culture.* Some experiments have been performed *in vitro* with pituitary clonal cells derived from the epithelium of Rathke's pouch (Ishikawa *et al.*, 1977; Shiino *et al.*, 1977). Ohtsuka *et al.* (1971, 1972) isolated free chromophobe cells from fresh anterior pituitaries of adult rat and cultivated them in a chemically defined medium suitably supplemented with various substances.

B. Presumptive Factors Influencing Cytodifferentiation

1. *Substances of Central Nervous System Origin*

To test whether central nervous structures, including hypothalamus, influence cytological differentiation of the pituitary gland, investigations with immunocytological methods were performed on anencephalic infants, on encephalectomized fetuses of laboratory animals, and on pituitary implants or cultures.

a. *Information Provided by Anencephaly Study.* In the hypophysis of human anencephalic fetuses at term, corticomelanotropic cells were scarce and hypoplastic (Dubois *et al.*, 1975; Osamura, 1977). Yet Hatakeyama (1969) and Salazar *et al.* (1969) observed various types of secretory cells in the pituitary gland of anencephals, and Satow *et al.* (1972) reported few ACTH cells (identified by ultrastructural features) in those fetuses. More recently, the use of antibodies to $\propto$-(17–39)-ACTH, β-(1–24)-ACTH, β-LPH, and α- and β-endorphin allowed observation of the same cells in anencephalic as well as in normal fetuses (Bégeot *et al.*, 1978a,b,c; Dubois *et al.*, 1979). However, in the former, immunoreactive cells were scarce and smaller than in the latter (Bégeot *et al.*, 1978b,c) (Fig. 7). Bégeot *et al.* (1976, 1977) studied the evolution of the cytoplasmic area of corticotropic cells in normal and anencephalic fetuses collected at different stages of gestation. Cytoplasmic areas were always statistically lower (30 to 40%) in anencephalic fetuses than in normal fetuses of the same age; moreover, the cytoplasmic area increased very slightly (10%) from 23 weeks to term (Fig. 8). At birth, cytoplasmic area was similar to that observed in normal fetuses ranging from 12 to 16 weeks (Fig. 8). Similar observations have also been reported for somatotropic cells (Bégeot *et al.*, 1976, 1977). From these data

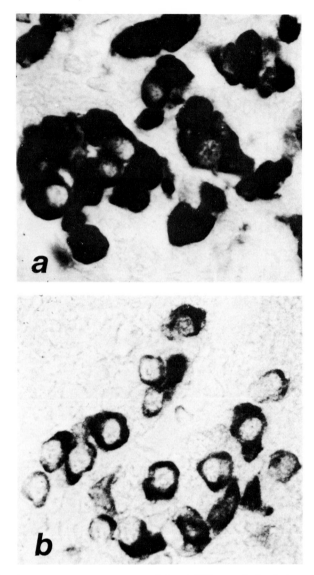

Fig. 7.   Corticotrophs revealed with anti-α-(17–39)-ACTH serum in the pituitary of a 23-week-old normal fetus (a) and of a human anencephalic fetus (b) at the same age. ×1300. (From Bégeot *et al.*, 1977.)

it can be speculated first that the appearance of corticotrophs occurs even in the complete absence of hypothalamic tissue and second that "hypothetic hypothalamic factors are needed for the normal development of cells" containing ACTH and growth hormone (Bégeot *et al.*, 1977).

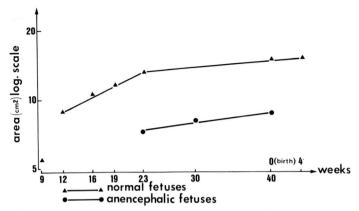

FIG. 8. Evolution of the cytoplasmic area of the corticotropic cells in normal and anencephalic human fetuses at different stages of gestation and newborns. (From Bégeot *et al.*, 1977.)

Such influence of anencephaly on ACTH-containing cells can be correlated with the poor ACTH pituitary content (7 ng/mg gland versus 1.1 μg/mg gland in adult) (Allen *et al.*, 1974), the low plasma ACTH concentration (Allen *et al.*, 1973, 1974; Hayek *et al.*, 1973; Winters *et al.*, 1974), and the striking underdevelopment of the adrenal cortex (Tuchmann-Duplessis and Larroche, 1958; Jost *et al.*, 1970; Jost, 1971; Eneroth *et al.*, 1972; Allen *et al.*, 1974). According to adrenal weight values presented in the literature (Jost *et al.*, 1970; Jost, 1971), the growth of the adrenals in anencephalic human fetuses was practically stopped after the fourth lunar month of pregnancy. The ACTH release was strongly reduced in anencephalic fetuses and neonates but could be stimulated readily with exogenous vasopressin (Hayek *et al.*, 1973; Allen *et al.*, 1974).

In human fetuses and neonates, normal adrenocorticotropin synthesis and secretion are under the control of the central nervous system.

Experimental fetal anencephaly was produced in several kinds of laboratory animals. Different types of granule-containing cells were present in the adenohypophysis of rats encephalectomized on the sixteenth day of gestation and sacrificed at term (Daikoku *et al.*, 1973). The secretory granules in the cells of the operated animals were smaller and fewer than those of the intact ones. However, as they have not identified the secretory cells by classic ultrastructural features, we reinvestigated with immunocytological methods the differentiation of the hypophysis in fetuses with complete absence of hypothalamic tissue from day 16 to term. We demonstrated that the development of the corticotrophs, melanotrophs (Chatelain *et al.*, 1976), and lipotrophs (Chatelain *et al.*, 1979; Dupouy *et al.*, 1979) did not require the presence of the fetal hypothalamus or other central nervous structures after day 16. These cells, revealed with specific antibodies, were, at term, as numerous in the encephalectomized fetuses on day 16 as in the littermate controls (Figs. 9 and 10); however, pituitary ACTH content was slightly reduced (Table II).

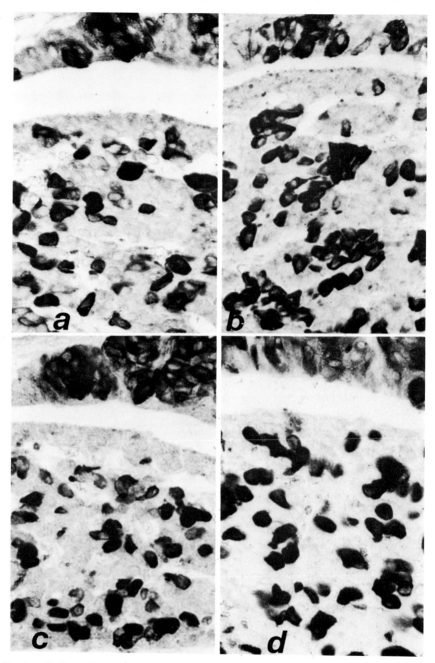

FIG. 9. Corticotrophs revealed with anti-α-(17–39)-ACTH serum (a,c) and anti-β-(1–24)-ACTH serum (b,d) in the pituitary gland of 21-day-old rat fetuses. (a,b) Control fetus; (c,d) littermate fetus, encephalectomized on day 16 of pregnancy. Immunoreactive cells can be observed in both pars distalis and pars intermedia. ×500. (Courtesy of A. Chatelain.)

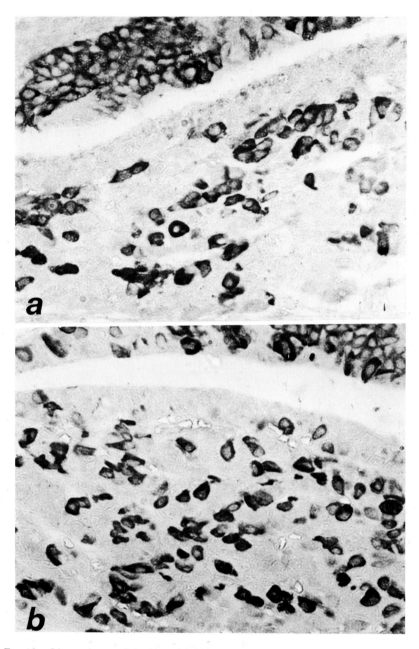

Fig. 10. Lipotrophs revealed with anti-β-LPH serum in the pituitary gland of a 21-day-old rat fetus. (a) Control fetus; (b) littermate fetus, encephalectomized on day 16 of pregnancy. Immunoreactive cells can be observed in both pars distalis and pars intermedia. ×500. (Courtesy of A. Chatelain.)

TABLE II

Effects of Fetal Encephalectomy and Hypophysectomy on Pituitary and Plasma ACTH Levels and Adrenal Weight in the Rat

Fetuses[a]	Pituitary ACTH[c,d,f] (pg/µg protein)	Plasma ACTH[c,d,f] (pg/ml)	Adrenal weight [e,f] (mg)
Controls	202 ± 74 (6)	208 ± 30 (21)	2.91 ± 0.17 (28)
Encephalectomized	124 ± 61 (7)	132 ± 19 (12)	1.45 ± 0.20 (16)
Hypophysectomized[b]		<30 (3)	1.26 ± 0.10 (10)

Pituitary ACTH: Controls vs Encephalectomized NS. Plasma ACTH: Controls vs Encephalectomized ***. Adrenal weight: Controls vs Encephalectomized ***; Encephalectomized vs Hypophysectomized NS; ***.

[a] Fetuses were operated on day 16 and sacrificed on day 21 of pregnancy.

[b] Hypophysectomy was performed by fetal decapitation according to Jost's technique (1947).

[c] A. Chatelain and J. P. Dupouy (unpublished data).

[d] ACTH was assayed by radioimmunoassay using synthetic human (1–39) ACTH as reference standard (Chatelain et al., 1980).

[e] A. Chatelain (personal communication).

[f] Values are expressed as means ± confidence intervals ($p = 0.05$). Number of animals or assays in parentheses. Mean's comparisons: NS, not significant difference ($p > 0.05$); ***, very highly significant difference ($p < 0.001$).

Osamura (1977) treated 15-day-old rat fetuses with methylazoxymethanol, which was known to damage parenchymal cells of the fetal central nervous system, and obtained microcephaly. Nevertheless, immunohistochemical stain showed no significant changes in ACTH cell development.

In all of the investigated species, experimental encephalectomy produced severe adrenal gland atrophy (Novy et al., 1977, in Rhesus monkey; Jost, 1966; Jost et al., 1966, 1970; Dupouy and Jost, 1970; Fujita et al., 1970; Cohen et al., 1971; Mitskevich and Rumyantseva, 1972; Eguchi et al., 1973; Coffigny and Dupouy, 1978, in rat; Mitskevich and Rumyantseva, 1972, in rabbit and guinea pig) and hypoactivity (Dupouy and Jost, 1970; Coffigny and Dupouy, 1978) correlated with reduced plasma ACTH concentration (Chatelain and Dupouy, unpublished data, in rat) (Table II).

These data indicated that the fetal hypothalamus was not required for the development of ACTH-, MSH-, and LPH-containing cells in the hypophysis of several mammals but, in its absence, the corticostimulating activity of the pituitary gland was very reduced.

However, since the experimental encephalectomy in the rat was performed relatively late in gestation (16 days), one cannot exclude any inductive influence of the central nervous system in early fetal development; the action of circulating substances in such in vivo experiments should be considered.

In the tadpole of Rana pipiens, the extirpation of the posterior hypothalamic anlage prevented the differentiation of the intermediate lobe but not that of the pars distalis (Hanaoka, 1967).

b. *Information Provided by Implants and Cultures.* Implants and cultures of Rathke's pouch or isolated cell lines established from pituitary primordium or chromophobe cells have been used to investigate whether pituitary cells present the capacity of self-differentiation or whether they need exogenous influences (Table III).

Watanabe *et al.* (1973) cultivated Rathke's pouch isolated from 12-, 14-, and 15–day-old rat fetuses and examined the ultrastructure of these explants 9, 6, and 5 days later, respectively. Since many adenohypophyseal cells contained electron-dense granules, the authors postulated that granule formation, which is a reliable sign of cytodifferentiation, can start in the absence of hypothalamic neurosecretory substances. More recently, the presence of ACTH-containing cells was demonstrated by immunohistochemical methods applied to cultivated adenohypophyseal primordia (Nemeskery *et al.*, 1976; Watanabe and Daikoku, 1976; Bégeot *et al.*, 1979). LPH- and endorphin-containing cells were also

TABLE III

LIST OF PUBLICATIONS ON PITUITARY CELL DIFFERENTIATION IN ORGAN OR CELL CULTURE
(EXPERIMENTS ON RATS)

Time of Rathke's pouch or pituitary primordium isolation (days)	Culture duration (days)	Culture medium	Antisera used for immunocytological investigation	References
(Adult pituitary chromophobes)	12	NCTC-109 + nucleic acids + thyroxine + colchicine + ovine CRF		Ohtsuka *et al.* (1971, 1972)
12, 14, 15	9, 6, 5	Synthetic medium 199 + 10% calf serum		Watanabe *et al.* (1973)
13 to 15	2 to 11	Parker 199 synthetic medium	p-ACTH α-(17–34)-ACTH	Nemeskery *et al.* (1976)
12	9	Synthetic medium 199 + 10–15% calf or rat serum	p-(1–39)-ACTH p-(17–39)-ACTH	Watanabe and Daikoku (1976)
11 to 13	14	85% Ham's F10 medium + 15% fetal calf serum		Ishikawa *et al.* (1977) Shiino *et al.* (1977)
12.5 to 15.5	9, 8, 7, 6	RPMI 16–40 medium + 20% fetal calf serum	β-(1–24)-ACTH α-(17–39)-ACTH β-LPH α-endorphin β-endorphin	Begeot *et al.* (1979)

revealed in explants (Bégeot *et al.*, 1979), while indirect evidence for the presence of melanotrophs was reported (Watanabe and Daikoku, 1976). The extent of cytodifferentiation *in vitro* depended largely on the age of the fetuses furnishing the explants. Indeed, the cultivated pituitaries removed from 15-day-old embryos showed more immunoreactive cells than those explanted from 12- or 13-day-old embryos (Nemeskery *et al.*, 1976; Bégeot *et al.*, 1979) (Figs. 11 and 12); moreover, their cytoplasmic areas were greater (Bégeot, personal communication) (Fig. 11). From these data it appeared that the capacity of self-differentiation was greater on day 15 that on days 12 and 13 for fetal rat pituitary. However, as most of the authors used synthetic medium supplemented with calf serum for culture, one might ask if substances present in the sera could influence the *in vitro* adenohypophyseal cell differentiation. Moreover, as connective tissue or mesenchymal elements were reported always to be present around explanted primordia when collecting the material (Watanabe *et al.*, 1973; Nemeskery *et al.*, 1976; Bégeot *et al.*, 1979), some influence of the mesenchyme on the cytodifferentiation cannot be excluded; such a hypothesis will be strengthened by further data on bird embryos.

Gash *et al.* (1975) studied the fate of pituitary anlage excised from 11–15-day-old rat embryos and transplanted into the median eminence of the hypothalamus of hypophysectomized adult females for 4 weeks. Only Rathke's pouch tissues from rat fetuses 13 days of age or older showed the ability to develop into good histological adenohypophyseal features; anlage from 12-day embryos almost invariably produced neoplasms. Adrenal enlargement in host animals indicated ACTH secretion. After such observations, Gash *et al.* (1975) postulated that the embryonic nervous system could exert inductive influences on presumptive pituitary and that critical changes seemed to occur between the thirteenth and fourteenth day of fetal life in rat.

Ishikawa *et al.* (1977) and Shiino *et al.* (1977) have successfully established clonal strains of rat pituitary anlage cells derived from the epithelium of the Rathke's pouch of 11- to 13-day-old rat fetuses. By supplementing the growth medium (85% Ham's F-10 medium and 15% fetal calf serum) with fetal brain extract (FBE) from fetuses younger than 16 days, Ishikawa *et al.* (1977) have been able to establish cell lines. In contrast, FBE from 16-day-old or older rats, as well as adult rat median eminence extract (MEE), were not effective in

---

FIG. 11.   Rat pituitary anlage in organ culture. (Culture medium RPMI 16–40, supplemented with 20% fetal calf serum and various antibiotics.) (a,b) Cells revealed with anti-α-(17–39)-ACTH serum. Adenohypophyseal primordia were explanted on day 12 (a) or day 14 (b) of gestation and kept in culture for 9 days (a) and 7 days (b). A greater number of immunoreactive cells can be observed in the adenohypophysis last explanted. $\times 400$. (c,d) Cells revealed with anti-β-LPH serum. Adenohypophyseal primordia were explanted on day 12 (c) or day 15 (d) of gestation and kept in culture for 9 days (c) and 6 days (d). The cytoplasmic area of the immunoreactive cells is more developed in the adenohypophysis last explanted. $\times 1000$. (Courtesy of M. Bégeot.)

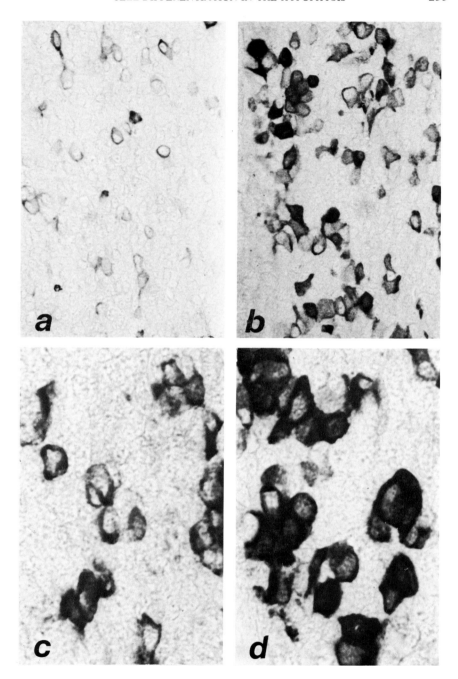

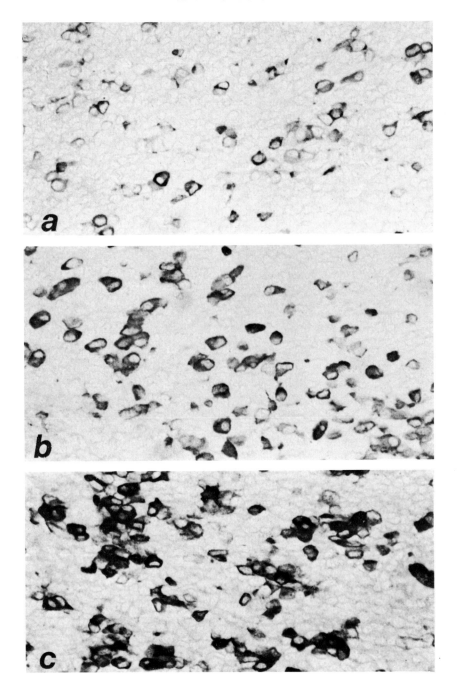

promoting the growth of pituitary anlage cells. After establishment of clonal strains, adult MEE was required for cell proliferation, cytodifferentiation, and production of one to six anterior pituitary hormones including ACTH. Ishikawa *et al.* (1977) postulated that at the early stage of development (11–13 days) some factor(s) present only in fetal brain extract were necessary for the differentiation of anterior pituitary cell lines and the development of receptors to adult neural substances. Shiino *et al.* (1977) reported ACTH, GH, and prolactin secretion by clonal cells derived from the epithelium of Rathke's pouch and grown *in vitro*. In contrast, glycoprotein hormone secretion was not detected. When cells of this clone were implanted in the hypophysiotropic areas of the hypothalamus (in the preoptic area or median eminence) of hypophysectomized female rats, the grafts were cytodifferentiated into various types of anterior pituitary cells (prolactin cells, somatotrophs, thyrotrophs, gonadotrophs, and follicular cells) with a rich vascularization.

Ohtsuka *et al.* (1971) reported that isolated chromophobes from adult rat pituitaries differentiated into acidophils and produced mainly ACTH when cultured in a chemically defined medium (NCTC-109 supplemented with nucleic acids and thyroxine) to which were added various amounts of CRF (extracted from ovine stalk median eminence) and colchicine. Small secretory granules (100 to 150 nm in diameter) were collected from the explants grown in the control medium without CRF; they were regarded as the storage carriers responsible for six trophic hormones: ACTH, TSH, FSH, LH, prolactin, and GH (Ohtsuka *et al.,* 1972). In the cells differentiating in CRF-supplemented medium (from chromophobes to acidophils), three different kinds of granules (classified by dimension) were successfully isolated; all contained ACTH activity alone regardless of their size (Ohtsuka *et al.,* 1972). Bioactive α-MSH was detected in the explants but not in the medium except temporarily and only in the earlier phase of differentiation of chromophobes (Ohtsuka *et al.,* 1972).

All these experiments performed on rat material provide evidence that some unknown factor(s) of central nervous system origin might produce an inductive influence at a critical phase of early embryonic development.

Mouse pituitary anlage, isolated at the eleventh day of gestation with mesenchymal tissue, was able to differentiate into acidophils and cyanophils when transplanted into the coelomic cavity of 3-day-old chick embryo (Ferrand and Nanot, 1968).

Dubois *et al.* (1976) put into culture pieces of adenohypophysis removed from

---

FIG. 12.   Rat pituitary anlage in organ culture. Cells revealed with anti-β-endorphin. Adenohypophyseal primordia were explanted on days 12 (a), 13 (b), or 15 (c) of gestation and kept in culture for, respectively, 9 days (a), 8 days (b), and 6 days (c). The abundance of differentiated cells depends largely on the age of the embryos furnishing the explants. ×400. (Courtesy of M. Bégeot.)

7- to 40-week-old human fetuses; several weeks later all the explants showed differentiated cells corresponding to various adenohypophyseal cell types.

Human and mouse pituitary primordia can differentiate without the presence of hypothalamus from an early stage of fetal development (11 days in mouse; 7 weeks in man).

In chick embryos, the Rathke's pouch appeared at stage 17 somites (Le Douarin *et al.*, 1967a; Ferrand, 1969a) and some inductive influence of the diencephalic floor on the cytodifferentiation of the epithelial primordium of the hypophysis has been demonstrated by these authors. Indeed, pituitary anlage that was removed from chick embryos at stages 17 to 24 somites and further associated with various mesenchymal tissues was unable to differentiate into adenohypophyseal tissue unless nervous tissue from the encephalic floor was added (Ferrand, 1969a, 1970a, 1972). After the stage 25 somites, the differentiation of the pituitary primordium occurred without the presence of hypothalamic elements but always needed the presence of mesenchymal tissue (Le Douarin *et al.*, 1967a,b; Ferrand, 1970a,b, 1972). The inductive influence of the ventral wall of the forebrain on the pituitary anlage was specific and not performed by any other nervous structures (Ferrand, 1969a, 1970a, 1972). In contrast, no specificity was observed for mesenchymal tissue (Ferrand and Le Douarin, 1968; Ferrand, 1969b, 1970a, 1972).

From these experiments, it appears that the hypothalamus exerts an inductive influence on the cytodifferentiation of chick pituitary gland during an early critical period of the incubation time. Investigations of Mikami *et al.* (1973) and Daikoku *et al.* (1974) support Ferrand's observations.

This is also the case in amphibians. An inductive influence of the forebrain was demonstrated during initial development of the pituitary anlage (Etkin, 1958a), whereas later, at the tail bud stage, self-differentiation of the pituitary primordium was evident (Atwell, 1935, 1937; Etkin, 1958b).

2. *Mesenchymal Tissue*

Mesenchymal cells may be involved in inducing cytodifferentiation of the hypophysis. Presumptive influence of the mesenchymal tissue was based at first on histological observations of the developing pituitary gland and afterward on *in vitro* experiments.

The appearance of the earliest granulated or hormone-containing cells has been reported to occur near the mesenchyme surrounding or invading the pituitary anlage in man (Conklin, 1968; Andersen *et al.*, 1970, 1971; Bugnon *et al.*, 1974, 1976a,b), rabbit (Schechter, 1970, 1971; Chatterjee, 1975), and rat (Svalander, 1974); such observations suggested but did not prove a mesenchymal effect on cytodifferentiation. That influence was particularly well demonstrated in bird embryos.

The epithelial primordium of the hypophysis isolated at an early stage of embryonic development (after the stage 25 somites or the third day of incubation for chick) was unable to differentiate into adenohypophyseal tissue unless mesenchymal tissue from various origins was added to the implant (Le Douarin *et al.*, 1967a,b; Le Douarin and Ferrand, 1968; Ferrand and Le Douarin, 1968; Ferrand, 1970a,b, 1971a, 1972). Similar observations were made with quail (Ferrand, 1969b, 1971a) and mouse embryos (Ferrand and Nanot, 1968).

Some mesenchymal cells, which were probably isolated with pituitary primordium of man (Dubois *et al.*, 1976) and rat (Watanabe *et al.*, 1973; Watanabe and Daikoku, 1976; Nemeskery *et al.*, 1976; Bégeot *et al.*, 1979), certainly influenced the cytodifferentiation of pituitary anlage in organ culture.

3. Vascularization

A great number of authors reported that the earliest differentiation of the adenohypophyseal secretory cells occurred at a time when primordial cell cords were seen penetrating the vascularized mesenchyme in man (Conklin, 1968; Pavlova *et al.*, 1968; Andersen *et al.*, 1971; Dubois *et al.*, 1973; Bugnon *et al.*, 1974, 1976a,b; Baker and Jaffe, 1975; Bégeot *et al.*, 1975; Celio, 1979), rabbit (Schechter, 1970, 1971; Chatterjee, 1975), rat (Svalander, 1974), mouse (Sano and Sasaki, 1969; Dearden and Holmes, 1976; Stoeckel *et al.*, 1979), hamster (Thompson *et al.*, 1976), birds (Ferrand, 1969b; 1971a,b; Fellman *et al.*, 1975), and amphibians (Doerr-Schott and Dubois, 1973; Rémy and Dubois, 1974). However, the vascularization was probably of variable importance for the cytodifferentiation of the different pituitary lobes; the pars intermedia of the fetal rat was nearly avascular (Negm, 1971), and in that of the fetal rabbit, cytodifferentiation was beginning before the establishment of vascular supply (Chatterjee, 1975).

VI. Localization of Pituitary Hormone-Like Substances in Extrapituitary Areas

Biochemical and biological investigations revealed the presence of pituitary hormone-like substances in the brain of several species. Observations have been reported for peptides related to ACTH, α- and β-MSH, β-LPH, and endorphins (see Table IV)

High-molecular-weight precursors to ACTH (Orwoll *et al.*, 1979) and corticotropin/β-endorphin-like material (Liotta *et al.*, 1979) have been demonstrated to be present in the brain.

Recently, endogenous peptides with opiate activity (methionine-enkephalin and leucine-enkephalin) have been found initially in brain extracts and subsequently in pituitaries from several species (Hughes, 1975; Hughes *et al.*, 1975; Elde *et al.*, 1976; Lazarus *et al.*, 1976; Simantov and Snyder, 1976a,b, Siman-

TABLE IV

LIST OF PUBLICATIONS ON THE DETECTION OF PITUITARY HORMONE-LIKE PEPTIDES IN THE BRAIN OF VARIOUS SPECIES

Peptides	Species	Biological detection	Immunocytology[a]	Radioimmuno- or radioreceptor assay	References
ACTH	Hog	+			Guillemin et al. (1962)
	Dog	+			Schally et al. (1962)
	Rat			+	Krieger et al. (1977a,b)
	Bovine			+	Krieger et al. (1977c)
	Rat		+ (Anti-purified-ACTH)		Larsson (1977)
	Various mammals (especially guinea pig)		+ (Anti-(17–39)-ACTH)		Tramu et al. (1977)
	Human		+ (Anti-(1–24)-ACTH and (17–39)-ACTH)		Bloch et al. (1978a,c)
	Various mammals			+	Moldow and Yalow (1978)
	Rat		+ (Anti-(17–39)-ACTH)		Toubeau (1978)
	Rat		+ (Anti-(1–24)-ACTH)		Watson et al. (1978a,b)
	Human (adult, fetus)		+ (Anti-(1–24)-ACTH and (17–39)-ACTH)		Bugnon et al. (1979a,b)
	Rat			+	Krieger et al. (1979)
	Bovine			+	Liotta et al. (1979)
	Ewe		+ (Anti-synthetic-h.-ACTH)		Nilaver et al. (1979)
	Rat, cat, monkey	+		+	Orwoll et al. (1979)
	Rat		+ (Anti-p-ACTH)		Pelletier (1979)
	Human		+ (Anti-p-ACTH)		Pelletier and Desy (1979)
	Rat		+ (Anti-(1–24)-ACTH)		Sofroniew (1979)
α, β-MSH	Rat	+			Mialhe-Voloss and Stutinsky (1953)
	Hog	+			Guillemin et al. (1962)
	Dog	+			Schally et al. (1962)

	Species					Reference
	Bovine, simian, human	+				Rudman et al. (1973)
	Rodent	+	+			Rudman et al. (1974)
	Rat	+	+		+ (α-MSH)	Barnea et al. (1977)
	Rat			+ (Anti-α-MSH)		Pelletier and Dube (1977)
	Rat			+ (Anti-α-MSH)		Swaab and Fisser (1977)
	Human			+ (Anti-α- and β-MSH)		Bloch et al. (1978a)
	Rat			+ (Anti-α-MSH)		Dube et al. (1978)
	Rat				+ (α-MSH)	Oliver and Porter (1978)
	Rat				+ (α-MSH)	Vaudry et al. (1978)
	Human (adult, fetus)	+		+ (Anti-α- and β-MSH)		Bugnon et al. (1979a,b)
	Rat				+ (α-MSH)	Orwoll et al. (1979)
β-LPH	Bovine					Krieger et al. (1977c)
	Bovine					Labella et al. (1977)
	Rat			+		Watson et al. (1977b, 1978b)
	Human			+		Bloch et al. (1978a,c)
	Human			+		Pelletier et al. (1978)
	Sheep, ox			+		Zimmerman et al. (1978)
	Human (adult, fetus)			+		Bugnon et al. (1979a,b)
	Ewe			+		Nilaver et al. (1979)
	Rat			+		Pelletier (1979)
	Rat				Electrophoresis	Scherrer et al. (1979)
Endorphins	Various species (general review)	+			+	Goldstein (1976)
	Pig	+ (α-end)				Guillemin et al. (1976)
	Pig				+ (α-, γ-end)	Ling et al. (1976)
	Bovine				+	Queen et al. (1976)

(continued)

TABLE IV (continued)

Peptides	Species	Biological detection	Immunocytology[a]	Radioimmuno- or radioreceptor assay	References
	Carp, goldfish		+ (α-end)		Follenius and Dubois (1977, 1978)
	Bovine			+ (β-end)	Krieger et al. (1977c)
	Bovine			+ (α-end)	Labella et al. (1977)
	Rat			+ (β-end)	Rossier et al. (1977)
	Human		+ (α-, β-end)		Bloch et al. (1978a,b,c)
	Rat		+ (β-end)		Bloom et al. (1978)
	Rat			+ (β-end)	Hollt et al. (1978)
	Rat		+ (β-end)		Watson et al. (1978b)
	Human (adult, fetus)		+ (α-, β-end)		Bugnon et al. (1979a,b)
	Rat			+ (β-end)	Krieger et al. (1979)
	Rat		+ (β-end)		Sofroniew (1979)

[a] h, Human; p, porcine.

tov *et al.*, 1976a,b; Austen *et al.*, 1977; Fredrickson, 1977; Hökfelt *et al.*, 1977; Rossier *et al.*, 1977; Watson *et al.*, 1977a; Tramu and Leonardelli, 1979). Met-enkephalin is a pentapeptide with the sequence Tyr-Gly-Gly-Phe-Met, which is the same as the NH_2-terminus of β-endorphin, whereas in Leu-enkephalin, the COOH-terminal amino acid methionine is replaced by leucine. It is noteworthy that the morphinomimetic activity of endorphins, Met-enkephalin, and Leu-enkephalin is associated with the fragment (61–65) of the β-LPH (Ling and Guillemin, 1976).

In the human fetus, Bugnon *et al.* (1979 a,b) recently revealed neurons with anti-β-endorphin or anti-(17–39)-ACTH sera, as early as the eleventh week of fetal life, in the mediobasal hypothalamus, immunoreactive cells presented an immature neuroblastic aspect (Bugnon *et al.*, 1979b).

In the human placenta, the presence of an ACTH-like substance was reported (review in Saxena, 1971; Genazzani *et al.*, 1974, 1975; Rees *et al.*, 1975; Liotta *et al.*, 1977), as well as immunoreactive β-lipotrophin, β-endorphin, and possibly their precursors (Nakai *et al.*, 1978).

Recently Dubois (1980) observed the presence of immunoreactive material in the retina of several species (pig, sheep, chick, and trout) when anti-α-MSH, anti-(1–24)-ACTH, anti-α and anti-β-endorphin, and anti-β-LPH sera were used.

When and how differentiation of these cells containing pituitary hormone-like substances takes place remains to be investigated.

VII. Concluding Remarks

In the embryonic and fetal hypophysis of several species, the differentiation of the hormone-containing cells related to ACTH, MSH, LPH, and endorphin appears as a self-determining event. However, factors of different origin (brain, blood, mesenchyme) seem to have an early inductive influence during the initial formation of the pituitary primordium. Its growth proceeds by the multiplication of both undifferentiated and differentiated cells; mitotic divisions have been observed in granulated cells of several species including sheep (Alexander *et al.*, 1973a), pig (Liwska, 1978), rat (Nakane *et al.*, 1973; Svalander, 1974; Setalo and Nakane, 1976), hamster (Thompson and Trimble, 1976), and toad (Mira-Moser, 1972) as well as in chromophobic cells (Sano and Sasaki, 1969; Dearden and Holmes, 1976; Stoeckel *et al.*, 1979). The latter cell type could form a pool of reserve cells from which differentiation may proceed along various lines.

In the early stages of development, one and the same cell may elaborate several kinds of structurally related hormones (ACTH, MSH, LPH, and endorphins). Moriarty and Garner (1977b) reported immunocytochemical evidence that some cells of the adult rat hypophysis contained both ACTH and FSH; such findings were surprising and difficult to explain, since there is no known chemical homology between the two hormones. More recently, immunohistochemical staining demonstrated the presence of enkephalins in somatotrophs (Weber *et al.*, 1978b), thyrotrophs, and gonadotrophs (Tramu and Leonardelli, 1979) of the adult hypophysis of the rat and guinea pig.

The precise relationships between enkephalins and pituitary hormones related to GH, TSH, and LH are yet to be found.

In the adult hypophysis, the polypeptide secretion of the pars distalis differs from that of the pars intermedia cells. Whether such differences exist during embryonic and fetal life has to be determined. One can speculate that the two kinds of cells produce different proteolytic enzymes that cleave ACTH–LPH precursor molecules and subsequently ACTH and β-LPH to form smaller peptides. How selective synthesis of such proteolytic enzymes occurs during embryological development and why there are differences between pars distalis and pars intermedia cells are questions that have not been answered at present.

The data relative to hormonal content and secretion of the pituitary gland during early development are too scarce to be precise in all species about when pituitary starts to function. Moreover, the biological significance of all the peptides identified in the same cell of embryonic and adult hypophysis is now only partly known.

Considerable progress in the knowledge of pituitary gland cytodifferentiation has been realized during the last 15 years, due mainly to the advent and the development of immunological and biochemical techniques of investigation The recent discovery of different kinds of hormone precursor molecules and of new smaller polypeptides raises great interest in the pituitary gland. There is no doubt that important discoveries can be expected in the future.

ACKNOWLEDGMENTS

The author is indebted to P. M. Dubois (Laboratoire d'Histologie et Embryologie, U. E. R. Médicale Lyon Sud-Ouest) for reading and commenting on the manuscript and M. P. Dubois (INRA, Station de Physiologie de la Reproduction, Tours) for the donation of antisera. M. Bégeot and J. Y. Li (Laboratoire d'Histologie et Embryologie, UER médicale Lyon-Sud-Ouest), A. Chatelain (Laboratoire de Physiologie Animale, Faculté des Sciences d'Amiens), J. Doerr-Schott (Laboratoire de Zoologie et d'embryologie expérimentale, Université Louis Pasteur, Strasbourg), and M. E. Stoeckel (Laboratoire de Physiologie générale, Université Louis Pasteur, Strasbourg) are gratefully acknowledged for the generous production and gift of original documents. The author also wishes to thank N. Delatte for typing the manuscript and J. B. Zazac for technical assistance.

REFERENCES

Abe, K., Nicholson, W. E., Liddle, G. W., Island D. P., and Orth, D. N. (1967). *J. Clin. Invest.* **46**, 1609.

Abe, K., Nicholson, W. E., Liddle, G. W., Orth, D. N., and Island D. P. (1969). *J. Clin. Invest.* **48,** 1580.

Alexander, D. P., Britton, H. G., Forsling, M. L., Nixon, D. A., and Ratcliffe, J. G. (1971). *J. Physiol.* **213,** 31P.

Alexander, D. P., Britton, H. G., Cameron, E., Foster, C. L., and Nixon, D. A. (1973a). *J. Physiol.* **230,** 10P.

Alexander, D. P., Britton, H. G., Forsling, M. L., Nixon, D. A., and Ratcliffe, J. G. 1973b). *In* "Endocrinology of Pregnancy and Parturition" (C. G. Pierrepoint, ed.), pp. 112–115. Alpha Publ., Cardiff.

Allen, J. P., Cook, D. M., Kendall, J. W., and McGilvra, R. (1973). *J. Clin. Endocrinol. Metab.* **37,** 230.

Allen, J. P., Greer, M. A., McGilvra, R., Castro, A., and Fisher, D. A. (1974). *J. Clin. Endocrinol. Metab.* **38,** 94.

Allen, R. G., Herbert, E., Hinman, M. Shibuya, H., and Pert, C. B. (1978). *Proc. Natl. Acad. Sci. U.S.A.* **75,** 4972.

Alluchon-Gerard, M. J. (1970). *C.R. Acad. Sci. (Paris)* **271,** 1195.

Alluchon-Gerard, M. J. (1971). *Z. Zellforsch.* **120,** 525.

Andersen, H., Mollgard, K., and Von Bulow, F. A. (1970). *Histochemie* **22,** 362.

Andersen, H., Von Bulow, F. A., and Mollgard, K. (1971). *Z. Anat. Entwicklungs.* **135,** 117.

Atwell, W. J. (1935) *Proc. Soc. Exp. Biol.* **33,** 224.

Atweel, W. J. (1937). *Anat. Rec.* **68,** 431.

Austen, B. M., Smyth, D. G., and Snell, C. R. (1977). *Nature (London)* **269,** 619.

Bachelot, I. G., Wolfsen, A. R., and Odell, W. D. (1976). *Clin. Res.* **24,** 130 A.

Bachelot, I. G., Wolfsen, A. R., and Odell, W. D. (1977). *J. Clin. Endocrinol. Metab.* **44,** 939.

Baker, B. L., and Jaffe, R. B. (1975). *Am. J. Anat.* **143,** 137.

Barnea, A., Oliver, C., and Porter, J. C. (1977). *J. Neurochem.* **29,** 619.

Bégeot, M., Chatelain, A., Dubois, M. P., Dubois, P. M. and Dupouy, J. P. (1975). *J. Physiol. (Paris)* **70,** 25B.

Bégeot, M., Dubois, M. P., and Dubois, P. M. (1976). *J. Physiol. (Paris)* **72,** 8B.

Bégeot, M., Dubois, M. P., and Dubois, P. M. (1977). *Neuroendocrinology* **24,** 208.

Bégeot, M., Dubois, M. P., and Dubois, P. M. (1978a). *Ann. Endocrinol.* **39,** 235.

Bégeot, M., Dubois, M. P., and Dubois, P. M. (1978b). *C. R. Acad. Sci. (Paris)* **286,** 213.

Bégeot, M., Dubois, M. P., and Dubois, P. M. (1978c). *Cell Tissue Res.* **193,** 413.

Bégeot, M., Dubois, M. P., and Dubois, P. M. (1979). *J. Physiol. (Paris)* **75,** 27.

Bertagna, X., and Girard, F. (1978). *Ann. Endocrinol.* **39,** 201.

Bertagna, X. Y., Nicholson, W. E., Sorenson, G. D., Pettengill, O. S., Mount, C. D., and Orth, D. N. (1978). *Proc. Natl. Acad. Sci. U.S.A.* **75,** 5160.

Bielánska-Osuchowska, Z., and Liwska, J. (1975). *Folia Morphol.* **2,** 143.

Bloch, B., Franco, N., Fellmann, D., and Hatier, R. (1975). *J. Physiol. (Paris)* **70,** 26B.

Bloch, B., Bugnon, C., Fellmann, D., and Lenys, D. (1978a). *C. R. Acad. Sci.* (Paris) **287,** 1019.

Bloch, B., Bugnon, C., Lenys, D., and Fellman, D. (1978b). *C. R. Acad. Sci. (Paris)* **287,** 309.

Bloch, B., Bugnon, C., Fellman, D., and Lenys, D. (1978c). *Neurosci. Lett.* **10,** 147.

Bloom, F., Battenberg, E., Rossier, J., Ling, N., and Guillemin, R. (1978). *Proc. Natl. Acad. Sci. U.S.A.* **75,** 1591.

Bloomfield, G. A., and Scott, A. P. (1974). *Proc. R. Soc. Med.* **67,** 748.

Bloomfield, G. A., Scott, A. P., Lowry, P. J., Gilkes, J. J. H., and Rees, L. H. (1974). *Nature (London)* **252,** 492.

Boddy, K., Jones, C. T., Mantell, C., Ratcliffe, J. G., and Robinson, J. S. (1974). *Endocrinology* **94,** 588.

Bugnon, C., Lenys, D., Bloch, B., and Fellman, D. (1974). *C. R. Soc. Biol.* **168,** 460.

Bugnon, C., Lenys, D., Bloch, B., and Fellman, D. (1975). *C. R. Soc. Biol.* **169,** 1271.

Bugnon, C., Lenys, D., Bloch, B., and Fellmann, D. (1976a). *C. R. Soc. Biol.* **170**, 975.
Bugnon, C., Lenys, D., Bloch, B., and Fellman, D. (1976b). *J. Microsc. Biol. Cell* **26**, 31a.
Bugnon, C., Bloch, B., Lenys, D., and Fellman, D. (1979a). *J. Physiol. (Paris)* **75**, 67.
Bugnon, C., Bloch, B., Lenys, D., and Fellmann, D. (1979b). *Cell Tissue Res.* **199**, 177.
Celio, M. R. (1979). *J. Histochem. Cytochem.* **27**, 1215.
Chatelain, A., and Dupouy, J. P. (1980). *Coll. Soc. Neuroendocrinol. Exp., 10th, Lyon, J. Physiol. (Paris)* (in press).
Chatelain, A., Dubois, M. P., and Dupouy, J. P. (1976). *Cell Tissue Res.* **169**, 335.
Chatelain, A., Dupouy, J. P., and Dubois, M. P. (1979). *Cell Tissue Res.* **196**, 409.
Chatelain, A., Dupouy, J. P., and Allaume, P. (1980). *Endocrinology* **106**, 1297.
Chatterjee, P. (1974). *Cell Tissue Res.* **152**, 113.
Chatterjee, P. (1975). *Cell Tissue Res.* **164**, 481.
Chrétien, M., Lis, M., and Gilardeau, C. (1973). *Union Med. Can.* **102**, 890.
Coffigny, H., and Dupouy, J. P. (1978). *Gen. Comp. Endocrinol.* **34**, 312.
Cohen, A., Dupouy, J. P., and Jost, A. (1971). *C.R. Acad. Sci. (Paris)* **273**, 883.
Conklin, J. L. (1968). *Anat. Rec.* **160**, 79.
Costoff, A., and McShan, W. H. (1969). *J. Cell Biol.* **43**, 564.
Crine, P., Gianoulakis, C., Seidah, N. G., Gossard, F., Pezalla, P. D., Lis, M., and Chrétien, M. (1978). *Proc. Natl. Acad. Sci. U.S.A.* **75**, 4719.
Daikoku, S., Kinutani, M., and Watanabe, Y. G. (1973). *Neuroendocrinology* **11**, 284.
Daikoku S., Ikeuchi, C., and Nakagawa, H. (1974). *Gen. Comp. Endocrinol.* **23**, 256.
Dearden, N. M., and Holmes, R. L. (1976). *J. Anat.* **121**, 551.
Dent, J. N., and Gupta, B. L. (1967). *Gen. Comp. Endocrinol.* **8**, 273.
Doerr-Schott, J. (1968). *Z. Zellforsch. Mikrosk. Anat.* **90**, 616.
Doerr-Schott, J. (1974). *Fortschr. Zool.* **22**, 245.
Doerr-Schott, J., and Dubois, M. P. (1973). *Z. Zellforsch.* **142**, 571.
Dube, D., Lissitzky, J. C., Leclerc, R., and Pelletier, G. (1978). *Endocrinology* **102**, 1283.
Dubois, M. P. (1977). *Bull. Soc. Zool. France* **102**, Suppl. 1, 63.
Dubois, M. P. (1980). *Coll. Neuroendocrinol. Exp., 10th, Lyon, J. Physiol. (Paris)* (in press).
Dubois, P. M. (1968). *C. R. Soc. Biol.* **162**, 689.
Dubois, P. M. (1971). *Z. Anat. Entwicklungs, g.* **133**, 318.
Dubois, P., Vargues-Regairaz, H., and Dubois, P. M. (1973). *Z. Zellforsch.* **145**, 131.
Dubois, P. M., Bethenod, M., Gilly, R., and Dubois, M. P. (1975). *Arch. Fr. Pediatr.* **32**, 647.
Dubois, P. M., Li, J. Y., and Bégeot, M. (1976). *J. Microsc. Biol. Cell.* **26**, 11a.
Dubois, P. M., Bégeot, M., Paulin, C., Alizon, E., and Dubois, M. P. (1979). *J. Physiol. (Paris)* **75**, 55.
Dupouy, J. P. (1976). *C.R. Acad. Sci. (Paris)* **282**, 211.
Dupouy, J. P., and Dubois, M. P. (1975). *Cell Tissue Res.* **161**, 373.
Dupouy, J. P., and Jost, A. (1970). *C. R. Soc. Biol.* **164**, 2422.
Dupouy, J. P., Chatelain A., and Dubois, M. P. (1979). *Proc. Int. Symp. Neuroendocrinol. Regul. Mech. Serb. Acad. Sci. Arts, Belgrade* p. 219.
Eguchi, Y., Hirai, O., Morikawa, Y., and Hashimoto, Y. (1973). *Endocrinology* **93**, 1.
Eipper, B. A., and Mains, R. E. (1977). *J. Biol. Chem.* **252**, 882.
Eipper, B. A., and Mains, R. E. (1978a). *J. Biol. Chem.* **253**, 5732.
Eipper, B. A., and Mains, R. E. (1978b). *J. Supramol. Struc.* **8**, 247.
Eipper, B. A., Mains, R. E., and Guenzi, D. (1976). *J. Biol. Chem.* **251**, 4121.
Elde, R., Hökfelt, T., Johansson, O., and Terenius, L. (1976). *Neuroscience* **1**, 349.
Enemar, A. (1963). *Ark. Zool. II* **16**, 169.
Eneroth, P., Ferngren, H., Gustafsson, J. A., Ivemark, B., and Stenberg, A. (1972). *Acta Endocrinol.* **70**, 113.

Etkin, W. (1941). *Proc. Soc. Exp. Biol.* **47**, 425.

Etkin, W. (1958a). *Anat. Rec.* **131**, 548.

Etkin, W. (1958b). *Proc. Soc. Exp. Biol.* **97**, 388.

Eurenius, L., and Jarskär, R. (1975). *Cell Tissue Res.* **164**, 11.

Falin, L. I. (1961). *Acta Anat.* **44**, 188.

Fellmann, D., Franco, N., Bloch, B., and Hatier, R. (1975). *Bull. Assoc. Anat.* **59**, 631.

Ferrand, R. (1969a). *C.R. Acad. Sci. (Paris)* **268**, 550.

Ferrand, R. (1969b). *C. R. Soc. Biol.* **163**, 2669.

Ferrand, R. (1970a). *Ann. Biol.* **9**, 357.

Ferrand, R. (1970b). *C. R. Acad. Sci. (Paris)* **270**, 1480.

Ferrand, R. (1971a). *Bull. Assoc. Anat.* **151**, 328.

Ferrand, R. (1971b). *C.R. Soc. Biol.* **165**, 392.

Ferrand, R. (1972). *Arch. Biol. (Liège)* **83**, 297.

Ferrand, R., and Le Douarin, N. (1968). *C.R. Soc. Biol.* **162**, 2215.

Ferrand, R., and Nanot, J. (1968). *C.R. Soc. Biol.* **162**, 983.

Ferrand, R., Pearse, A. G. E., Polak, J. M., and Le Douarin, N. M. (1974). *Histochemistry* **38**, 133.

Ferrand, R., Le Douarin, N. M., Polak, J. M., and Pearse, A. G. E. (1975). *Experientia* **31**, 1096.

Fink, G., and Smith, G. (1971). *Z. Zellf. Mikrosk. Anat.* **119**, 208.

Follenius, E., and Dubois, M. P. (1977). *C.R. Acad. Sci. (Paris)* **285**, 1065.

Follenius, E., and Dubois, M. P. (1978). *C.R. Acad. Sci. (Paris)* **286**, 1383.

Fredrickson, R. C. A. (1977). *Life Sci.* **21**, 23.

Fujita, T., Eguchi, Y., Morikawa, Y., and Hashimoto, Y. (1970). *Anat. Rec.* **166**, 651.

Fukuda, T. (1973). *Virchows Arch. A Pathol. Anat.* **359**, 19.

Gash, D. M., Roos, T. B., and Chambers, W. F. (1975). *Neuroendocrinology* **19**, 214.

Genazzani, A. R., Hurlimann, J., Fioretti, P., and Felber, J. P. (1974). *Experientia* **30**, 430.

Genazzani, A. R., Fraioli, F., Hurlimann, J., Fioretti, P., and Felber, J. P. (1975). *Clin. Endocrinol.* **4**, 1.

Gilkes, J. J. H., Bloomfield, G. A., Scott, A. P., Lowry, P. J., Ratcliffe, J. G., Landon, J., and Rees, L. H. (1975). *J. Clin. Endocrinol. Metab.* **40**, 450.

Girod, C. (1977). *Bull. Assoc. Anat.* **62**, 417.

Giroud, A., and Martinet, M. (1957). *Arch. Anat. Microsc. Morphol. Exp.* **46**, 247.

Goldstein, A. (1976). *Science* **193**, 1081.

Grieshaber, C. K., and Hymer, W. C. (1968). *Proc. Soc. Exp. Biol. Med.* **128**, 459.

Guillemin, R., Schally, A. V., Lipscomb, H. S., Andersen, R. N., and Long, J. M. (1962). *Endocrinology* **70**, 471.

Guillemin, R., Ling, N., and Burgus, R. (1976). *C.R. Acad. Sci. (Paris)* **282**, 783.

Guillemin, R., Vargo, T., Rossier, J., Minick, S., Ling, N., Rivier, C., Vale, W., and Bloom, F. (1977). *Science* **197**, 1367.

Hanaoka, Y. (1967). *Gen. Comp. Endocrinol* **8**, 417.

Hanström, B. (1966). *In* "The Pituitary Gland" (G. W. Harris and B. T. Donovan, eds.), Vol. 1. Butterworths, London.

Harris, J. I. (1959). *Nature (London)* **184**, 167.

Harris, J. I., and Lerner, A. B. (1957). *Nature (London)* **179**, 1346.

Hatakeyama, S. (1969). *Endocrinol. Jpn.* **16**, 187.

Hayek, A., Driscoll, S. G., and Warshaw, J. B. (1973). *J. Clin. Invest.* **52**, 1636.

Hirata, Y., Sakamoto, N., Matsukura, S., and Imura, H. (1975). *J. Clin. Endocrinol. Metab.* **41**, 1092.

Hirata, Y., Matsukura, S., Imura, H., Nakamura, M., and Tanaka, A. (1976). *J. Clin. Endocrinol. Metab.* **42**, 33.

Hökfelt, T., Elde, R., Johansson, O., Terenius, L., and Stein, L. (1977). *Neurosci. Lett.* **5**, 25.

Hollt, V., Przewlocki, R., and Herz, A. (1978). *Life Sci.* **23,** 1057.

Hughes, J. (1975). *Brain Res.* **88,** 295.

Hughes, J., Smith, T. W., Kosterlitz, H. W., Fothergill, L. A., Morgan, B. A., and Morris, H. R. (1975). *Nature (London)* **258,** 577.

Hymer, W. C., and McShan, W. H. (1963). *J. Cell Biol.* **17,** 67.

Ishikawa, H., Shiino, M., and Rennels, E. G. (1977). *Proc. Soc. Exp. Biol. Med.* **155,** 511.

Jacob, J. (1977). *Anna. Pharmac. Fr.* **35,** 399.

Jeffcoate, W. J., Rees, L. H., Lowry, P. J., Hope, J., and Besser, G. M. (1978). *J. Endocrinol.* **77,** 27P.

Jones, C. T., and Roebuck, M. M. (1979). *Int. Symp. J. Steroid Bioch. 4th Paris* p. 35P.

Jost, A. (1947). *C.R. Acad. Sci. (Paris)* **225,** 322.

Jost, A. (1966). *Rec. Progr. Horm. Res.* **22,** 541.

Jost, A. (1971). *In* "Hormones in Development" (M. Hamburgh and E. J. W. Barrington, eds) pp. 1–18. Appleton, New York.

Jost, A., and Picon, L. (1970). *Adv. Metab. Disorders* **4,** 123.

Jost, A., Dupouy, J. P., and Monchamp, A. (1966). *C.R. Acad. Sci. (Paris)* **262,** 147.

Jost, A., Dupouy, J. P., and Geloso-Meyer, A. (1970). *In* "The Hypothalamus" (L. Martini, M. Motta, and F. Fraschini, eds.), pp. 605–615. Academic Press, New York.

Kastin, A. J., Gennser, G., Arimura, A., Miller, M. C., and Schally, A. V. (1968). *Acta Endocrinol. (Kbh)* **58,** 6.

Krieger, D. T., Liotta, A., and Brownstein, M. J. (1977a). *Proc. Natl. Acad. Sci. U.S.A.* **74,** 648.

Krieger, D. T., Liotta, A., and Brownstein, M. J. (1977b). *Brain Res.* **128,** 575.

Krieger, D. T., Liotta, A., Suda, T., Palkovits, M., and Brownstein, M. J. (1977c). *Biochem. Biophys. Res. Commun.* **76,** 930.

Krieger, D. T., Liotta, A. S., Nicholsen, G., and Kizer, J. S. (1979). *Nature (London)* **278,** 562.

Labella, F., Queen, G., Senyshyn, J., Lis, M., and Chretien, M. (1977). *Biochem. Biophys. Res. Commun.* **75,** 350.

Larsson, L. I. (1977). *Lancet* **II** 1321.

Lazarus, L. H., Ling, N., and Guillemin, R. (1976). *Proc. Natl. Acad. Sci. U.S.A.* **73,** 2156.

Le Douarin, G., and Ferrand, R. (1968). *C.R. Acad. Sci. (Paris)* **266,** 697.

Le Douarin, N., Ferrand, R., and Le Douarin, G. (1967a). *C.R. Soc. Biol.* **161,** 1807.

Le Douarin, N., Ferrand, R., and Le Douarin, G. (1967b). *C. R. Acad. Sci. (Paris)* **264,** 3027.

Levina, S. E. (1968). *Gen. Comp. Endocrinol.* **11,** 151.

Levina, S. E., Ivanova, E. A., and Sergeenkova, G. P. (1968). *Dokl. Akad. Nauk.* **172,** 246.

Li, C. H., and Chung, D. (1976). *Nature (London)* **260,** 622.

Li, J. Y., Dubois, M. P., and Dubois, P. M. (1976). *J. Microsc. Biol. Cell.* **26,** 18a.

Li, J. Y., Dubois, M. P., and Dubois, P. M. (1979). *Cell Tissue Res.* **204,** 37.

Ling, C. H., and Guillemin, R. (1976). *Proc. Natl. Acad. Sci. U.S.A.* **73,** 3308.

Ling, N., Burgus, R., and Guillemin, R. (1976). *Proc. Natl. Acad. Sci. U.S.A.* **73,** 3942.

Liotta, A. S., Osathanondh, R., Ryan, K. J., and Krieger, D. T. (1977). *Endocrinology* **101,** 1552.

Liotta, A. S., Gildersleeve, D., Brownstein, M. J., and Krieger, D. T. (1979). *Proc. Natl. Acad. Sci. U.S.A.* **76,** 1448.

Liwska, J. (1975). *Folia Morphol.* **3,** 211.

Liwska, J. (1978). *Folia Histochem.* **16,** 307.

Loh, Y. P. (1979). *Proc. Natl. Acad. Sci. U.S.A.* **76,** 796.

Loh, Y. P., and Gainer, H. (1978). *FEBS Lett.* **96,** 269.

Lowry, P. J., and Scott, A. P. (1975). *Gen. Comp. Endocrinol.* **26,** 16.

Maillard, M. (1963). *J. Microsc.* **2,** 81.

Mains, R. E., Eipper, B. A., and Ling, N. (1977). *Proc. Natl. Acad. Sci. U.S.A.* **74,** 3014.

Martin, R., Weber, E., and Voigt, K. H. (1979). *Cell Tissue Res.* **196,** 307.
Mialhe-Voloss, C., and Stutinsky, F. (1953). *Ann. Endocrinol.* **14,** 681.
Mikami, S., Hashikawa, T., and Farner, D. S. (1973). *Z. Zellforsch. Mikrosk. Anat.* **138,** 299.
Mira-Moser, F. (1972). *Z. Zellforsch.* **125,** 88.
Mitskevich, M. S., and Rumyantseva, O. N. (1972). *Ontogenesis* **3,** 376.
Moldow, R., and Yalow, R. S. (1978). *Proc. Natl. Acad. Sci. U.S.A.* **75,** 994.
Moriarty, G. C. (1973). *J. Histochem. Cytochem.* **21,** 855.
Moriarty, G. C., and Garner, L. L. (1977a). *Front. Horm. Res.* **4,** 26.
Moriarty, G. C., and Garner, L. L. (1977b). *Nature (London)* **265,** 356.
Moriarty, G. C., and Halmi, N. S. (1972a). *J. Histochem. Cytochem.* **20,** 590.
Moriarty, G. C., and Halmi, N. S. (1972b). *Z. Zellforsch. Mikrosk. Anat.* **132,** 1.
Moriarty, G. C., and Moriarty, C. M. (1973). *Anat. Rec.* **175,** 393.
Moriarty, G. C., Moriarty, C. M., and Sternberger, L. A. (1973). *J. Histochem. Cytochem.* **21,** 825.
Nakai, Y., Nakao, K., Oki, S., and Imura, H. (1978). *Life Sci.* **23,** 2013.
Nakane, P. K., and Pierce, G. B. (1966). *J. Histochem. Cytochem.* **14,** 929.
Nakanishi, S., Inoue, A., Taii, S., and Numa, S. (1977). *FEBS Lett.* **84,** 105.
Nakanishi, S., Inoue, A., Kita, T., Nakamura, M., Chang, A. C. Y., Cohen, S. N., and Numa, S. (1979). *Nature (London)* **278,** 423.
Nakano, K. K. (1973). *Dev. Med. Child Neurol.* **15,** 383.
Negm, I. M. (1970). *Acta Anat.* **77,** 422.
Negm, I. M. (1971). *Acta Anat.* **80,** 604.
Negm, I. M. (1972). *Acta Anat.* **83,** 95.
Nemeskery, A., Nemeth, A., Setalo, G., Vigh, S., and Halasz, B. (1976). *Cell Tissue Res.* **170,** 263.
Nilaver, G., Zimmerman, E. A., Defendini, R., Liotta, A. S., Krieger, D. T., and Brownstein, M. J. (1979). *J. Cell Biol.* **81,** 50.
Novy, M. J., Walsh, S. W., and Kittinger, G. W. (1977). *J. Clin. Endocrinol. Metab.* **45,** 1031.
Nyholm, N. E. I. (1969). *Gen. Comp. Endocrinol.* **13,** 523.
Nyholm, N. E. I. (1972). *Gen. Comp. Endocrinol.* **18,** 113.
Nyholm, N. E. I., and Doerr-Schott, J. (1977). *Cell Tissue Res.* **180,** 231.
Ohtsuka, Y., Ishikawa, H., Omoto, T., Takasaki, Y., and Yoshimura, F. (1971). *Endocrinol. Jpn.* **18,** 33.
Ohtsuka, Y., Ishikawa, H., Watanabe, T., and Yoshimura, F. (1972). *Endocrinol. Jpn.* **19,** 237.
Oliver, C., and Porter, J. C. (1978). *Endocrinology* **102,** 697.
Orth, D. N., Nicholson, W. E., Mitchell, W. M., Island, D. P., Shapiro, M., and Byyny, R. L. (1973). *Endocrinology* **92,** 385.
Orth, D. N., Guillemin, R., Ling, N., and Nicholson, W. E. (1978). *J. Clin. Endocrinol. Metab.* **46,** 849.
Orwoll, E., Kendall, J. W., Lamorena, L., and McGilvra, R. (1979). *Endocrinology* **104,** 1845.
Osamura, R. Y. (1977). *Acta Pathol. Jpn.* **27,** 495.
Pavlova, E. B., Pronina, T. S., and Skebelskaya, Y. B. (1968). *Gen. Comp. Endocrinol* **10,** 269.
Pedernera, E. (1972). *Gen. Comp. Endocrinol.* **19,** 589.
Pehlemann, F. W. (1965). *Gen. Comp. Endocrinol.* **5,** 704.
Pelletier, G. (1979). *J. Histochem. Cytochem.* **27,** 1046.
Pelletier, G., and Desy, L. (1979). *Cell Tissue Res.* **196,** 525.
Pelletier, G., and Dube, D. (1977). *Am. J. Anat.* **150,** 201.
Pelletier, G., Leclerc, R., Labrie, F., Cote, J., Chretien, M., and Lis, M. (1977). *Endocrinology* **100,** 770.
Pelletier, G., Robert, F., and Hardy, J. (1978). *J. Clin. Endocrinol. Metab.* **46,** 534.

Perdue, J. F., and McShan, W. H. (1966). *Endocrinology* **78**, 406.

Pezalla, P. D., Seidah, N. G., Benjannet, S., Crine, P., Lis, M., and Chrétien, M. (1978). *Life Sci.* **23**, 2281.

Queen, G., Pinsky, C., and Labella, F. (1976). *Biochem. Biophys. Res. Commun.* **72**, 1021.

Rees, L. H., Burke, C. W., Chard, T., Evans, S. W., and Letchworth, A. T. (1975). *Nature (London)* **254**, 620.

Remy, C., and Dubois, M. P. (1974). *C.R. Soc. Biol.* **168**, 1275.

Roberts, J. L., and Herbert, E. (1977a). *Proc. Natl. Acad. Sci. U.S.A.* **74**, 4826.

Roberts, J. L., and Herbert, E. (1977b). *Proc. Natl. Acad. Sci. U.S.A.* **74**, 5300.

Roberts, J. L., Phillips, M., Rosa, P. A. and Herbert, E. (1978). *Biochemistry* **17**, 3609.

Rossier, J., Vargo, T. M., Minick, S., Ling, N., Bloom, F. E., and Guillemin, R. (1977). *Proc. Natl. Acad. Sci. U.S.A.* **74**, 5162.

Rubinstein, M., Stein, S., and Udenfriend, S. (1978). *Proc. Natl. Acad. Sci. U.S.A.* **75**, 669.

Rudman, D., Del Rio, A. E., Hollins, B. M., Houser, D. H., Keeling, M. E., Sutin, J., Scott, J. W., Sears, R. A., and Rosenberg, M. Z. (1973). *Endocrinology* **92**, 372.

Rudman, D., Scott, J. W., Del Rio, A. E., Houser, D. H., and Sheen, S. (1974). *Am. J. Physiol* **226**, 682.

Salazar, H., MacAulay, M. A., Charles, D., and Pardo, M. (1969). *Arch. Pathol.* **87**, 201.

Sano, M., and Sasaki, F. (1969). *Z. Anat. Entwicklungsg.* **129**, 195.

Satow, Y., Miyabara, S., Ueno, T., Ikeda, T., and Okamoto, N. (1971). *Hirosh. J. Med. Sci.* **20**, 223.

Satow, Y., Okamoto, N., Ikeda, T., Ueno, T., and Miyabara, S. (1972). *J. Electron Microsc.* **21**, 29.

Saxena, B. N. (1971). *Vitam. Horm.* **29**, 95.

Schally, A. V., Lipscomb, H. S., Long, J. H., Dear, W. E., and Guillemin, R. (1962). *Endocrinology* **70**, 478.

Schechter, J. (1970). *Gen. Comp. Endocrinol.* **14**, 53.

Schechter, J. (1971). *Gen. Comp. Endocrinol.* **16**, 1.

Scherrer, H., Benjannet, S., Pezalla, P. D., Bourassa, M., Seidah, N. G., Lis, M., and Chrétien, M. (1978). *FEBS Lett.* **90**, 353.

Scherrer, H., Seidah, N. G., Benjannet, S., Crine, Ph., and Chrétien, M. (1979). *C.R. Acad. Sci. (Paris)* **288**, 543.

Schwind, J. (1928). *Am. J. Anat.* **41**, 295.

Scott, A. P., and Lowry, P. J. (1974). *Biochem. J.* **139**, 593.

Scott, A. P., Ratcliffe, J. G., Ress, L. H., Landon, J., Bennett, H. P. J. Lowry, P. J., and McMartin, C. (1973). *Nature (London) New Biol.* **244**, 65.

Setalo, G., and Nakane, P. K. (1972). *Anat. Rec.* **172**, 403.

Setalo, G., and Nakane, P. K. (1976). *Endocrinol. Exp.* **10**, 155.

Shiino, M., Ishikawa, H., and Rennels, E. (1977). *Cell Tissue Res.* **181**, 473.

Silman, R. E. (1979). *J. Endocrinol.* **80**, 1P.

Silman, R. E., Chard, T., Lowry, P. J., Smith, I., and Young, I. M. (1976). *Nature (London)* **260**, 716.

Silman, R. E., Holland, D., Chard, T., Lowry, P. J., Hope, J., Rees, L. H., Thomas, A., and Nathanielsz, P. (1979). *J. Endocrinol* **81**, 19.

Simantov, R., and Snyder, S. H. (1976a). *Life Sci.* **18**, 781.

Simantov, R., and Snyder, S. H. (1976b). *Proc. Natl. Acad. Sci. U.S.A.,* **73**, 2515.

Simantov, R., Goodman, R., Aposhian, D., and Snyder, S. H. (1976a). *Brain Res.* **111**, 204.

Simantov, R., Kuhar, M. J., Pasternak, G. W., and Snyder, S. H. (1976b). *Brain Res.* **106**, 189.

Skebelskaya, Y. B. (1965). *Probl. Endocrinol. Horm. Ther.* **4**, 77.

Sofroniew, M. V. (1979). *Am. J. Anat.* **154**, 283.

Stoeckel, M. E., Hindelang-Gertner, C., Porte, A., Dellmann, H. D. and Stutinsky, F. (1973a). *C. R. Acad. Sci. (Paris)* **277**, 97.

Stoeckel, M. E., Porte, A., Hindelang-Gertner, C., and Dellmann, H. D. (1973b). *Z. Zellf. Mikrosk. Anat.* **142**, 347.

Stoeckel, M. E., Hindelang-Gertner, C., and Porte, A. (1979). *Cell Tissue Res.* **198**, 465.

Svalander, C. (1974). *Acta Endocrinol. Suppl.* **188**, 1.

Swaab, D. F., and Fisser, B. (1977). *Neurosci. Lett.* **7**, 313.

Székely, G., Endröczi, E., and Szentagothai, J. (1958). *Acta Biol. Hung.* **8**, 283.

Tanaka, K., Mount, C. R., and Orth, D. H. (1976). *Proc. Meet. Endocr. Soc. 58th, San Francisco* (Abstr.) p. 129.

Thompson, S. A., and Trimble, J. J. (1976). *Anat. Embryol.* **150**, 7.

Toubeau, G. (1978). *C.R. Acad. Sci. (Paris)* **287**, 499.

Tramu, G., and Leonardelli, J. (1979). *Brain Res.* **168**, 457.

Tramu, G., Leonardelli, J., and Dubois, M. P. (1977). *Neurosci. Lett.* **6**, 305.

Tuchmann-Duplessis, H., and Larroche, J. (1958). *C.R. Soc. Biol.* **152**, 300.

Vaudry, H., Tonon, M. C., Delarue, C., Vaillant, R., and Kraicer, J. (1978). *Neuroendocrinology* **27**, 9.

Watanabe, Y. G., and Daikoku, S. (1976). *Cell Tissue Res.* **166**, 407.

Watanabe, Y. G., and Daikoku, S. (1979). *Dev. Biol.* **68**, 557.

Watanabe, Y. G., Matsumura, H., and Daikoku, S. (1973). *Z. Zellforsch.* **146**, 453.

Watson, S. J., Akil, H., Sullivan, S., and Barchas, J. D. (1977a). *Life Sci.* **21**, 733.

Watson, S. J., Barchas, J. D., and Li, C. H. (1977b). *Proc. Natl. Acad. Sci. U.S.A.* **74**, 5155.

Watson, S. J., Richard, C. W., and Barchas, J. D. (1978a). *Science* **200**, 1180.

Watson, S. J., Akil, H., Richard, C. W., and Barchas, J. D. (1978b). *Nature (London)* **275**, 226.

Weber, E., Voigt, K. H., and Martin, R. (1978a). *Endocrinology* **102**, 1466.

Weber, E., Voigt, K. H., and Martin, R. (1978b). *Proc. Natl. Acad. Sci. U.S.A.* **75**, 6134.

Wingstrand, K. G. (1966). *In* "The Pituitary Gland" (G. W. Harris and B. T. Donovan, eds.), Vol. 1, pp. 1–23., Butterworths, London.

Winters, A. J., Oliver, C., Colston, C., Mac Donald, P. C., and Porter, J. C. (1974). *J. Clin. Endocrinol. Metab.* **39**, 269.

Yoshida, Y. (1966). *In* "Methods and Achievements in Experimental Pathology" (E. Bajusz and G. Jasmin, eds.), Vol. 1, p. 439., Karger, Basel.

Yoshimi, H., Matsukura, S., Sueoka, S., Fukase, M., Yokota, M., Hirata, Y., and Imura, H. (1978). *Life Sci.* **22**, 2189.

Yoshimura, F., Harumiya, K., and Kiyama, H. (1970). *Arch. Histol. Jpn.* **31**, 333.

Zimmerman, E. A., Liotta, A., and Krieger, D. T. (1978). *Cell Tissue Res.* **186**, 393.

INTERNATIONAL REVIEW OF CYTOLOGY, VOL. 68

Cell Death: The Significance of Apoptosis

A. H. Wyllie

Department of Pathology, University of Edinburgh, Edinburgh, Scotland

J. F. R. Kerr

Department of Pathology, University of Queensland, Brisbane, Australia

A. R. Currie

Department of Pathology, University of Edinburgh, Edinburgh, Scotland

I. Introduction

Death of cells has aroused far less interest than other basic cellular processes such as proliferation and differentiation. The relative neglect probably results at least in part from the wide prevalence of an unjustifiably circumscribed and restricted notion of cell death as a degenerative phenomenon produced by injury. This concept evolved early in the history of cellular pathology (Virchow, 1858) and has tended to dominate thinking about both the incidence and the mechanisms of cell death ever since. Thus, the fact that cell death can occur as a controlled event in healthy animals has not gained general recognition even though the vital role of focal cell death in the morphogenesis of normal embryos was clearly defined more than thirty years ago (for review, see Glücksmann, 1951). Moreover, the possibility that such "normal cell death" might involve active self destruction rather than passive degeneration has been virtually ignored; most regard the changes that take place in all dying cells as being akin to

251

the postmortem autolysis of a corpse, a process hardly likely to stimulate en-
thusiastic investigation.

Classification of cell death can be based on morphological or biochemical
criteria or on the circumstances of its occurrence. In this chapter, we have
adopted a morphological approach for two reasons. First, irreversible structural
alteration provides, at present, the only unequivocal evidence of death; biochemical indicators of cell death that are universally applicable have still to be precisely defined, and studies of cell function or of reproductive capacity do not
necessarily differentiate between death and dormant states from which recovery
may be possible. Second, it has proved feasible to categorize most if not all dying
cells into one or the other of two discrete and distinctive patterns of
morphological change, which have, in general, been found to occur under dispa-
rate but individually characteristic circumstances. One of these patterns is the
familiar swelling proceeding to rupture of plasma and organelle membranes and
dissolution of organized structure, which has long been termed "coagulative"
necrosis. It classically results from injury by agents such as toxins and ischemia,
affects cells in groups rather than singly, and evokes exudative inflammation
when it develops *in vivo*. The other morphological pattern is characterized by
condensation of the cell with maintenance of organelle integrity and the formation of surface protuberances that separate as membrane-bounded globules; in
tissues, these are phagocytosed and digested by resident cells, there being no
associated inflammation. The occurrence of this second mode of cell death has
been shown in many situations to be under the control of physiological stimuli.
The name *shrinkage necrosis* was originally used for it on morphological
grounds (Kerr, 1965, 1971), but when its widespread incidence and its kinetic
significance in health and disease were recognized by us, we proposed the name
apoptosis (Kerr *et al.*, 1972).

We shall discuss first, the historical approaches to the study of cell death;
second, the morphology of necrosis and apoptosis; third, their incidence; and
last, what is known of their mechanisms. The main thrust of the chapter will be
directed toward apoptosis since it is much less well known than necrosis, which
has been the subject of many reviews in the past. However, some consideration
of necrosis is necessary, since the significance of apoptosis can be fully ap-
preciated only when it is set in the context of cell death as a whole.

II. Historical Approaches

There have been three major approaches to the study of cell death, which have
been determined by the circumstances of its occurrence, the availability of suitable models, and the observational techniques available.

First, experiments have been designed to provide information about the cell death that develops in natural diseases in which tissue injury is produced by defined noxious agents. In essence, these experiments involve exposing relatively homogeneous groups of previously healthy cells to violent and non-physiological changes in their environment (for reviews, see Majno *et al.,* 1960; McLean *et al.,* 1965; Trump and Ginn, 1969; Trump and Mergner, 1974; Jennings *et al.,* 1975). Precise control of the severity and timing of the lethal stimulus is possible, especially when *in vitro* methods are used, and, since contiguous cells tend to be uniformly affected, morphological and biochemical studies representative of the population as a whole can be performed on samples of convenient size. Nevertheless, the cells of many tissues do show variability in their response, which may depend on normal biochemical heterogeneity (Novikoff and Essner, 1960), and the development of inflammation with migration of leukocytes into an area of injury may complicate interpretation of biochemical studies of cell death produced *in vivo* (Tappel *et al.,* 1963). On the whole, a good correlation between morphological and biochemical changes in this type of cell death has been attained (Trump and Ginn, 1969; Trump and Mergner, 1974). Early in the history of these studies, however, Bessis (1964) pointed out that such "accidental" death might be very different from the "natural" cell death that occurs in physiological processes.

The second major approach has been by biologists studying normal metamorphosis and embryonic development, where processes such as the elimination of larval organs and phylogenetic vestiges and the fashioning of limbs and organs involve massive deletion of cells (for reviews, see Glücksmann, 1951; Saunders, 1966; Saunders and Fallon, 1966; Menkes *et al.,* 1970; Lockshin and Beaulaton, 1974). Here the cell death occurs spontaneously and predictably under physiological conditions and has been shown to be subject to control by genetic, hormonal, or local tissue factors. Historically, the death was first detected by direct morphological observation; subsequently, biochemical studies were undertaken, but their interpretation has been rendered difficult by the presence of many viable cells intimately intermixed with the dying ones.

In the third major approach, the occurrence of cell death has been inferred indirectly from kinetic studies; the disparity between the rate of change in size of a tissue and the rate of proliferation of its component cells provides a quantitative measure of cell loss. Most attention has been devoted to neoplasms, where continuous and spontaneous cell loss has been shown to be a major though variable determinant of growth (Iversen, 1967; Steel, 1967, 1968; Laird, 1969; Dethlefsen, 1971; Terz *et al.,* 1971; Cooper *et al.,* 1975; Tubiana and Malaise, 1976). However, there is evidence that cells are also regularly lost from healthy adult tissues that remain constant in size (Cameron, 1971). Involution such as that occurring in endocrine-dependent tissues in response to changes in the blood concentration of trophic hormones must also be accompanied by cell loss. In all these situations, the kinetic data provide no clue to the relative contributions of *in situ* death of cells and loss of cells from the tissue by exfoliation or migration; attempts to directly observe cell death have been surprisingly few (Mendelsohn, 1960; Kerr, 1971; Kerr and Searle, 1972a; Kerr *et al.*, 1972; Wyllie *et al.*, 1973a,b).

These three approaches have little in common and have led to divergent conclusions. It is probable that the processes studied—although all called cell death—represent fundamentally different phenomena. Historically, however, there has been no satisfactory means of recognizing and classifying different types of death: early attempts were based either on the circumstances in which death occurred (Bessis, 1964) or on subjective interpretation of its biological function (Glücksmann, 1951). There is need for a new means of classification based on changes in the affected cell itself. It is our view that study of morphology provides the essentials of such a classification.

As indicated in the introduction to this article, two distinct patterns of morphological change that embrace most, and possibly all, cell death found in higher animals under natural and experimental conditions have been identified. Further, a strong correlation has emerged between the circumstances of occurrence and putative biological significance of cell death on the one hand and the morphological type of cell death found to be present on the other. In synopsis, gross environmental perturbation usually produces necrosis, whereas both naturally occurring morphogenetic death in the embryo and death involved in the regulation of tissue size in postnatal life display the features of apoptosis. There are a few apparent exceptions to this general statement, the implications of which will be discussed later.

III. Morphology

A. NECROSIS

There is good agreement on the ultrastructural features of necrosis in a wide range of cells (McLean *et al.*, 1965; Trump and Ginn, 1969; Buckley, 1972; Trump *et al.*, 1973; Kloner *et al.*, 1974; Ganote *et al.*, 1975; Jennings *et al.*, 1975; Laiho and Trump, 1975). Early abnormalities, which are compatible either with recovery of an injured cell or with progression to necrosis, include marginal clumping of loosely textured nuclear chromatin, dilatation of the endoplasmic reticulum and mild dispersal of ribosomes. A finding that strongly suggests irreversibility is gross swelling of the matrix of mitochondria. The subsequent evolution of necrosis (Fig. 1) is accompanied by rupture of nuclear, organelle, and plasma membranes, the appearance of flocculent and sometimes also granular matrix densities in mitochondria, and the dissolution of ribosomes and

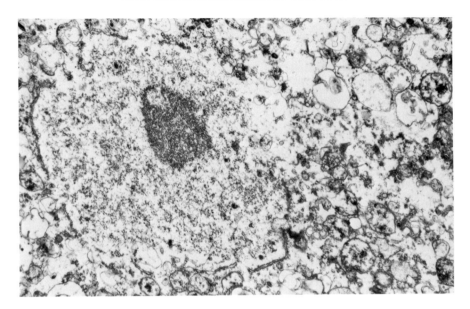

FIG. 1. Ischemic necrosis in center of solid nodule of mouse sarcoma 180. Note flocculent matrix densities in grossly swollen mitochondria and loss of chromatin from disintegrating nucleus. ×9800.

lysosomes; as the nucleus swells, the masses of clumped chromatin may become slightly dispersed, but they soon disappear altogether.

The histological appearances (Fig. 2A) are well described in standard pathology texts. The chromatin often initially appears fairly uniformly compacted (*pyknosis*), but with swelling of the nucleus and rupture of its membrane, the marginated chromatin masses may become evident as small discrete masses (*karyorrhexis*). Whatever the early changes, all basophilia is then lost leaving a faintly stained nuclear "ghost" (*karyolysis*). The swollen cytoplasm also loses its basophilia, and cell boundaries become indistinct (Fig. 2A). Typically a number of contiguous cells are affected, and exudative inflammation develops in the adjoining viable tissue. The debris is eventually ingested and degraded by specialized phagocytic cells.

B. APOPTOSIS

The account that follows is based mainly on our own studies, which cover a wide variety of cell types (Kerr, 1965, 1971, 1973; Crawford *et al.*, 1972; Kerr and Searle, 1972a,b, 1973, 1980; Kerr *et al.*, 1972, 1974; Searle *et al.*, 1973, 1975, 1977; Wyllie *et al.*, 1973a,b; Bird *et al.*, 1976; Don *et al.*, 1977; Robertson *et al.*, 1978; Waddell *et al.*, 1979; Weedon *et al.*, 1979). Difficulty

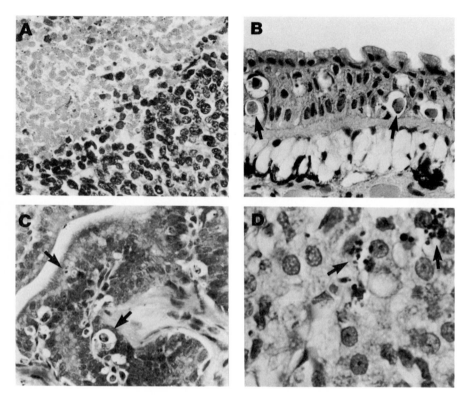

Fɪɢ. 2.   Light microscopic appearances of cell death. (A) Focal ischemic necrosis in human ovarian carcinoma; (B) apoptosis in skin of tail of tadpole during metamorphosis; (C) apoptosis in epithelium of rat ventral prostate 3 days after castration; (D) apoptosis in adrenal cortex of 20-day rat fetus following hypophysectomy effected by decapitation *in utero* on day 18 of gestation. In B–D, apoptotic bodies with nuclear fragments are indicated by arrows.

Fɪɢ. 3.   Stylized diagram illustrating the formation and fate of apoptotic bodies in a glandular epithelium. (1) Early apoptosis showing nuclear changes and cytoplasmic condensation; (2) cluster of intercellular apoptotic bodies; (3) extrusion into lumen with closing of ranks of remaining viable cells; (4) phagocytosis by adjacent epithelial cells; (5) phagocytosis by macrophage; (6) degeneration and lysosomal digestion. The macrophage illustrated (stippled) is within the epithelium, as in rat prostate and hamster endometrium.

Fɪɢ. 4.   Early nuclear changes of apoptosis in acinar epithelium of rat ventral prostate 2 days after castration. ×12,200. (From Kerr and Searle, 1973.)

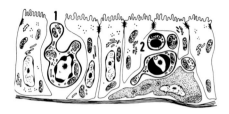

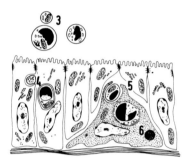

Fig. 3.

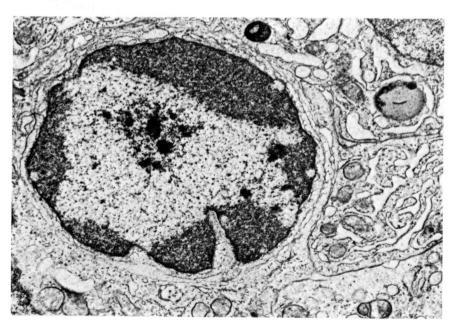

Fig. 4.

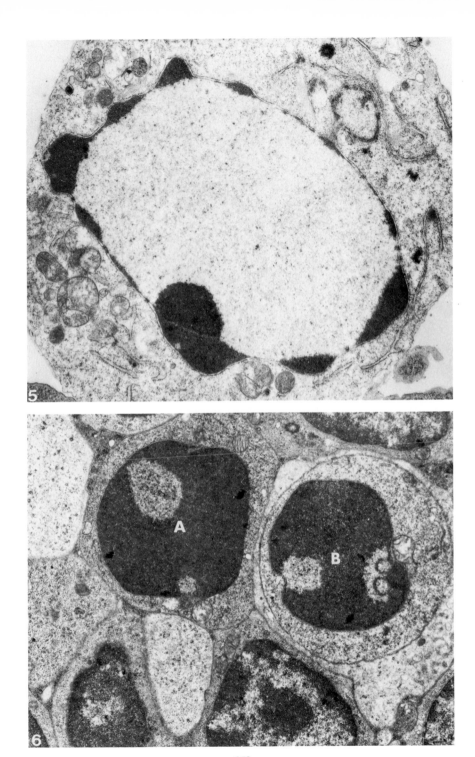

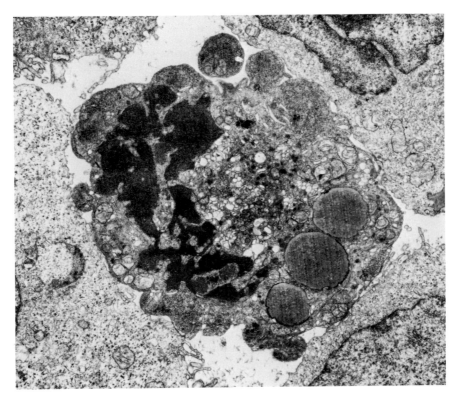

Fig. 7. Apoptosis occurring spontaneously in mouse sarcoma 180 growing as solid tumor. Early stage showing convoluted cell surface, cytoplasmic vacuoles, and chromatin condensation. ×7200.

arises from the fact that certain of the morphological manifestations of apoptosis have been described under a diversity of other names, sometimes apparently without awareness of their place in the sequence that we regard as characteristic of the process. We shall attempt a rationalization of nomenclature toward the end of this section and make much more comprehensive reference to the descriptions of other authors in the section dealing with incidence.

1. *Morphological Sequence*

Apoptosis characteristically affects scattered single cells rather than tracts of contiguous cells (Fig. 2B–D). The cardinal morphological events as revealed by electron microscopy are shown diagramatically in Fig. 3.

In the earliest stage so far recognized (Figs. 4–6), most of the chromatin has aggregated in large compact granular masses that abut on the nuclear membrane.

Fig. 5. Early nuclear changes of apoptosis in L5178 YC3 mouse lymphoma cell incubated for 30 hours in tissue culture medium containing 10^{-5} *M* methylprednisolone. ×12,000.

Fig. 6. Thymic cortex of juvenile rat 1 hour after intravenous injection of cortisol in dose of 50 mg per kg body weight. (A) Early apoptosis of thymocyte; (B) phagocytosed apoptotic cell. ×11,000.

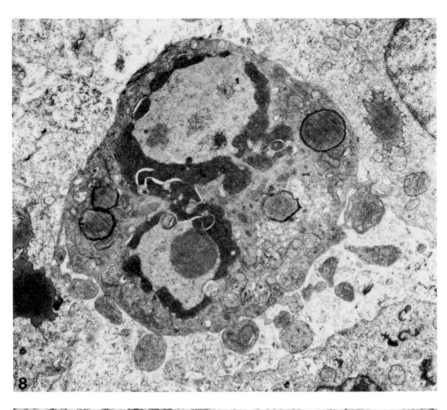

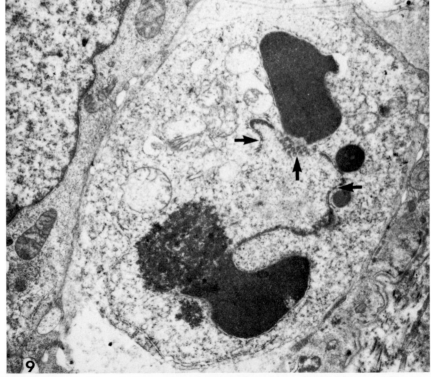

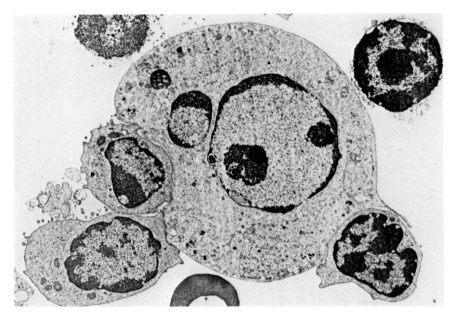

FIG. 10. Early nuclear changes of apoptosis in mouse mastocytoma cell following incubation with specifically allergized T lymphocytes for 30 minutes. Three lymphocytes are closely applied to its surface. Two other lymphocytes, physically disrupted during their extraction from the spleen, show changes of necrosis. ×4500. (From Don *et al.*, 1977.)

The nuclear outline is usually abnormally convoluted (Figs. 4 and 5); at a slightly later stage, it often becomes grossly indented (Figs. 7–9), and apparently discrete nuclear fragments may be present (Fig. 10). The nucleolus is enlarged, its granules being coarse and abnormally scattered (Fig. 4). Compact round masses of fine granules of uncertain origin are seen near the center of the nucleus in some sections (Fig. 8). Nuclear pores have not been detected with certainty over the marginated chromatin masses, but appear intact in parts of the membrane that adjoin dispersed chromatin (Fig. 9). Concomitant with these early nuclear changes, the cytoplasm has usually started to condense, microvilli (if originally present) have disappeared, and blunt protuberances have often formed on the cell surface (Figs. 7 and 8). In solid tissues, affected cells separate from their neighbors at about this stage, and, in the case of epithelial cells, the desmosomal

FIG. 8. Spontaneous apoptosis in mouse sarcoma 180. In addition to previously noted features, this cell shows round intranuclear masses of fine granular material. ×7000.

FIG. 9. Human breast lobular epithelial cell undergoing apoptosis. The material was obtained from an apparently normal region in a mastectomy specimen. The nuclear membrane is deeply convoluted and focally underlain by condensed chromatin. Nuclear pores, plentiful in some sites (arrows), are apparently less frequent overlying the condensed chromatin. ×11,000. (By courtesy of D. Ferguson.)

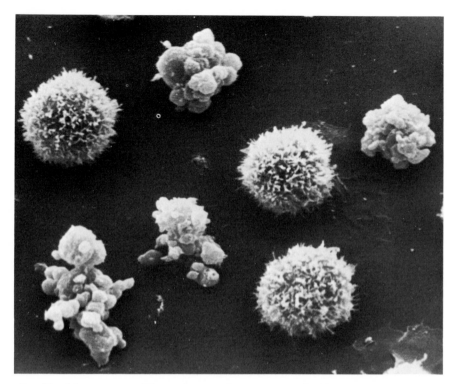

Fɪɢ. 11.   Mouse sarcoma 180 growing as ascites tumor treated with actinomycin D 18 hours previously. The surface of intact tumor cells is covered with microvilli, whereas cells undergoing apoptosis show relatively smooth surface protuberances of various sizes. (By courtesy of C. Bishop.)

attachments break down. Progression of the cytoplasmic condensation then results in marked crowding of organelles, which characteristically retain their integrity, and, in many cell types, the focal surface protrusion becomes very pronounced; this is particularly well seen when cells are examined with the scanning electron microscope (Figs. 11–14; see also Mullinger and Johnson, 1976). The overall compaction of the cytoplasm is frequently associated with the development of translucent cytoplasmic vacuoles (Fig. 7). Estimates of the degree of compaction suggest that it is enough to accommodate the increase in surface-to-volume ratio that must accompany the surface convolution. In the

Fɪɢ. 12.   Mouse sarcoma 180 growing as ascites tumor treated with actinomycin D 18 hours previously. (A) Scanning electron micrograph of cell proven by prior light microscopy to have a fragmented nucleus; (B) transmission electron micrograph of surface protuberances of similar cell. (By courtesy of C. Bishop.)

Fɪɢ. 13.   Apoptosis occurring 4 hours after methylprednisolone treatment of BLA1 cell line established from a patient with lymphoblastic leukemia. (From Robertson et al., 1978.)

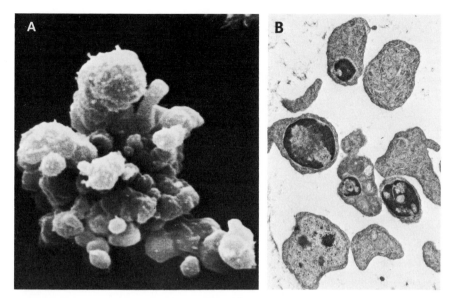

FIG. 12.

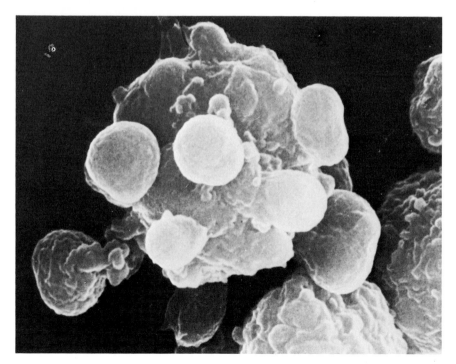

FIG. 13.

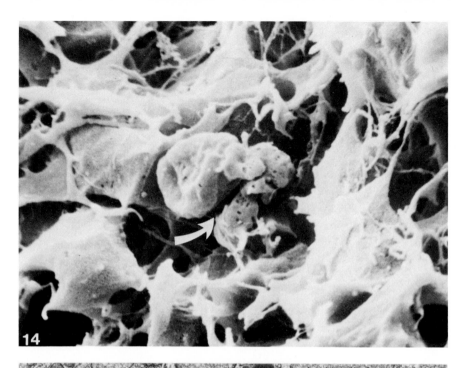

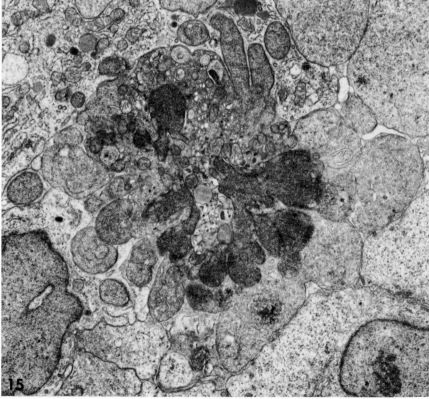

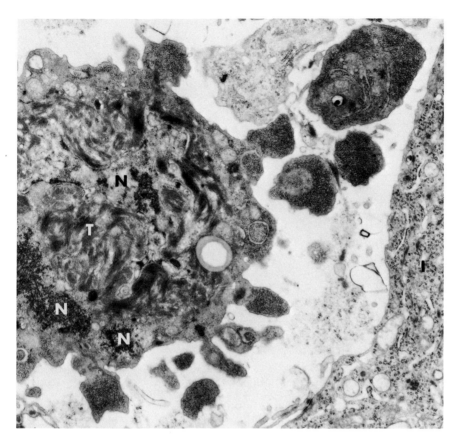

Fig. 16. Early apoptosis of keratinocyte in human epidermis in lichen planus. T, Closely packed tonofilaments; N, nuclear fragment; I, intact keratinocyte. ×11,000. (From Weedon *et al.*, 1979. Copyright by Masson Publishing USA, Inc., New York.)

meantime, the nucleus has usually broken up into a number of discrete fragments. Characteristically, these show condensed chromatin occupying their entire cut surface or arranged in crescentic caps (Figs. 12B and 15). Some of the nuclear fragments are surrounded by double membranes, but no membrane can be detected around others (Fig. 16). Finally, the protuberances on the cell surface separate with plasmalemmal sealing to produce membrane-bounded apoptotic bodies of roughly spherical or ovoid shape (Figs. 17–21). The number, size, and

---

Fig. 14. Apoptosis of mesenchymal cell (arrow) in "posterior necrotic zone" of wingbud of normal chick embryo. (From Hurle and Hinchliffe, 1978.)

Fig. 15. Apoptosis in solid nodule of mouse sarcoma 180 treated with 1000 rads X rays 6 hours previously. The nucleus has fragmented, and several discrete apoptotic bodies have formed. ×6400.

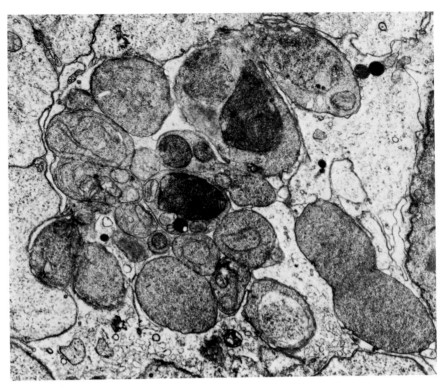

Fig. 17.   Cluster of apoptotic bodies in solid nodule of mouse sarcoma 180 treated with 1000 rads X rays 6 hours previously. A variety of organelles is included in the different bodies. ×8000.

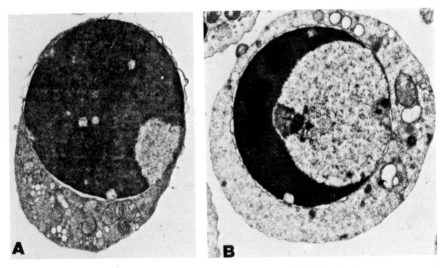

Fig. 18.   Apoptotic bodies formed from mouse mastocytoma cells during incubation with specifically allergized T lymphocytes for 1 hour. A, ×8800. B, ×7800. (From Don *et al.*, 1977.)

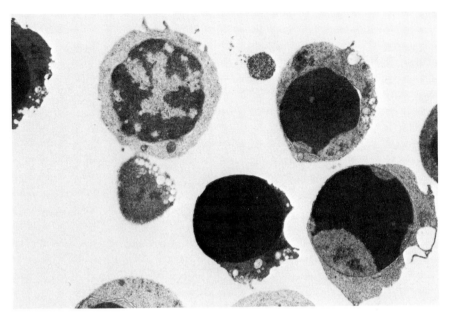

FIG. 19.   Apoptotic bodies formed from juvenile rat cortical thymocytes during incubation for 5 hours *in vitro* with 10^{-5} *M* methylprednisolone. One normal thymocyte is also present. ×6200.

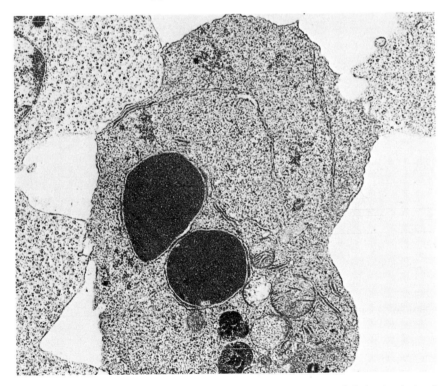

FIG. 20.   Apoptotic body formed from L5178 YC3 mouse lymphoma cell during incubation for 30 hours *in vitro* with 10^{-5} *M* methylprednisolone. ×13,500.

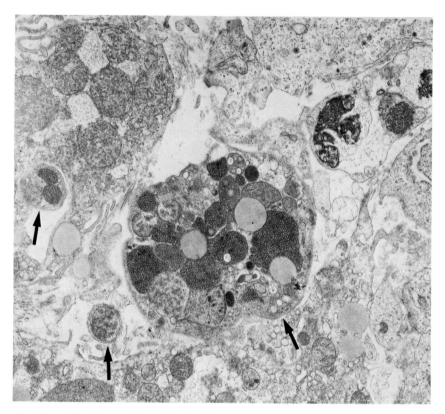

FIG. 21.   Adrenal cortex of normal 2-day-old rat. Apoptotic bodies of various sizes (arrows) lie in the intercellular space. Many of their closely packed mitochondria are condensed. ×8000.

composition of apoptotic bodies deriving from a cell may vary widely. Some contain one or more nuclear fragments in addition to condensed cytoplasm (Figs. 17–20), whereas others contain cytoplasmic elements alone (Figs. 17 and 21). The possession of nuclear components by a particular apoptotic body is not consistently related to its size.

The membrane-bounded apoptotic bodies are dispersed in the intercellular tissue spaces (Figs. 17 and 21), and are either extruded into an adjacent lumen or, more commonly, phagocytosed by resident tissue cells. Some are ingested by cells of the mononuclear phagocytic system (Fig. 22), but many of those formed in epithelia are taken up by the adjacent epithelial cells (Figs. 23–25). In our experience, the majority of apoptotic bodies formed deep within tumor nodules are ingested by neoplastic cells.

Neutrophil leukocytes are not involved in the phagocytosis of apoptotic bodies, nor is there other evidence of exudative inflammation such as accompanies necrosis.

Phagosomes containing apoptotic bodies acquire digestive enzymes by fusing with primary and secondary lysosomes (Fig. 26A). The phagocytosed bodies initially undergo a change that resembles necrosis of whole cells (Figs. 26B and 27); this is then followed by rapid degradation (Figs. 22–24), and soon all that remains of a deleted cell is a small amount of nondigestable material in the secondary lysosomes of the ingesting cells. There is evidence that the occurrence of extensive apoptosis in a tissue increases synthesis of lysosomal enzymes in the remaining viable cells (Ballard and Holt, 1968; Kerr and Searle, 1972b). It is emphasized that it is the lysosomal enzymes of the engulfing cells—not of the dying cells—that are involved in the digestion of apoptotic bodies. Further, there is nothing to suggest that lysosomes play any part in the early stages of apoptosis when the cell is already irreversibly committed to destruction. Indeed, electron microscopy shows that the lysosomes of recently formed apoptotic bodies are intact, like their other organelles, and this has been confirmed histochemically (Kerr, 1965; Searle et al., 1973).

Although apoptotic bodies within tissues usually undergo rapid phagocytosis,

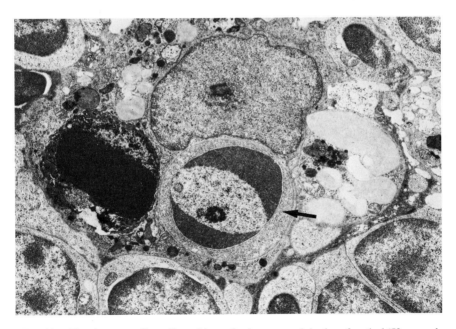

FIG. 22. Thymic cortex of juvenile rat 1 hour after intravenous injection of cortisol (50 mg per kg body weight). One apoptotic body within a macrophage is well preserved (arrow); a second is partly degraded. ×4600.

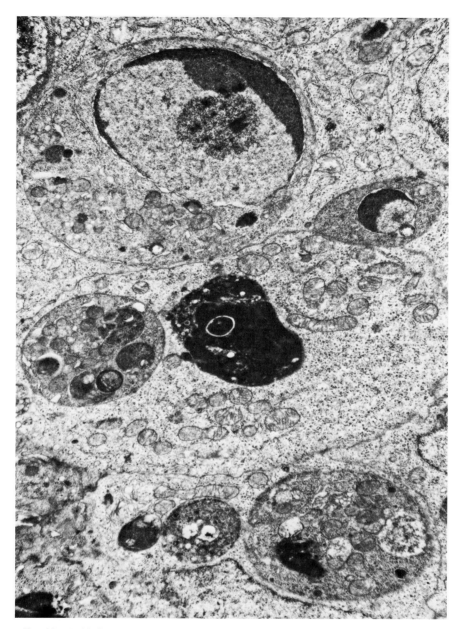

Fig. 23.   Epithelium of mouse small intestinal crypt 2 hours after intraperitoneal injection of cytosine arabinoside in dose of 250 mg per kg body weight. Apoptotic bodies have been phagocytosed by epithelial cells, and one shows advanced degradation. ×7700. (From Searle *et al.*, 1975.)

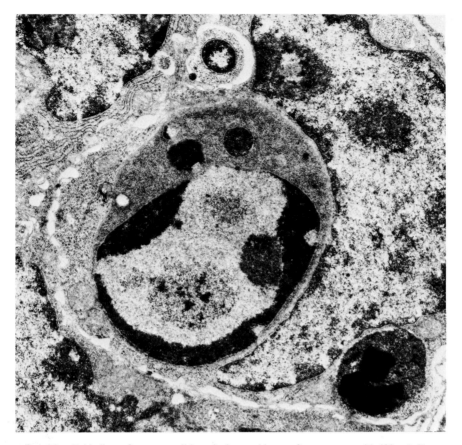

FIG. 24. Epithelium of mouse small intestinal crypt 4 hours after treatment with 400 rads X rays. One small apoptotic body lies in the intercellular space; a second has been phagocytosed but is still well preserved; a third shows advanced degradation. ×9000.

they may escape ingestion when dispersed in fluid, as occurs in suspension cultures (Fig. 28; Don *et al.*, 1977; Robertson *et al.*, 1978), in ascites tumors (Searle *et al.*, 1975), and after their extrusion into a gland lumen (Kerr and Searle, 1973). Such free-floating bodies eventually undergo spontaneous degeneration with swelling and membrane rupture (Fig. 28), a process morphologically identical with necrosis. This "secondary necrosis" of apoptotic bodies can be distinguished from cellular necrosis occurring *ab initio* by the small size of most of the degenerate cell fragments, their spherical or ovoid shape, and the presence of nuclear fragments of typical morphology in some of them.

Light microscopy of apoptosis shows that intensely basophilic, Feulgen-positive material corresponds in distribution with the condensed chromatin ob-

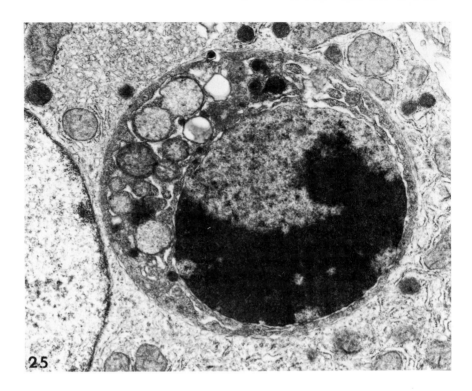

25

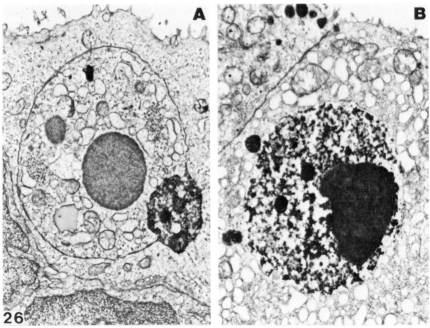

26

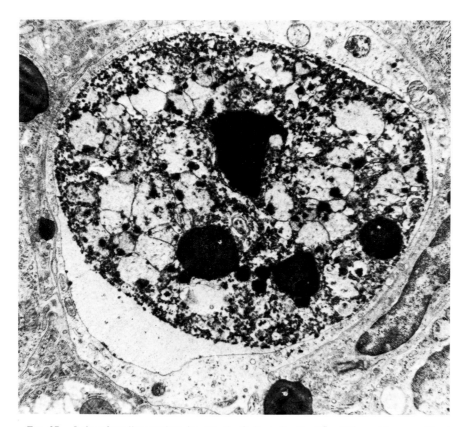

FIG. 27. Lobe of rat liver undergoing atrophy 3 days after ligation of its portal venous blood supply. An apoptotic body of parenchymal epithelial cell origin lies within a phagosome in a macrophage. Note nuclear fragments and flocculent densities in swollen mitochondria. ×5600. (From Kerr, 1971.)

FIG. 25. Rejecting pig liver allograft 1 week after transplantation. An apoptotic body containing a nuclear fragment lies with a phagosome in a parenchymal epithelial cell. ×12,000.

FIG. 26. Acinar epithelium of rat ventral prostate 2 days after castration. (A) Secondary lysosome fusing with phagosome containing apoptotic body with nuclear fragments. ×11,400. (B) Degeneration of phagocytosed apoptotic body containing a nuclear fragment. ×7600. (From Kerr and Searle, 1973.)

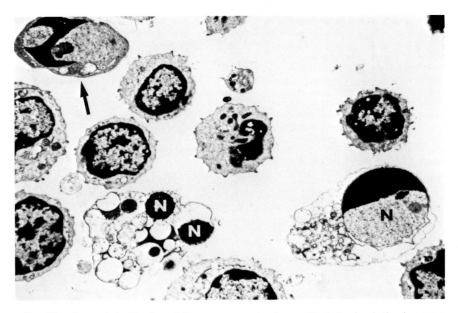

<figure>FIG. 28.   Apoptotic bodies formed from mouse mastocytoma cells during incubation in suspension culture with specifically allergized T lymphocytes for 2 hours. One (arrow) is well preserved; two others, which can be recognized by the presence of remnants of characteristic nuclear fragments (N), show "secondary necrosis." ×3700. (From Don *et al.*, 1977.)</figure>

served by electron microscopy. Thus, at an early stage, basophilic masses are found around the margins of nuclei in most cell types, but in certain cells such as thymocytes, where condensed chromatin occupies much of the nuclear volume (Fig. 6), the entire nucleus appears deeply basophilic (*pyknosis*). Later in the process, the nuclear fragments present in the apoptotic bodies appear as discrete basophilic masses; when numerous, they give rise to the histological appearance of *karyorrhexis*. While large apoptotic bodies can be readily identified in tissues as spherical or roughly ovoid acidophilic globules, irrespective of the presence within them of a nuclear component (Figs. 2B and 2C), the smaller ones are difficult to detect unless they contain a basophilic chromatin mass (Fig. 2D). The bodies occur both singly and in small clusters and are sometimes surrounded by clear halos. Their exact spatial relationship to adjoining intact tissue cells is often impossible to define.

The need for ultrastructural confirmation of light microscopic identification of dying cells will be evident. Despite the difference between the processes of apoptosis and necrosis, the terms pyknosis and karyorrhexis can be legitimately used to describe the appearances in the light microscope of certain phases in the evolution of both.

2. Modifications in Particular Cell Types

Although apoptosis involves a stereotyped sequence [the cardinal morphological events being essentially the same in insects and vertebrates (Kerr *et al.*, 1972; Giorgi and Deri, 1976)], minor modifications of the process occur in certain cells, perhaps because of their particular structural features. For example, fragmentation of skeletal muscle fibers in tadpoles during metamorphosis (Kerr *et al.*, 1974) is achieved by dilatation and confluence of the sarcoplasmic reticulum, a process possibly akin to the lucent vacuole formation in other cell types; strutting of the fibers by closely packed myofilaments probably precludes fragmentation in the usual way. Likewise, the abundance of tonofilaments in human keratinocytes seems to impart a rigidity to much of their cytoplasm, and restricts surface convolution; the protuberances that do form are small and virtually devoid of tonofilaments (Fig. 16: Weedon *et al.*, 1979). An epidermal cell thus often gives rise to one or two unusually large apoptotic bodies, which have been assigned special names such as Civatte bodies in lichen planus and sunburn cells developing after exposure to ultraviolet light (see Weedon *et al.*, 1979). Relatively restricted fragmentation of both nucleus and cytoplasm is a feature of apoptosis of cortical thymocytes (Figs. 6 and 19; Wyllie, 1980a).

3. Time-scale and Kinetics

Phase contrast microscopy of cultured cells (Hurwitz and Tolmach, 1969; Stanek, 1970; Russell *et al.*, 1972; Mullinger and Johnson, 1976; Sanderson, 1976; Matter, 1979) shows that apoptosis is of sudden onset and that its early stages are swiftly effected. Previously substrate-attached cells round up. This is rapidly followed by cellular shrinkage and violent pulsation and blebbing of the membrane, which have been likened to "boiling of the cytoplasm" (Sanderson, 1976). Many of the surface protrusions then assume a spherical configuration, detach, and float away in the medium as discrete globules. This whole sequence is often completed within a few minutes. Over the next hour or two, the residual cell mass swells and takes on the appearance of an empty shell (Sanderson, 1976), a change that probably corresponds with the "secondary necrosis" described previously.

Analogous studies have not been conducted within tissues, but an attempt has been made to measure the average period for which an apoptotic body formed in a tissue remains visible with the light microscope (Wyllie, 1975). A wave of apoptosis was produced in the rat adrenal cortex by ACTH withdrawal (effected by injection of prednisolone), and then further initiation of apoptosis was inhibited by injection of ACTH. The "life-span" of identifiable bodies was found to lie between 12 and 18 hours, a figure that corresponds with the time taken for phagocytes to digest material of biological origin in other systems (Perkins and Makinodan, 1965; Gordon and Cohn, 1973).

Ultrastructural studies of several tissues harvested consecutively after initiation of a wave of apoptosis (Kerr, 1971; Kerr and Searle, 1973; Searle *et al.*, 1975) are in accord with the time-scale indicated by the preceding observations. Thus, condensing cells with convoluted outlines are only rarely encountered, as expected of an extremely transient event; apoptotic bodies within phagosomes invariably far outnumber those still lying free in the intercellular spaces, indicating that apoptotic bodies, once formed, are rapidly ingested; lastly, a significant proportion of phagocytosed bodies show evidence of advanced lysosomal digestion within several hours of initiation of the wave.

A corollary of the rapid disposal of apoptotic bodies is that the findings of moderate numbers on light microscopy of a tissue indicates that extensive cell death is occurring; this is well illustrated by the tail of the tadpole in metamorphosis (Fig. 2B), which, from comprising a large part of the animal, shrinks and disappears within several days, apparently due to apoptosis (Kerr *et al.*, 1974). As cells are deleted, the adjoining cells successively close ranks, and very rapid involution of a tissue may thus occur in an ordered fashion.

Finally, there is no simple answer to the question as to how quickly apoptosis may be initiated by a suitable stimulus. Different stimuli and different cell systems appear to have diverse response times. Moreover, the response time to a single stimulus appears to vary among cells in the same population, so that apoptosis is never completely synchronous. For example, a single small dose of ionizing radiation produces an extensive wave of apoptosis in mouse gut crypts that peaks at 3 to 6 hours (Potten, 1977; Potten *et al.*, 1978); the time between attachment of a specifically allergized T lymphocyte to its target cell and the onset of apoptosis of the target cell may vary from a few minutes to several hours (Sanderson, 1976); following continuous exposure of rat thymocytes *in vitro* to 10^{-7} M methylprednisolone, no morphological changes are evident for 1 hour, and thereafter the proportion of cells undergoing apoptosis rises rapidly to over 90% at 12 hours (Wyllie, 1980a); atrophy of part of the rat liver resulting from interruption of its portal blood supply is associated with a marked increase in apoptosis that peaks at 3 days and is sustained for over a week (Kerr, 1971).

4. *Problems of Recognition and Nomenclature*

Several factors appear to have led to underrecognition and frequent misidentification of apoptosis in the past, including the nature of its morphology, confusion with other processes, and a plethora of redundant terminology.

a. *The Nature of the Process.* The nature of the process itself promotes underrecognition. Thus the involvement of only scattered single cells, the evanescence of the early stages, the small size of most apoptotic bodies, and the rapid dispersal, phagocytosis, and digestion of bodies in tissues without any inflammatory response all militate against detection and even more so against quantification. Further, apoptotic bodies formed in monolayer cultures tend to detach from the substrate and should be sought in the supernatant as well as among the surviving cells.

b. *Confusion with Other Processes.* Histologically, apoptosis has frequently not been distinguished from necrosis. Nuclear condensation (pyknosis) and fragmentation (karyorrhexis) have long been correctly accepted as histological criteria of cell death, but the fact that these terms can be applied to the nuclear changes found at certain stages in both necrosis (see Section III,A) and apoptosis (see Section III, B,1)has tended to conceal the incidence of the latter. While the two processes can usually be distinguished with the light microscope, the differences are unequivocal in the electron microscope.

Ultrastructurally, apoptotic bodies have often been misidentified as au-

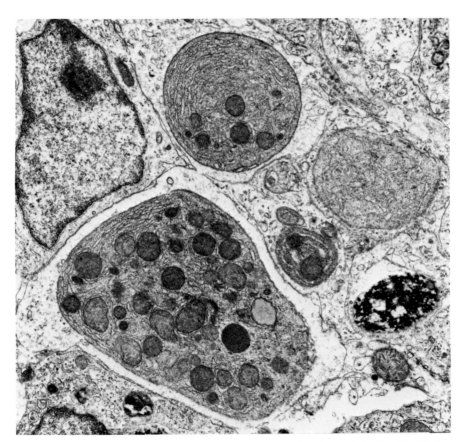

Fig. 29.   Rejecting pig liver allograft 1 week after transplantation. Apoptotic bodies, identified as being of parenchymal epithelial cell origin by the structure of their ergastoplasm, have been phagocytosed by infiltrating mononuclear cells. ×7500.

tophagic vacuoles (for example, see Gona, 1969; Helminen and Ericsson, 1971; Zeligs *et al.*, 1975; Morgan, 1976; Sadler and Kochhar, 1976). Certain identification of phagocytosed apoptotic bodies rests on the presence either of a recognizable nuclear fragment (Figs. 22–27, 30) or of organelles alien to the ingesting cell type (Figs. 29 and 30). However, suspicion that the observed process is apoptosis may be aroused by the sheer size of the inclusion.

Lastly, the presence of what we consider to be phagocytosed apoptotic bodies in viable cells has been claimed by some to represent an early stage in a process leading to the death and dissolution of these cells; the cytolysosomes, interpreted as autophagic vacuoles, are said to increase in number and then to rupture with release of digestive enzymes and consequent disintegration of the cells in which

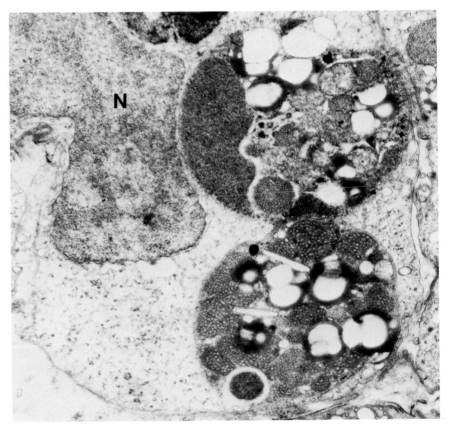

FIG. 30.   Adrenal cortex of fetal rat killed on the twentieth day of gestation after ACTH suppression *in utero* by maternal injection with prednisolone. Within phagosomes of a histiocyte (nucleus marked N) there lie apoptotic bodies identified as of cortical epithelial origin by their characteristic mictochondria and lipid vacuoles. ×14,000.

they lie (Schweichel and Merker, 1973; Sadler and Kochhar, 1976). In a similar category is the suggestion of Menkes *et al.* (1970), based on light microscopy, that Feulgen-positive basophilic "necrospherules" in the cytoplasm of cells might represent an early sign of their degeneration.

    c. *Nomenclature.* Sanderson (1976) and Sanderson and Glauert (1977) used the term *zeiosis* for death induced by cell-mediated immune attack; the morphology is identical with apoptosis. The word "zeiosis" was originally coined by Costero and Pomerat (1951) for blebbing of the surface of cultured neurones and has the disadvantage that it has often been applied to reversible phenomena (Cooper *et al.*, 1975; Godman *et al.*, 1975); its use for cell death clearly causes ambiguity. Russell *et al.* (1972) suggested the picturesque term *popcorn-type cytolysis* for cell death produced by sensitized T lymphocytes, but did not relate their findings to existing descriptions of the same process occurring in other contexts. Johnson *et al.* (1978) now consider that the phenomenon they described under the name *extrusion subdivision* with the production of *minisegregants* (Johnson *et al.*, 1975; Mullinger and Johnson, 1976) is the same as apoptosis, a conclusion with which we concur. *Necrobiosis* has sometimes been used as a general term for physiological cell death, including processes such as the destruction of senescent erythrocytes and flaking off of keratinized squames (Walter and Israel, 1974), and Yasuzumi *et al.* (1960) and Weinberger and Banfield (1965) have figured ultrastructurally typical apoptosis occurring spontaneously in malignant tumors as necrobiosis. However, this term is also used in a vague way to designate incipient necrosis (Müller, 1955) and "degeneration" of collagen in skin diseases such as necrobiosis lipoidica diabeticorum (Allen, 1977), and its meaning is so ambiguous that it is best avoided. Lastly, apoptosis in the normal embryo has sometimes been referred to as *necrosis* (Saunders, 1966; Hammer and Mottet, 1971; Schweichel and Merker, 1973; Hurle *et al.*, 1977), a practice that is clearly highly undesirable.

    Apoptotic bodies have been given various names by light microscopists, often in the belief that they were describing a morphological change unique to the particular circumstances under consideration. Such redundant names include *acidophilic* and *Councilman bodies* in the liver (see Kerr, 1971; Kerr *et al.*, 1972; Edmondson and Peters, 1977), *tingible bodies* in lymphoid germinal centers (see Swartzendruber and Congdon, 1963; Kerr *et al.*, 1972; Searle *et al.*, 1975), *karyolytic bodies* in gut crypts after X-irradiation (see Kerr and Searle, 1980), *sarcolytes* (Weber, 1969) derived from striated muscle fibers in the regressing tadpole tail (see Kerr *et al.*, 1974), and *Civatte bodies, sunburn cells,* and *satellite cell dyskeratosis* in the skin (see Weedon *et al.*, 1979).

    Biological realities may be obscured or illuminated by terminology. *Apoptosis* has no ambiguous historical connotations and was proposed by us expressly to emphasize the widely ranging occurrence and the functional implications of a distinctive morphological entity.

Our review of the morphology of necrosis and apoptosis indicates that they are discrete and distinct phenomena. The salient differences are summarized in Table I.

TABLE I

COMPARISON OF MORPHOLOGICAL FEATURES OF NECROSIS AND APOPTOSIS

Feature	Necrosis	Apoptosis
Histological appearances	Usually affects tracts of contiguous cells	Characteristically affects scattered individual cells
	Eosinophilic "ghosting" of entire cells	Affected cell represented by one or more roughly spherical cytoplasmic masses, some also containing basophilic particles of condensed chromatin. They lie both in the intercellular space and within tissue cells
	Exudative inflammation usually present	Exudative inflammation absent
Ultrastructural changes		
Chromatin	Marginates in small, loosely textured aggregates; disappears eventually, when nuclear membrane destroyed	Marginates in condensed, coarsely granular aggregates; confluent over entire nucleus or localized to large crescentic caps
Nucleolus	Evident as compact body until cytoplasmic degradation advanced	Disperses to shower of granules while cytoplasm structurally intact
Nuclear membrane	Retains pore structures until cytoplasmic degradation advanced; eventually destroyed with other organelles	Progressive convolution; resulting protuberances eventually separate. Pores retained adjacent to euchromatin, but lost next condensed chromatin. Eventually becomes discontinuous, so that dense chromatin masses lie among cytoplasmic organelles
Cytoplasm	Swelling of all compartments followed by rupture of membranes and destruction of organelles. Mitochondrial matrix densities characteristic	Condensation of cytosol (endoplasmic reticulum may dilate focally); structurally intact mitochondria and other organelles compacted together; protuberances from cell surface separate to form apoptotic bodies

IV. Incidence

The purpose of surveying the incidence of necrosis and apoptosis is 3-fold. First, it will become apparent that the two types of cell death differ markedly in the contexts in which they develop, supporting the conclusion based on morphology that they are fundamentally different processes. Second, the circumstances associated with their occurrence permit inferences to be drawn about the intracellular mechanisms that underlie them. Third, the relative significance of necrosis and apoptosis can be judged; apoptosis clearly emerges as the major mode of cell death in living tissues.

A. Necrosis

The occurrence of necrosis is invariably associated with a gross departure from physiological conditions such as severe hypoxia and ischemia (Kloner *et al.*, 1974; Ganote *et al.*, 1975; Jennings *et al.*, 1975), major changes in environmental temperature (Buckley, 1972), disruption of cell membranes by complement (Hawkins *et al.*, 1972), and exposure to toxins (McLean *et al.*, 1965; McDowell, 1972; Evan and Dail, 1974). Significantly, it is also the mode of cell death in autolysis *in vitro* (Trump *et al.*, 1965).

By contrast, there is no evidence implicating necrosis in cell population kinetics or in normal embryonic development and metamorphosis. The focal necrosis sometimes seen in the centers of tumor nodules (Figs. 1 and 2A) is almost certainly the result of hypoxia and ischemia (Thomlinson and Gray, 1955; Schatten, 1962; Schatten and Burson, 1965).

Thus, the occurrence of necrosis is determined, not by factors intrinsic to the cell itself, but by environmental perturbation, which must be violent.

B. Apoptosis

Apoptosis is implicated in the steady-state kinetics of healthy adult tissues and accounts for focal deletion of cells during normal embryonic development and metamorphosis. It is enhanced in several types of atrophy, including that produced in endocrine-dependent tissues by changes in the blood concentrations of trophic hormones. It occurs spontaneously in growing neoplasms, and it is increased in both neoplasms and rapidly proliferating normal cell populations by certain radiomimetic cancer-chemotherapeutic agents and ionizing radiation. It is induced in target cells by attachment of specifically allergized T lymphocytes. Its relationship to aging needs further study. These varied circumstances of occurrence will now be dealt with in more detail.

1. Cell Turnover in Healthy Tissues

Cell death with the morphology of apoptosis has been described in a number of healthy mature mammalian tissues, including lymphoid germinal centers (Swartzendruber and Congdon, 1963), thymic cortex (Lundin and Schelin, 1965; van Haelst, 1967a), liver (Kerr, 1965, 1971), adrenal cortex (Wyllie *et al.*, 1973a), prostate (Kerr and Searle, 1973), and small gut crypts (Searle *et al.*, 1975; Potten, 1977). However, the kinetic significance of the cell death was not assessed in any of these studies.

In rat adrenal cortex, an attempt has been made to quantify cell loss by apoptosis (Wyllie, 1975). Despite inaccuracies inherent in the morphometric methods used, the results indicated that apoptosis is quantitatively capable of balancing most or all of the cell gain by mitosis. Further, it was found that acute ACTH withdrawal resulted in a 9-fold increase in the number of cells undergoing apoptosis without any immediate change in the number of mitotic figures (Table II). ACTH withdrawal is known to occur and to enhance adrenal apoptosis under physiological conditions in the neonatal period (Figs. 21 and 31; Wyllie *et al.*, 1973b). The inference may be drawn that apoptosis is a significant factor in normal tissue homeostatic regulation. In the rat adrenal cortex at least, its frequency is of the right order to balance that of mitosis, and it responds swiftly to alterations in trophic hormone levels.

2. Embryonic Development

Published electron micrographs of the cell death that occurs during normal embryonic development show typical apoptosis. It has been illustrated in the preimplantation mouse blastocyst (El-Shershaby and Hinchliffe, 1974) and in very early avian embryos (Bellairs, 1961). It is prominent during the early formation of the neural tube in the mouse (Schlüter, 1973), and later, during the development of the chick spinal cord when cervical motor neurones that are part of an abortive visceral system are eliminated (O'Connor and Wyttenbach, 1974).

TABLE II

MITOSIS AND APOPTOSIS IN THE RAT ADRENAL CORTEX

	Normal	Prednisolone-treated
Mitotic figures per section (mean ± 1 SE)	1.7 ± 0.4	1.9 ± 0.6
Apoptotic bodies per section (mean ± 1 SE)	5.7 ± 0.9	55.7 ± 12.2
Mitotic rate per apoptotic rate in whole gland[a]	≤ 1.6	≤ 0.2

[a] Mitotic and apoptotoc rates indicate number of cells entering each process per unit of time. The ratio was derived from the observed numbers of mitotic and apoptotic cells (given above) and was corrected to allow for their different size distribution and observable life-spans and for the geometry of the whole gland. (After Wyllie, 1975.)

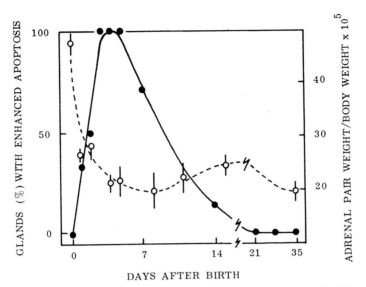

FIG. 31.   Apoptosis in the normal neonatal adrenal cortex. "Enhanced apoptosis," derived from semiquantitative assessment on a 3-point scale (Wyllie *et al.*, 1973b), is plotted against time after birth. On the same graph is shown adrenal weight relative to total body weight. Enhancement of apoptosis and decline in adrenal weight are both attributable to the postnatal physiological drop in ACTH secretion. (After Wyllie *et al.*, 1973b.)

It is involved in the morphogenesis of the heart (Manasek, 1969; Pexieder, 1972; Ojeda and Hurle, 1975; Hurle *et al.*, 1977) and in the deletion of redundant epithelial cells that follows fusion of the palatine processes in mammals (Farbman, 1968; Smiley and Dixon, 1968; Matthiessen and Anderson, 1972). It is extensive in the duodenal mucosa of the rat fetus during formation of the villi, when stratified epithelium is transformed into a single layer of columnar cells (Behnke, 1963). Focally massive apoptosis is responsible for the elimination of interdigital webs in reptiles (Fallon and Cameron, 1977), birds (Webster and Gross, 1970; Hammar and Mottet, 1971; Mottet and Hammar, 1972), and mammals (Ballard and Holt, 1968) and is also seen in the so-called "posterior necrotic zone" of the chick wingbud (Fig. 14; see also Saunders, 1966; Saunders and Fallon, 1966; Hurle and Hinchliffe, 1978). A genetically determined wingless malformation in the chick results from abnormal and excessive deletion of wingbud cells by ultrastructurally typical apoptosis (Hinchliffe and Ede, 1973).

In all these circumstances, apoptosis occurs at predictable sites and times in development. Further, it has been demonstrated (Saunders, 1966; Saunders and Fallon, 1966; Fallon and Saunders, 1968) that, after explantation of small fragments of the appropriate tissue, apoptosis occurs "on schedule" *in vitro;* the "death sentence" may, however, sometimes be commuted by culturing the explanted cells in association with tissue from a different part of the embryo, even when separated from it by a millipore filter. The inference is that the focal apoptosis involved in embryonic morphogenesis is under precise local and perhaps intracellular control.

3. Metamorphosis

Cells with the ultrastructural features of apoptosis have been described in metamorphosis of insects (Goldsmith, 1966) and amphibia (Kerr *et al.*, 1974; Decker, 1976). However, the morphological changes accompanying cell death during metamorphosis have also been subject to various other interpretations (Weber, 1964; Gona, 1969; Michaels *et al.*, 1971; Hourdry, 1972; Schiaffino *et al.*, 1974; Fox, 1975). We believe that the electron micrographs illustrating the papers just cited are all consistent with apoptosis; indeed, we feel that they exemplify the tendency to miss the morphologically unique early stages and to misinterpret the later stages, to which we drew attention in the section dealing with morphology. Hence the reports of unequivocal apoptosis in metamorphosis, though few in number, are highly significant; we can find nothing in the literature to negate the conclusion that apoptosis is the usual mode of cell death involved.

Death of cells in metamorphosis is known to be controlled by specific hormones. It often affects several different types of cells simultaneously, yet similar cell types outside the regressing region are not affected and may show an anabolic response to the hormone (Saunders, 1966). Facile explanations for the regional localization of the cell death (for example, altered blood supply) can be excluded by experiments in which tadpole tail regression stimulated by thyroid hormone has been faithfully reproduced *in vitro* (Tata, 1966; Frieden, 1967). By inference, the basis for the selectivity of the apoptosis occurring in metamorphosis resides within the affected cells themselves.

4. Endocrine-Dependent Atrophy and Hyperplasia

The atrophy of endocrine-dependent tissues that follows lowering of the blood level of trophic hormones has long been known to involve a diminution in the

---

FIG. 32. Hyperplastic adrenal cortical cells do not show the usual enhancement of apoptosis on ACTH withdrawal. The graphs show: (A) progressive changes in adrenocortical apoptosis induced by a single injection of methylprednisolone to suppress ACTH, (B) adrenal gland weight, and (C) plasma corticosterone, with (▼) and without (●) ACTH suppression by prednisolone, during regression of hyperplasia. The hyperplasia was induced by daily ACTH injections over the 7 days immediately preceding. Apoptosis, which is normally enhanced after suppression of endogenous ACTH, remains at low levels until the fifteenth day of regeneration (*p*) values indicate significance of difference from nonhyperplastic glands). Adrenal gland weight shows a similar slow return to normal. By contrast, basal plasma corticosterone is normal and shows normal reduction after ACTH suppression by day 4. Early corticosterone values are low, probably because of hypothalamic suppression by the injected ACTH used to produce adrenal hyperplasia. All points are means (± 1 SE) from at least 5 animals. The control values (1 SE on either side of the mean) obtained from nonhyperplastic glands are indicated by the hatched bars. (After Wyllie, 1975.)

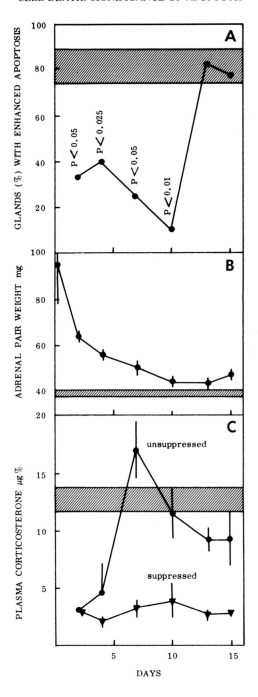

size of their constituent cells (Ingle *et al.*, 1938; Helminen and Ericsson, 1971). However, extensive cell deletion by apoptosis also occurs. The tissues studied include rat adrenal cortex after ACTH withdrawal (Fig. 2D; Wyllie *et al.*, 1973a), rat ventral prostate after orchidectomy (Figs. 2C, 4, and 26; Kerr and Searle, 1973), Huggins rat mammary tumor after oophorectomy (Kerr *et al.*, 1972), endometrial epithelium after estrogen withdrawal in mice (Martin *et al.*, 1973), hamsters (Sandow *et al.*, 1979), and man (Hopwood and Levison, 1976), sheep theca interna during follicular atresia (O'Shea *et al.*, 1978), and vascular endothelium within regressing ovine corpora lutea (O'Shea *et al.*, 1977).

The preceding examples implicate *decrease* in trophic hormone stimulation in the initiation of apoptosis. Apoptosis may also be enhanced by the appropriate hormone stimulation, as in progesterone-stimulated hamster endometrium (West *et al.*, 1978; Sandow *et al.*, 1979). Similarly, glucocorticoid hormones cause atrophy of lymphoid tissues of certain species (Claman, 1972), much of it due to a direct lethal effect of the hormone on the lymphoid cells, which can be reproduced fin cell suspensions *in vitro* (Munck, 1971). Light and electron microscopic studies of death of lymphocytes and cortical thymocytes after glucocorticoid treatment have consistently shown it to have the morphology of apoptosis (Figs. 6, 19, and 22; see also Dougherty and White, 1945; Makman *et al.*, 1967; van Haelst, 1967b; Whitfield *et al.*, 1968; La Pushin and de Harven, 1971; Waddell *et al.*, 1979; Wyllie, 1980a), and similar changes are observed in lymphoid cell lines (Figs. 5, 13, and 20; Robertson *et al.*, 1978).

In contrast, there is some evidence that endocrine-induced hyperplasia is associated with a loss of the ability of trophic hormone withdrawal to enhance apoptosis (Wyllie, 1975). Apoptosis could not be induced by this means in adrenal glands rendered grossly hyperplastic by previous ACTH administration until the cortical mass and morphology had almost reverted to normal (Fig. 32).

These observations on endocrine-dependent tissues support the inference drawn earlier that apoptosis is a regulated phenomenon of significance in tissue homeostasis.

5. *Other Types of Atrophy*

Tissue mass may decrease as a result of stimuli far less well understood than the trophic hormones previously discussed. Mild ischemia produces atrophy that is associated with reduction in cell numbers but with conservation of basic tissue organization: it is of interest that apoptosis is the type of cell death involved. Chronic mild ischemia was induced in the rat liver by interrupting the portal venous supply to the left and median lobes; rapid atrophy of these lobes ensued, with concomitant hyperplasia of the rest of the organ. By the eighth day, the mass of the ischemic lobes had decreased to one-sixth normal size and thereafter remained constant; the lobules were very small, but their basic histological architecture was preserved. The evolution of the atrophy was associated with massive deletion of hepatocytes by apoptosis (Fig. 27; Kerr, 1971); hepatocyte necrosis was absent, provided the arterial blood flow was not compromised.

Regrettably, we do not know what role apoptosis plays in the tissue atrophy of aging. Even for cell senescence *in vitro,* there are remarkably few detailed morphological studies (see Brock and Hay, 1971).

6. *Cell Death Caused by Radiation, Cytotoxic Cancer-Chemotherapeutic Agents, and Hyperthermia*

This section is concerned primarily with the effects of certain agents commonly used in the treatment of neoplasms.

Ionizing radiation and cytotoxic cancer-chemotherapeutic agents have long been known to cause death of cells as well as to inhibit mitosis in rapidly proliferating cell populations. Review of the literature concords with our own studies in showing that, in doses similar to those employed therapeutically, these agents cause death by apoptosis.

Thus cytotoxic drugs have been shown to produce massive cell death with the morphology of apoptosis in the embryo (Webster and Gross, 1970; Schweichel and Merker, 1973; Sadler and Kochhar, 1976; Bannigan and Langman, 1979); in rapidly proliferating adult cell populations such as gut crypt epithelium (Fig. 23; see also Trier, 1962; Millington, 1965; Lieberman *et al.,* 1970; Verbin *et al.,* 1972, 1973; Philips and Sternberg, 1975; Searle *et al.,* 1975), lymphoid germinal centers (Searle *et al.,* 1975), and spermatogonia (Harrison, 1975); and in several malignant neoplasms (Figs. 11 and 12, see also Heine *et al.,* 1966; Schwartz *et al.,* 1966; Uzman *et al.,* 1966; Krishan and Frei, 1975; Searle *et al.,* 1975; Kerr and Searle, 1980). Likewise, moderate doses of various types of radiation have been found to cause increased apoptosis, not necrosis, in these and similar cell populations (Figs. 15, 17, and 24; see also Montagna and Wilson, 1955; Hugon *et al.,* 1965; Hugon and Borgers, 1966; Trowell, 1966; Jordan, 1967; Ghidoni and Campbell, 1969; Hurwitz and Tolmach, 1969; Lucas and Peakman, 1969; Pratt and Sodicoff, 1972; Strange and Murphree, 1972; Cheng and Leblond, 1974; Olson and Everett, 1975; Aleksandrova and Guljaev, 1976; Eisenbrandt and Phemister, 1977; Potten, 1977; Searle and Potten, 1977; Potten *et al.,* 1978; Weedon *et al.,* 1979; Kerr and Searle, 1980). It may be significant that radiomimetic cytotoxic drugs cause extensive death among cells at the base of gut crypts, in the basal layer of squamous epithelium, and in lymph follicles: all populations that normally proliferate continuously. By contrast, the same agents exert a much lesser lethal effect on normally quiescent populations such as liver, accessory sex gland, and salivary gland, even when these are stimulated to proliferate (Farber and Baserga, 1969; Searle *et al.,* 1975; Alison and Wright, 1979). Radiomimetic cytotoxic drugs and ionizing radiation both exert their major effects on nuclear DNA, and it may be speculated that tissues whose normal function involves continuous proliferation possess mechanisms to selectively eliminate progenitors with acquired genetic abnormalities (Searle *et al.*, 1975; Potten, 1977). The validity of this hypothesis is not necessarily negated by the fact that other chemicals, including certain teratogens and carcinogens, also enhance apoptosis in a variety of tissues (Kerr, 1969, 1972; Menkes *et al.*, 1970; Crawford *et al.*, 1972). The mode of action of many of these latter substances is not precisely known, but it is perhaps significant that some of them bind to DNA.

In spite of the recent surge of interest in the use of hyperthermia in tumor therapy, there have been few ultrastructural studies of the cell deletion that must occur when the treatment proves successful (Overgaard, 1976). However, following transient mild hyperthermia, ultrastructurally typical apoptosis has been observed in the brains of guinea-pig fetuses (Wanner *et al.*, 1976), the leg and eye-antennal discs of larvae of a heat-sensitive mutant of *Drosophila* (Clark and Russell, 1977), and the seminiferous tubules of rat testes (Harrison, 1975).

7. *Induction of Apoptosis by Cell-Mediated Immunity*

Death of cells in mammalian allografts is frequently the result of ischemia, which may, in turn, be related either to technical failure of the grafting procedure or to immunologically induced vascular lesions. However, in pig liver allografts during rejection, vascular lesions are mild, and the extent of hepatocyte death after technically successful transplantation correlates closely with the degree of lymphoid infiltration of the parenchyma (Searle *et al.*, 1977). The only type of cell death detectable in such rejecting liver grafts is apoptosis (Figs. 25 and 29; Battersby *et al.*, 1974; Searle *et al.*, 1977), and hepatocytes showing early ultrastructural changes of apoptosis in their nuclei are found to have lymphoid cells closely applied to their surfaces (Searle *et al.*, 1977).

Cell death with the morphology of apoptosis has been demonstrated in various tissues in graft-versus-host reactions (Woodruff *et al.*, 1969; Salvin and Woodruff, 1974; De Dobbeleer *et al.*, 1975; Grogan *et al.*, 1975; Gallucci *et al.*, 1979), and close proximity of infiltrating lymphocytes to the dying cells is a conspicuous feature of the lesions.

Such observations clearly suggest that cell-mediated immune attack causes apoptosis rather than necrosis, but conclusions drawn from *in vivo* studies cannot be more than tentative because of the complexity of immune reactions in intact animals. However, cell death with the morphology of apoptosis developing *in vitro* after attachment to specifically allergized T lymphocytes (Figs. 10, 18, 28, and 33) has now been reported by a number of investigators (see Russell *et al.*, 1972; Koren *et al.*, 1973; Firket and Degiovanni, 1975; Sanderson, 1976; Don *et al.*, 1977; Liepins *et al.*, 1977; Sanderson and Glauert, 1977; Matter, 1979), and similar changes have been recorded in cultured cells treated with lymphotoxin (Russell *et al.*, 1972; Rosenau and Tsoukas, 1976). Further, recent *in vitro* experiments (Kerr *et al.*, 1979) suggest that antibody-dependent, cell-mediated immune attack (K cell activity) also induces apoptosis of target cells. In contrast,

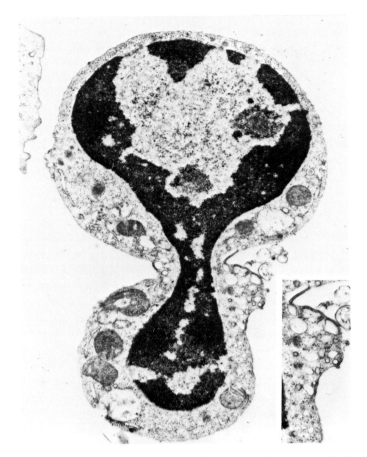

FIG. 33.   Apoptosis of mouse mastocytoma cell following incubation with specifically allergized
T lymphocytes for 1 hour. Note typical nuclear changes and subplasmalemmal densities in the
cleavage furrows. ×8800; inset ×12,500. (By courtesy of W. Clouston.)

antibody-dependent, complement-mediated cell death has the ultrastructural features of necrosis (Goldberg and Green, 1959; Hawkins *et al.*, 1972).

In accord with these findings are observations of increased apoptosis in human
diseases in which there is evidence for cell-mediated immune reactions. For
example, enhanced apoptosis occurs in association with lymphoid infiltration of
the epidermis in fixed drug eruptions and lichen planus (Fig. 16; see also
Hashimoto, 1976; Weedon *et al.*, 1979), and apoptosis, not necrosis, is seen in
the lesion known as "piecemeal necrosis" (Kerr *et al.*, 1979), which charac-
terizes chronic active hepatitis. Extensive apoptosis accompanied by a lymphoid
infiltrate has been recorded in spontaneously regressing plane warts (Weedon *et*

al., 1979) and in foci of spontaneous regression in cutaneous basal cell carcinomas (Curson and Weedon, 1979; Weedon *et al.*, 1979). Lastly, it is possible that apoptosis of hepatocytes in acute viral hepatitis, long recognized as a characteristic feature of the disease under the terms Councilman or acidophilic bodies (Klion and Schaffner, 1966; Edmondson and Peters, 1977), may be another example of immunological induction (Edgington and Chisari, 1975; Alberti *et al.*, 1976; Kerr *et al.*, 1979).

Selective cell deletion through cell-mediated immunity is of great biological importance, perhaps in surveillance (Burnet, 1970), certainly in preventing and containing many cytopathic and oncogenic virus infections (Allison, 1974; Blanden, 1974). Since apoptosis is involved, this cell killing may be due to activation of preexisting mechanisms within the target cell (see Don *et al.*, 1977; Matter, 1979).

8. *Spontaneous Occurrence in Neoplasms*

Pathologists have long been aware of the disparity between the slow growth of certain human tumors and their high mitotic rates evident in sections (Willis, 1953), a paradox for which an obvious histological explanation such as the development of ischemic necrosis is often lacking. It was the application of autoradiographic methods, however, that demonstrated the widely ranging occurrence of spontaneous cell loss in tumors and its frequently massive numerical dimensions (Iversen, 1967; Refsum and Berdal, 1967; Steel, 1967, 1968; Frindel *et al.*, 1968; Laird, 1969; Weinstein and Frost, 1970; Dethlefsen, 1971; Terz *et al.*, 1971; Lala, 1972; Tubiana and Malaise, 1976); in many human tumors, the rate of cell loss falls only fractionally behind the rate of cell gain (Table III). While exfoliation and migration do sometimes contribute to cell loss *in vivo*, they cannot account for the loss observed *in vitro*, and there is no doubt that continual *in situ* cell death is a major parameter in neoplastic growth. It is thus

TABLE III
CELL LOSS FROM HUMAN TUMORS[a]

	Cases	Cell loss factor[b] (%)
Undifferentiated bronchial carcinoma	23	97
Colorectal carcinoma	12	96
Malignant melanoma	25	73

[a] Reproduced from Steel (1977) with permission.

[b] The cell loss factor measures the disparity between the observed growth rate of the tumor and the rate expected from knowledge of cell proliferation kinetics alone. It has a value of 0% when cell loss is absent and 100% when loss and gain exactly balance.

significant that scattered cells displaying the morphological features of apoptosis have now been recorded in tumors of many different types (Figs. 7 and 8; see also Mendelsohn, 1960; Yasuzumi *et al.*, 1960; Weinberger and Banfield, 1965; Kerr and Searle, 1972a, b, 1980; Kerr *et al.*, 1972; Lala, 1972; Kerr, 1973; Searle *et al.*, 1973, 1975; Donald *et al.*, 1974; Huggins *et al.*, 1974; Wyllie, 1974; Cooper *et al.*, 1975; Bird *et al.*, 1976; Sanderson, 1976; Don *et al.*, 1977; Robertson *et al.*, 1978; Weedon *et al.*, 1979), whereas necrosis is often absent. Detailed comparisons between the rate of apoptosis and the "cell loss factor" calculated from kinetic data are not yet available for various individual tumors, but the extent of the apoptosis observed in histological sections of some tumors indicates that its kinetic significance may be considerable (Kerr and Searle, 1972a).

The causes of apoptosis in neoplasms is largely unknown. There is little doubt that some results from ischemia (Kerr *et al.*, 1972) and that some may be the result of immunological attack (Curson and Weedon, 1979). However, the tumor tissue may also retain intrinsic homeostatic regulatory mechanisms, albeit inadequate to arrest its progressive growth.

9. Conclusions

Although apoptosis is apparently not observed in cell deletion in planaria (Bowen and Ryder, 1974), it occurs very widely among phylogenetically higher animals. Its constant morphology suggests—but manifestly does not prove—that similar intracellular mechanisms may underlie cell death in these diverse circumstances. Where apoptosis occurs physiologically, it is clearly a controlled process involved in tissue homeostasis. Apoptosis also occurs following injury; a homeostatic role is here less apparent, but on teleological grounds it is possible that it represents selective deletion of cells whose survival would prejudice that of the organism as a whole. The causes of its occurrence in neoplasms are complex and incompletely understood, but of obvious importance.

V. Mechanisms

A. Necrosis

There is much evidence to support the view that the critical event leading to the development of necrosis is loss of cellular volume homeostasis. Specific evidence of altered membrane permeability has come from ultrastructural studies in which the influx of colloidal electron-dense material was correlated with morphology (Hoffstein *et al.*, 1975); from measurement of potassium loss, sodium entry, and fall in membrane potential (Prieto *et al.*, 1967; Saladino and Trump, 1968; Laiho and Trump, 1974; Whalen *et al.*, 1974; Jennings *et al.*,

1975); and from freeze–fracture studies demonstrating transmembrane defects (Ashraf and Halverson, 1977). Gross degrees of increased permeability of the plasmalemma are reflected in failure to exclude vital dyes such as nigrosine, trypan blue, and eosin Y, and in accelerated loss of previously cell-bound radioactive chromium; this forms the basis of widely used methods of measuring cell "death."

Although increase in membrane permeability appears to be the final critical event, different types of injury may produce it (Fig. 34). Thus, complement, (Hawkins *et al.*, 1972) or ouabain (Ginn *et al.*, 1968) affect membrane function

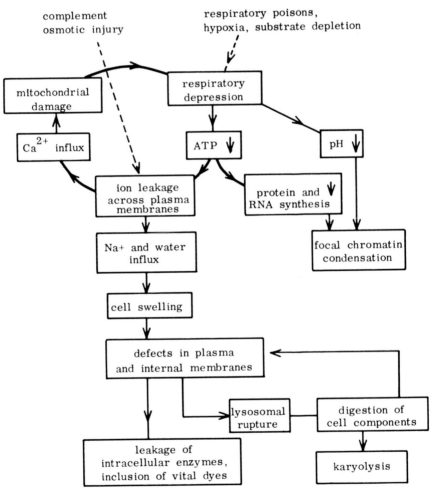

FIG. 34.   Scheme of mechanisms in necrosis. (After Trump and Mergner, 1974.)

directly, whereas certain toxins interfere first with the energy supply on which the selective control of fluxes across the plasma membrane depends (Trump and Mergner, 1974). More complex modifications of this model suggest that initial, potentially reversible alterations in membrane permeability permit increases in the intracellular concentrations of certain constituents such as calcium ion; local high concentrations of this ion within mitochondria may then cause irreversible damage to the respiratory apparatus (Whalen *et al.*, 1974; El-Mofty *et al.*, 1975; Farber and El-Mofty, 1975; Laiho and Trump, 1975; Chien *et al.*, 1977).

Loss of volume homeostasis, irreversible damage to mitochondria, altered ionic concentrations within the cell and falling energy supplies, are of course associated with collapse of all other vital functions, including macromolecular synthesis (Ginn *et al.*, 1968). Thus, most workers would agree that cells that show no loss of ATP or intracellular potassium ion, that evince no other evidence of increased membrane permeability, and in which macromolecular synthesis is conserved are unlikely to be undergoing necrosis. However, some aspects of necrosis are not satisfactorily explained by this scheme, such as the dependence on oxygen supply of some of the manifestations of osmotic failure (Kloner *et al.*, 1974; Ganote *et al.*, 1975, 1977).

B. APOPTOSIS

From the discussion of the incidence of apoptosis, it is clear that the initiating stimuli may vary widely, and it is obvious that the mechanisms involved in reception of these stimuli are also heterogeneous (see e.g., Gehring and Coffino, 1979). While it is possible that the effector mechanisms are likewise numerous, the stereotyped morphology of apoptosis suggests that a search for underlying intracellular events common to many different circumstances might be fruitful. This search is at an early stage and necessarily is restricted to a small number of experimental systems—those in which sufficiently large numbers of apoptotic cells can be generated to render biochemical studies feasible. With these provisos in mind, we attempt to answer three questions relating to the effector mechanisms of apoptosis: (1) Do apoptosis and necrosis differ in their mechanisms? (2) What molecular events mediate the cardinal morphological changes in apoptosis—chromatin condensation, cytoplasmic contraction, and early phagocytosis by resident tissue cells? (3) What are the ultimate intracellular controls of the effector mechanisms of apoptosis?

1. *Apoptosis Differs from Necrosis in Mechanism*

Unlike cells undergoing necrosis, apoptotic cells do not show evidence of increased membrane permeability, at least until after the characteristic morphological changes have appeared. Thus, there is no net sodium flux into the intracellular space in mammary tumors during their hormone-induced regression *in vivo* (Gullino and Lanzerotti, 1972); and in lymphotoxin-treated target cells, membrane ion fluxes remain normal *in vitro* (Walker and Lucas, 1972; Rosenau *et al.,* 1973). As noted earlier, the dying cells in all these circumstances show the morphology of apoptosis. In thymocytes treated with glucocorticoid *in vitro,* potassium ion concentration is within normal limits at a time when chormatin condensation is well developed (G. Boyd and A. H. Wyllie, unpublished), and vital dye exclusion remains normal for many hours later (Waddell *et al.,* 1979). This disparity between apoptosis and "death" scored on the basis of membrane permeability may contribute to the observed insensitivity of dye exclusion and chromium release in the assessment of lethal effects of cytotoxic cancer-chemotherapeutic drugs (Bhuyan *et al.,* 1976; Roper and Drewinko, 1976) or cell-mediated immune attack (Don *et al.,* 1977).

Content of energy-rich nucleotides also emphasizes differences between apoptotic and necrotic cells. Unlike necrosis, depletion of intracellular ATP in apoptosis does not regularly precede the onset of irreversible morphological changes; apoptosis can occur in cells with normal ATP content (Waddell *et al.,* 1980). Further, apoptosis occurring physiologically in the embryo may have an absolute requirement for energy-rich nucleotide: death in palatal shelf epithelium can be accelerated by artificially raising intracellular cyclic 3':5'-adenosine monophosphate (cAMP) (Pratt and Martin, 1975). In this system, death (which has the morphology of apoptosis) can be inhibited by epidermal growth factor, and this inhibition is thought to be mediated by depletion of intracellular cAMP (Hassell and Pratt, 1977). Similar dependence on cAMP is recorded for regression of hormone-sensitive mammary tumors (Cho-Chung and Clair, 1977; Cho-Chung, 1978), and cAMP itself has a lethal effect at low concentration on some lymphoid cells (Coffino *et al.,* 1975). It is extremely unlikely that cAMP would be available in cells sufficiently depleted in ATP (its major precursor) to cause failure of membrane ion pumps.

2. *Molecular Events Associated with Morphological Change*

a. *Chromatin Condensation.*   Changes in chromatin structure have been studied in rodent thymic cells after treatment with glucocorticoids *in vitro.* Altered chromatin configuration, reflected in the binding of intercalating dyes, can be detected with an hour (Alvarez and Truitt, 1977). From 1 hour onward synchronously with the morphologically demonstrable condensation of chromatin, endogenous endonuclease activity becomes apparent (Wyllie and Currie, 1979; Wyllie, 1980b). Chromatin fragments, consisting of well-organized chains of from one to more than ten nucleosomes are excised from the genome (Fig. 35). It seems unlikely that the nuclease activity merely represents one facet of generalized lysosomal activation within the apoptotic cell. We have already stressed that there is no evidence for lysosomal activation in the genesis of apoptosis. Further, electrophoresis of the DNA fragments from apoptotic cells

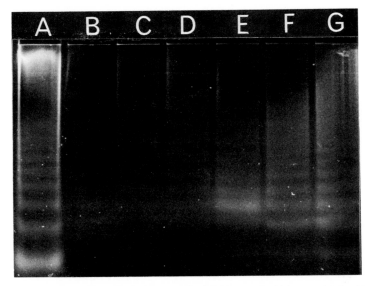

FIG. 35. Apoptosis of rat thymocytes treated with glucocorticoid is associated with endonucleolytic excision of well-organized nucleosome chains from nuclear chromatin. In this 2% agarose gel viewed under UV light after ethidium bromide staining to visualize DNA, the tracks show: a micrococcal nuclease digest of fresh thymocyte nuclei, as a marker (A); 27,000-G supernatants of lysates prepared from equal numbers of thymocytes, immediately after suspension *in vitro* (B), and 5 hours after incubation *in vitro* without steroid (C) or with methylprednisolone at concentrations of 10^{-8} M (D), 10^{-7} M (E), 10^{-6} M (F), or 10^{-5} M (G). The "ladder" pattern apparent in tracks E–G demonstrates endonucleolytic internucleosomal excision of chromatin fragments with the same well-ordered structure as normal whole chromatin (shown in the marker track A). This endonuclease activity becomes evident at concentrations of glucocorticoid close to those saturating the high-affinity cytoplasmic receptors, as does the morphological appearance of apoptosis.

suggests that they are the product of chromatin digestion by nuclease, without concurrent protease activity.

We know very little about the steroid-activated nuclease, and in particular there is complete ignorance as to whether its appearance represents new synthesis or activation of a preexisting molecule within the cell. There are, however, many nucleases within normal cells to which no physiological function has been attributed with certainty (Sierakowska and Shugar, 1977) but that would be capable of subserving a role in apoptosis. It is of interest that endogenous chromatin digestion, similar to that of apoptosis, occurs during the terminal differentiation of lens (Appleby and Modak, 1977) and erythroid cells (Williamson, 1970). Further, the activity of several of the endogenous nucleases is critically dependent upon local concentration of divalent cations (Hewish and Burgoyne, 1971; Ishida *et al.*, 1974): alteration of flux of these could provide a convenient method for "switching on" chromatin destruction. There exists some evidence that divalent cations do mediate death of lymphoid cells (Kaiser and Edelman, 1978a,b) although the relevance of this to the lethal effects of glucocorticoids has been questioned (Nicholson and Young, 1978a).

b. *Cytoplasmic Condensation and Apoptotic Body Formation.* Preliminary experiments using smooth-muscle antibodies suggest redistribution of cytoplasmic microfilaments in apoptosis (Clouston and Kerr, 1979), and electron micrographs sometimes show well-defined subplasmalemmal densities in the cleavage furrows of fragmenting cells (Fig. 33), which may represent contractile rings. Redistribution of cytoskeletal elements is likely to be linked to changes in the spatial organization and composition of the plasma membrane (Mullinger and Johnson, 1976), and this may underlie the redistribution of membrane particles in cells undergoing apoptosis after treatment with lymphotoxin observed by electron microscopy of freeze–fracture replicas (Friend and Rosenau, 1977).

c. *Phagocytosis of Apoptotic Cells. In vivo,* phagocytosis of apoptotic bodies is swift but selective. In an attempt to investigate the mechanism underlying this, macrophage monolayers were exposed to thymocyte populations rich in apoptotic cells or to their normal controls. Monolayers exposed to the apoptotic populations showed an increased incidence of phagocytosis, and this effect was found to depend on the presence of a heat-labile serum factor (Gouldesbrough *et al.,* 1979). Preliminary results suggest that this factor is a complement component.

3. *Ultimate Intracellular Control Mechanism*

A major inference that may be drawn from the data presented in this chapter is that apoptosis in at least some situations is the result of coordinated, actively initiated processes within the affected cell. This raises the question whether death is genetically controlled. There can be little doubt that genetic control of at least a remote character exists in some situations: in birds (Hinchliffe and Ede, 1973; Hinchliffe and Thorogood, 1974) and insects (Fristrom, 1969; Murphy, 1974), deficient or excessive apoptosis in mutant organisms follows well-defined patterns of inheritance. However, it is much more difficult to establish that the intracellular initiation of apoptosis is itself the result of new gene expression. Evidence that may relate to this hypothesis has been drawn from three sources: experiments with inhibitors, mammalian cell genetics, and direct biochemical search for new gene products. It is of interest to collate this evidence as far as possible, but it must be remembered that study so far has been restricted to only a small number of cell types; apoptosis in other systems may not share the same control mechanisms.

a. *Experiments with Inhibitors.* Inhibitors of protein synthesis suppress apoptosis in thyroxine-treated tadpole tails (Weber, 1965; Tata, 1966), glucocorticoid-treated thymocytes (Wyllie, 1980a), palatal shelf epithelium at the time of fusion (Pratt and Greene, 1976), and gut and bone marrow cells treated with radiation and cancer-chemotherapeutic agents (Lieberman *et al.*, 1970; Verbin *et al.*, 1973; Ben-Ishay and Farber, 1975). Inhibitors of transcription also suppress apoptosis in thyroxine-treated tadpole tails (Tata, 1966; Frieden, 1967) and glucocorticoid-treated thymocytes (Wyllie, 1980a). However, doses of actinomycin D sufficient to inhibit transcription did not affect palatal shelf apoptosis (Pratt and Greene, 1976). Further, regression of mammary tumors following endocrine ablation was unaffected by inhibitors of both translation and transcription (Gullino *et al.*, 1974).

There are many valid criticisms of experiments of this type. Actinomycin D (the agent most frequently used in these experiments to inhibit transcription) also inhibits protein synthesis (Cooper and Braverman, 1977). Similarly, all the inhibitors used may be capable of influencing apoptosis by mechanisms other than their well-known effects on macromolecular synthesis. The complexity of drug metabolism *in vivo* makes particularly difficult the interpretation of results of experiments that involve whole animals. Finally, inhibitor experiments merely indicate a requirement for continuing macromolecular synthesis and do not necessarily demonstrate dependence on the appearance of new species of molecules.

b. *Cellular Genetics.* A promising approach to the study of programmed cell death is the development *in vitro* of cell clones that are sensitive or resistant to the lethal action of hormones such as glucocorticoids (Sibley and Tomkins, 1974a; Bourgeois *et al.*, 1978). In theory, detailed biochemical comparison of sensitive and resistant cells should provide much information about the genetic controls of cell death, particularly if supplemented by cell hybridization studies. In practice, most of the information obtained from cell systems of this type has related to the genetic aspects of transduction of the initial environmental stimulus and not to the effector mechanisms of the death itself. Thus, defects in the steroid receptor mechanism are commonly observed in steroid-resistant clones (Sibley and Tomkins, 1974b; Gehring, 1979). Theoretically, resistance due to mutation influencing the effector mechanism would be associated with normal steroid receptors, and some early hybridization experiments sought to dissect the genetic control of death in mutants considered to be of this type (Gehring *et al.*, 1972). However, the basis—and indeed the existence—of the true "deathless" state is still in doubt.

A different type of comparison between cells sensitive and resistant to the lethal actions of glucocorticoids has been instituted by Nicholson *et al.* (1979). Using two-dimensional gel electrophoresis, they demonstrated that cells resistant to steroid synthesized a small number of proteins not present in genetically related sensitive cells. Surprisingly, similar proteins (as defined by size and charge) were associated with glucocorticoid resistance in P1789 lymphoma cell strains and rodent thymus subpopulations. Since the glucocorticoid-resistant cells were also known to be less vulnerable to osmotic damage (Nicholson and Young,

1978b), the hypothesis was advanced that glucocorticoid-induced changes in gene expression are lethal only when occurring in cells whose genetic constitution renders them particularly vulnerable to injury of many different types. The proteins that in this hypothesis protect the cell from such injury and hence confer resistance to lethal glucocorticoid action are at present defined solely by electrophoretic mobility. Their intracellular site and their function are at present speculative.

c. *New Gene Products.* The above approaches would be superseded if products of new gene expression could be demonstrated within apoptotic cells and shown to be essential for the initiation of the process. Search for such products has been largely unrewarding, and this is perhaps not surprising since specific new gene expression would be expected to involve small numbers of new protein species at low abundance. Thus Beckingham Smith and Tata (1976) found no new protein species at any stage in tadpole tail regression, and Young (1979), working with thymus cells, found no new RNA species shortly after glucocorticoid treatment. However, Kunitomi *et al.* (1975) demonstrated a substantial rise in the activity of RNA polymerase B and in polyadenylated mRNA in L-cells undergoing apoptosis following lymphotoxin treatment, and Borthwick

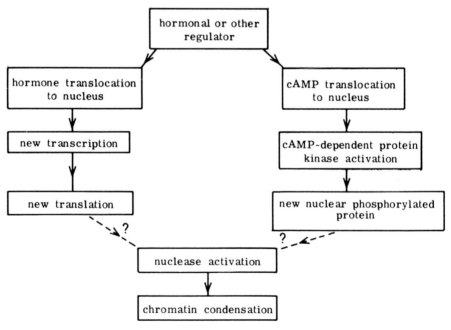

FIG. 36.   Postulated mechanisms in apoptosis.

and Bell (1978) showed a rise in RNA polymerase B (but not in other RNA polymerases) within 15 minutes of initiation of steroid treatment in thymocytes. Further, the claim has been made that a new protein species appears in such thymocytes within an hour of treatment (Borthwick, 1979). This work requires confirmation.

d. *Summary.* In summary, the ultimate control of the effector mechanisms of apoptosis is still obscure (see Fig. 36). The experiments outlined herein indicate that in some cell types there is genetic control over the reception and transduction of certain lethal stimuli. Many cell types apparently require intact capacity for protein synthesis for the execution of apoptosis. We do not know if these findings are representative of apoptosis in other circumstances. The hypothesis that new gene expression initiates apoptosis is neither refuted nor confirmed; it is equally possible that apoptosis, like fertilization and mitosis, is a process for which the genetic controls are remote, and the immediate triggers depend upon simpler events such as changes in the intracellular concentration of small molecules or ions.

VI. Conclusions and Summary

1. It is feasible to categorize most if not all dying cells into one or the other of two morphologically distinct modes: necrosis and apoptosis. What is known of the incidence and mechanism of these processes supports the conclusion that they are fundamentally different phenomena.

2. Necrosis involves cellular dissolution following collapse of internal homeostasis. The mechanisms underlying apoptosis are less clear but differ from those in necrosis in that (a) the process is frequently triggered by physiological stimuli, (b) depletion of cellular energy supply is not necessary, and (c) in the cell systems for which information is available, a step involving macromolecular synthesis is essential.

3. Because apoptosis occurs *in vivo* in single cells surrounded and outnumbered by viable neighbors, most studies have been restricted to its morphology. The process is of widespread incidence and apparently great biological significance, however, and merits more intensive investigation of its mechanisms.

4. Promising cell culture systems now exist in which apoptosis can be generated in numbers of cells sufficiently large for biochemical study. Such systems are likely to provide information on the requirement for new gene expression in initiation of apoptosis, the properties of the endonuclease whose activation is associated with the chromatin condensation, the mechanisms underlying the cytoplasmic contraction, and the recognition of apoptotic cells by phagocytes.

ACKNOWLEDGMENTS

Much of this work was supported by the Cancer Research Campaign, the Australian National Health and Medical Research Council, and the Queensland Cancer Fund.

REFERENCES

Alberti, A., Realdi, G., Tremolada, F., and Spina, G. P. (1976). *Clin. Exp. Immunol.* **25,** 396.
Aleksandrova, S. E., and Guljaev, V. A. (1976). *Radiobiol. Radiother. (Berlin)* **17,** 597.
Alison, M. R., and Wright, N. A. (1979). *Cell Tissue Kinet.* **12,** 477.
Allen, A. C. (1977). *In* "Pathology" (W. A. D. Anderson and J. M. Kissane, eds.), 7th ed., p. 1826. Mosby, St Louis, Missouri.
Allison, A. C. (1974). *Transplant. Rev.* **19,** 3.
Alvarez, M. R., and Truitt, A. J. (1977). *Exp. Cell Res.* **106,** 105.
Appleby, D. W., and Modak, S. P. (1977). *Proc. Natl. Acad. Sci. U.S.A.* **74,** 5579.
Ashraf, M., and Halverson, C. A. (1977). *Am. J. Pathol.* **88,** 583.
Ballard, K. J., and Holt, S. J. (1968). *J. Cell Sci.* **3,** 245.
Bannigan, J., and Langman, J. (1979). *J. Embryol. Exp. Morphol.* **50,** 123.
Battersby C., Egerton, W. S., Balderson, G., Kerr, J. F., and Burnett, W. (1974). *Surgery (St Louis)* **76,** 617.
Beckingham Smith, K., and Tata, J. R. (1976). *Exp. Cell Res.* **100,** 129.
Behnke, O. (1963). *J. Cell Biol.* **18,** 251.
Bellairs, R. (1961). *J. Anat.* **95,** 54.
Ben-Ishay, Z., and Farber, E. (1975). *Lab. Invest.* **33,** 478.
Bessis, M. (1964). *In* "Ciba Foundation Symposium on Cellular Injury" (A. V. S. de Reuck and J. Knight, eds.), p. 287. Churchill, London.
Bhuyan, B. K., Loughman, B. E., Fraser, T. J., and Day, K. J. (1976). *Exp. Cell Res.* **97,** 275.
Bird, C. C., Wyllie, A. H., and Currie, A. R. (1976). *In* "Scientific Foundations of Oncology" (T. Symington and R. L. Carter, eds.), p. 52. Heinemann, London.
Blanden, R. V. (1974). *Transplant. Rev.* **19,** 56.
Borthwick, N. M. (1979). *In* "Glucocorticoid Action and Leukaemia" (P. A. Bell and N. M. Borthwick, eds.), p. 41. Alpha Omega, Cardiff.
Borthwick, N. M., and Bell, P. A. (1978). *Molec. Cell. Endocrinol.* **9,** 269.
Bourgeois, S., Newby, R. F., and Huet, M. (1978). *Cancer Res.* **38,** 4279.
Bowen, I. D., and Ryder, T. A. (1974). *Cell Tissue Res.* **154,** 265.
Brock, M. A., and Hay, R. J. (1971). *J. Ultrastruct. Res.* **36,** 291.
Buckley, I. K. (1972). *Lab. Invest.* **26,** 201.
Burnet, F. M. (1970). "Immunological Surveillance," p. 161. Pergamon, Oxford.
Cameron, I. L. (1971). *In* "Cellular and Molecular Renewal in the Mammalian Body" (I. L. Cameron and J. D. Thrasher, eds.), p. 45. Academic Press, New York.
Cheng, H., and Leblond, C. P. (1974). *Am. J. Anat.* **141,** 537.
Chien, K. R., Abrams, J., Pfau, R. G., and Farber, J. L. (1977). *Am. J. Pathol.* **88,** 539.
Cho-Chung, Y. S. (1978). *Cancer Res.* **38,** 4071.
Cho-Chung, Y. S., and Clair, T. (1977). *Nature (London)* **265,** 452.
Claman, H. N. (1972). *New Engl. J. Med.* **287,** 388.
Clark, W. C., and Russell, M. A. (1977). *Dev. Biol.* **57,** 160.
Clouston, W. M., and Kerr, J. F. R. (1979). *Clin. Exp. Pharmacol. Physiol.* **6,** 451.
Coffino, P., Bourne, H. R., and Tomkins, G. M. (1975). *Am. J. Pathol.* **81,** 199.

Cooper, E. H., Bedford, A. J., and Kenny, T. E. (1975). *Adv. Cancer Res.* **21**, 59.

Cooper, H. L., and Braverman, R. (1977). *Nature (London)* **269**, 527.

Costero, I., and Pomerat, C. M. (1951). *Am. J. Anat.* **89**, 405.

Crawford, A. M., Kerr, J. F. R., and Currie, A. R. (1972). *Br. J. Cancer* **26**, 498.

Curson, C., and Weedon, D. (1979). *J. Cutan. Pathol.* **6**, 432.

Decker, R. S. (1976). *Dev. Biol.* **49**, 101.

De Dobbeleer, G. D., Ledoux-Corbusier, H. M., and Achten, G. A. (1975). *Arch. Dermatol.* **111**, 1597.

Dethlefsen, L. A. (1971). *Cell Tissue Kinet.* **4**, 123.

Don, M. M., Ablett, G., Bishop, C. J., Bundesen, P. G., Donald, K. J., Searle, J., and Kerr, J. F. R. (1977). *Aust. J. Exp. Biol. Med. Sci.* **55**, 407.

Donald, K. J., Van Griensen, L. J. L. D., and Van't Hull, E. (1974). *Pathology* **6**, 315.

Dougherty, T. F., and White, A. (1945). *Am. J. Anat.* **77**, 81.

Edgington, T. S., and Chisari, F. V. (1975). *Am. J. Med. Sci.* **270**, 213.

Edmondson, H. A., and Peters, R. L. (1977). *In* "Pathology" (W. A. D. Anderson and J. M. Kissane, eds.), 7th ed., p. 1329. Mosby, St Louis, Missouri.

Eisenbrandt, D. L., and Phemister, R. D. (1977). *Lab. Invest.* **37**, 437.

El-Mofty, S. K., Scrutton, M. C., Serroni, A., Nicolini, C., and Farber, J. L. (1975). *Am. J. Pathol.* **79**, 579.

El-Shershaby, A. M., and Hinchliffe, J. R. (1974). *J. Embryol. Exp. Morphol.* **31**, 643.

Evan, A. P., and Dail, W. G. (1974). *Lab. Invest.* **30**, 704.

Fallon, J. F., and Cameron, J. (1977). *J. Embryol. Exp. Morphol.* **40**, 285.

Fallon, J. F., and Saunders, J. W. (1968). *Dev. Biol.* **18**, 553.

Farber E., and Baserga, R. (1969). *Cancer Res.* **29**, 136.

Farber, J. L., and El-Mofty, S. K. (1975). *Am. J. Pathol.* **81**, 237.

Farbman, A. I. (1968). *Dev. Biol.* **18**, 93.

Firket, H., and Degiovanni, G. (1975). *Virchows Arch. B* **17**, 229.

Fox, H. (1975). *J. Embryol. Exp. Morphol.* **34**, 191.

Frieden, E. (1967). *Recent Prog. Horm. Res.* **23**, 139.

Friend, D. S., and Rosenau, W. (1977). *Am. J. Pathol.* **86**, 149.

Frindel, E., Malaise, E., and Tubiana, M. (1968). *Cancer (Philadelphia)* **22**, 611.

Fristrom, D. (1969). *Molec. Gen. Genet.* **103**, 363.

Gallucci, B. B., Shulman, H. M., Sale, G. E., Lerner, K. G., Caldwell, L. E., and Thomas, E. D. (1979). *Am. J. Pathol.* **95**, 643.

Ganote, C. E., Seabra-Gomes, R., Nayler, W. G., and Jennings, R. B. (1975). *Am. J. Pathol.* **80**, 419.

Ganote, C. E., Worstell, J., Iannotti, J. P., and Kaltenbach, J. P. (1977). *Am. J. Pathol.* **88**, 95.

Gehring, U. (1979). *In* "Glucorcorticoid Action and Leukaemia" (P. A. Bell and N. M. Borthwick, eds.), p. 99. Alpha Omega, Cardiff.

Gehring, U., and Coffino, P. (1977). *Nature (London)* **268**, 167.

Gehring, U., Mohit, B., and Tomkins, G. M. (1972). *Proc. Natl. Acad. Sci. U.S.A.* **69**, 3124.

Ghidoni, J. J. and Campbell, M. M. (1969). *Arch. Pathol.* **88**, 480.

Ginn, F. L., Shelburne, J. D., and Trump, B. F. (1968). *Am. J. Pathol.* **53**, 1041.

Giorgi, F., and Deri, P. (1976). *J. Embryol. Exp. Morphol.* **35**, 521.

Glücksmann, A. (1951). *Biol. Rev.* **26**, 59.

Godman, G. C., Miranda, A. F., Deitch, A. D., and Tanenbaum, S. W. (1975). *J. Cell Biol.* **64**, 644.

Goldberg, B., and Green, H. (1959). *J. Exp. Med.* **109**, 505.

Goldsmith, M. (1966). *J. Cell Biol.* **31**, 41A.

Gona, A. G. (1969). *Z. Zellforsch. Mikrosk. Anat.* **95**, 483.

Gordon, S., and Cohn, Z. A. (1973). *Int. Rev. Cytol.* **36**, 171.

Gouldesbrough, D., Murray, S., and Wyllie, A. H. (1979). *Proc. Pathol. Soc. G. B. Ireland 139th Meet.* p. 14.

Grogan, T. M., Broughton, D. D., and Doyle, W. F. (1975). *Arch. Pathol* **99**, 330.

Gullino, P. M., and Lanzerotti, R. H. (1972). *J. Natl. Cancer Inst.* **49**, 1349.

Gullino, P. M., Cho-Chung, Y. S., Losonczy, I., and Grantham, F. H. (1974). *Cancer Res.* **34**, 751.

Hammar, S. P., and Mottet, N. K. (1971). *J. Cell Sci.* **8**, 229.

Harrison, M. W. (1975). B. Med. Sc. Thesis, University of Queensland, Australia.

Hashimoto, K. (1976). *Acta Dermatovener.* **56**, 187.

Hassell, J. R., and Pratt, R. M. (1977). *Exp. Cell Res.* **106**, 55.

Hawkins, H. K., Ericsson, J. L. E., Biberfeld, P., and Trump, B. F. (1972). *Am. J. Pathol.* **68**, 255.

Heine, U., Langlois, A. J., and Beard, J. W. (1966). *Cancer Res.* **26**, 1847.

Helminen, H. J., and Ericsson, J. L. E. (1971). *J. Ultrastruct. Res.* **36**, 708.

Hewish, D. R., and Burgoyne, L. A. (1973). *Biochem. Biophys. Res. Commun.* **52**, 504.

Hinchliffe, J. R., and Ede, D. A. (1973). *J. Embryol. Exp. Morphol.* **30**, 753.

Hinchliffe, J. R., and Thorogood, P. V. (1974). *J. Embryol. Exp. Morphol.* **31**, 747.

Hoffstein, S., Gennaro, D. E., Fox, A. C., Hirsch, J., Streuli, F., and Weissmann, G. (1975). *Am. J. Pathol.* **79**, 207.

Hopwood, D., and Levison, D. A. (1976). *J. Pathol.* **119**, 159.

Hourdry, J. (1972). *J. Ultrastruct. Res.* **39**, 327.

Huggins, C. B., Yoshida, H., and Bird, C. C. (1974). *J. Natl. Cancer Inst.* **52**, 1301.

Hugon, J., and Borgers, M. (1966). *Lab. Invest.* **15**, 1528.

Hugon, J., Maisin, J. R., and Borgers, M. (1965). *Radiat. Res.* **25**, 489.

Hurle, J., and Hinchliffe, J. R. (1978). *J. Embryol. Exp. Morphol.* **43**, 123.

Hurle, J. M., Lafarga, M., and Ojeda, J. L. (1977). *J. Embryol. Exp. Morphol.* **41**, 161.

Hurwitz, C., and Tolmach, L. J. (1969). *Biophys. J.* **9**, 607.

Ingle, D. J., Higgins, G. M., and Kendall, E. C. (1938). *Anat. Rec.* **71**, 363.

Ishida, R., Akiyoshi, H., and Takahashi, T. (1974). *Biochem. Biophys. Res. Commun.* **56**, 703.

Iversen, O. H. (1967). *Eur. J. Cancer* **3**, 389.

Jennings, R. B., Ganote, C. E., and Reimer, K. A. (1975). *Am. J. Pathol.* **81**, 179.

Johnson, R. T., Mullinger, A. M., and Skaer, R. J. (1975). *Proc. R. Soc. London, B* **189**, 591.

Johnson, R. T., Mullinger, A. M., and Downes, C. S. (1978). *In* "Methods in Cell Biology" (D. M. Prescott, ed.), Vol. 20, p. 255. Academic Press, New York.

Jordan, S. W. (1967). *Exp. Molec. Pathol.* **6**, 156.

Kaiser, N., and Edelman, I. S. (1978a). *Cancer Res.* **38**, 3599.

Kaiser, N., and Edelman, I. S. (1978b). *Endocrinology* **103**, 936.

Kerr, J. F. R. (1965). *J. Pathol. Bacteriol.* **90**, 419.

Kerr, J. F. R. (1969). *J. Pathol.* **97**, 557.

Kerr, J. F. R. (1971). *J. Pathol.* **105**, 13.

Kerr, J. F. R. (1972). *J. Pathol.* **107**, 217.

Kerr, J. F. R. (1973). *In* "Lysosomes in Biology and Pathology" (J. T. Dingle, ed.), Vol. 3, p. 365. North-Holland Publ., Amsterdam.

Kerr, J. F. R., and Searle, J. (1972a). *J. Pathol.* **107**, 41.

Kerr, J. F. R., and Searle, J. (1972b). *J. Pathol.* **108**, 55.

Kerr, J. F. R., and Searle, J. (1973). *Virchows Arch. B* **13**, 87.

Kerr, J. F. R., and Searle, J. (1980). *In* "Radiation Biology in Cancer Research" (R. E. Meyn and H. R. Withers, eds.). Raven, New York (In press).

Kerr, J. F. R., Wyllie, A. H., and Currie, A. R. (1972). *Br. J. Cancer* **26**, 239.

Kerr, J. F. R., Harmon, B., and Searle, J. (1974). *J. Cell Sci.* **14**, 571.

Kerr, J. F. R., Cooksley, W. G. E., Searle, J., Halliday, J. W., Halliday, W. J., Holder, L., Roberts, I., Burnett, W., and Powell, L. W. (1979). *Lancet* **2**, 827.

Klion, F. M., and Schaffner, F. (1966). *Am. J. Pathol.* **48**, 755.

Kloner, R. A., Ganote, C. E., Whalen, D. A., and Jennings, R. B. (1974). *Am. J. Pathol.* **74**, 399.

Koren, H. S., Ax, W., and Freund-Moelbert, E. (1973). *Eur. J. Immunol.* **3**, 32.

Krishan, A., and Frei, E. (1975). *Cancer Res.* **35**, 497.

Kunitomi, G., Rosenau, W., Burke, G. C., and Goldberg, M. L. (1975). *Am. J. Pathol.* **80**, 249.

Laiho, K. U., and Trump, B. F. (1974). *Lab. Invest.* **31**, 207.

Laiho, K. U., and Trump, B. F. (1975). *Lab. Invest.* **32**, 163.

Laird, A. K. (1969). *In* "Human Tumor Cell Kinetics (National Cancer Institute Monograph 30)" (S. Perry, ed.), p. 15. National Cancer Institute, Bethesda, Maryland.

Lala, P. K. (1972). *Cancer (Philadelphia)* **29**, 261.

La Pushin, R. W., and de Harven, E. (1971). *J. Cell Biol.* **50**, 583.

Lieberman, M. W., Verbin, R. S., Landay, M., Liang, H., Farber, E., Lee, T.-N., and Starr, R. (1970). *Cancer Res.* **30**, 942.

Liepins, A., Faanes, R. B., Lifter, J., Choi, Y. S., and de Harven, E. (1977). *Cell. Immunol.* **28**, 109.

Lockshin, R. A., and Beaulaton, J. (1974). *Life Sci.* **15**, 1549.

Lucas, D. R., and Peakman, E. M. (1969). *J. Pathol.* **99**, 163.

Lundin, P. M., and Schelin, U. (1965). *Acta Pathol Microbiol. Scand.* **65**, 379.

McDowell, E. M. (1972). *J. Pathol.* **108**, 303.

McLean, A. E. M., McLean, E., and Judah, J. D. (1965). *Int. Rev. Exp. Pathol.* **4**, 127.

Majno, G., La Gattuta, M., and Thompson, T. E. (1960). *Virchows Arch. Pathol. Anat. Physiol.* **333**, 421.

Makman, M. H., Nakagawa, S., and White, A. (1967). *Recent Prog. Horm. Res.* **23**, 195.

Manasek, F. J. (1969). *J. Embryol. Exp. Morphol.* **21**, 271.

Martin, L., Finn, C. A., and Trinder, G. (1973). *J. Endocrinol.* **56**, 133.

Matter, A. (1979). *Immunology* **36**, 179.

Matthiessen, M., and Andersen, H. (1972). *Z. Anat. Entwicklungsg.* **137**, 153.

Mendelsohn, M. L. (1960). *J. Natl. Cancer Inst.* **25**, 485.

Menkes, B., Sandor, S., and Ilies, A. (1970). *Adv. Teratol.* **4**, 169.

Michaels, J. E., Albright, J. T., and Patt, D. I. (1971). *Am. J. Anat.* **132**, 301.

Millington, P. F. (1965). *Z. Zellforsch. Mikrosk. Anat.* **65**, 607.

Montagna, W., and Wilson, J. W. (1955). *J. Natl. Cancer Inst.* **15**, 1703.

Morgan, P. R. (1976). *Dev. Biol.* **51**, 225.

Mottet, N. K., and Hammar, S. P. (1972). *J. Cell Sci.* **11**, 403.

Müller, E. (1955). *In* "Handbuch der Allgemeinen Pathologie" (F. Büchner, E. Letterer and F. Roulet, eds.), Vol. 2, Pt. 1, p. 613. Springer-Verlag, Berlin and New York.

Mullinger, A. M., and Johnson, R. T. (1976). *J. Cell Sci.* **22**, 243.

Munck, A. (1971). *Persp. Biol. Med.* **14**, 265.

Murphy, C. (1974). *Dev. Biol.* **39**, 23.

Nicholson, M. L., and Young, D. A. (1978a). *J. Supramol. Struct.* 8 (Suppl. 2), 145.

Nicholson, M. L., and Young, D. A. (1978b). *Cancer Res.* **38**, 3673.

Nicholson, M. L., Voris, B. P., and Young, D. A. (1979). *Cancer Treat. Rep.* **63**, 1196.

Novikoff, A. B., and Essner, E. (1960). *Am. J. Med.* **29**, 102.

O'Connor, T. M., and Wyttenbach, C. R. (1974). *J. Cell Biol.* **60**, 448.

Ojeda, J. L., and Hurle, J. M. (1975). *J. Embryol. Exp. Morphol.* **33**, 523.

Olson, R. L., and Everett, M. E. (1975). *J. Cutan. Pathol.* **2**, 53.

O'Shea, J. D., Nightingale, M. G., and Chamley, W. A. (1977). *Biol. Reprod.* **17**, 162.

O'Shea, J. D., Hay, M. F., and Cran, D. G. (1978). *J. Reprod. Fertil.* **54**, 183.

Overgaard, J. (1976). *Cancer Res.* **36,** 983.
Perkins, E. H., and Makinodan, T. (1965). *J. Immunol.* **94,** 765.
Pexieder, T. (1972). *Z. Anat. Entwicklungsg.* **138,** 241.
Philips, F. S., and Sternberg, S. S. (1975). *Am. J. Pathol.* **81,** 205.
Potten, C. S. (1977). *Nature (London)* **269,** 518.
Potten, C. S., Al-Barwari, S. E., and Searle, J. (1978). *Cell Tissue Kinet.* **11,** 149.
Pratt, N. E., and Sodicoff, M. (1972). *Arch. Oral Biol.* **17,** 1177.
Pratt, R. M., and Greene, R. M. (1976). *Dev. Biol.* **54,** 135.
Pratt, R. M., and Martin, G. R. (1975). *Proc. Natl. Acad. Sci. U.S.A.* **72,** 874.
Prieto, A., Kornblith, P. L., and Pollen, D. A. (1967). *Science* **157,** 1185.
Refsum, S. B., and Berdal, P. (1967). *Eur. J. Cancer* **3,** 235.
Robertson, A. M. G., Bird, C. C., Waddell, A. W., and Currie, A. R. (1978). *J. Pathol.* **126,** 181.
Roper, P. R., and Drewinko, B. (1976). *Cancer Res.* **36,** 2182.
Rosenau, W., and Tsoukas, C. D. (1976). *Am. J. Pathol.* **84,** 580.
Rosenau, W., Goldberg, M. L., and Burke, G. C. (1973). *J. Immunol.* **111,** 1128.
Russell, S. W., Rosenau, W., and Lee, J. C. (1972). *Am. J. Pathol.* **69,** 103.
Sadler, T. W., and Kochhar, D. M. (1976). *Toxicol. Appl. Pharmacol.* **37,** 237.
Saladino, A. J., and Trump, B. F. (1968). *Am. J. Pathol.* **52,** 737.
Sanderson, C. J. (1976). *Proc. R. Soc. London, B* **192,** 241.
Sanderson, C. J., and Glauert, A. M. (1977). *Proc. R. Soc. London B* **198,** 315.
Sandow, B. A., West, N. B., Norman, R. L., and Brenner, R. M. (1979). *Am. J. Anat.* **156,** 15.
Saunders, J. W. (1966). *Science* **154,** 604.
Saunders, J. W., and Fallon, J. F. (1966). *In* "Major Problems in Developmental Biology" (M. Locke, ed.), p. 289. Academic Press, New York.
Schatten, W. E. (1962). *Cancer Res.* **22,** 286.
Schatten, W. E., and Burson, J. L. (1965). *Neoplasma (Bratislqva)* **12,** 435.
Schiaffino, S., Burighel, P., and Nunzi, M. G. (1974). *Cell Tissue Res.* **153,** 293.
Schlüter, G. (1973). *Z. Anat. Entwicklungsg.* **141,** 251.
Schwartz, H. S., Sodergren. J. E., Sternberg, S. S. and Philips, F. S. (1966). *Cancer Res.* **26,** 1873.
Schweichel, J. U., and Merker, H. J. (1973). *Teratology* **7,** 253.
Searle, J., and Potten, C. S. (1977). *Pathology* **9,** 73.
Searle, J., Collins, D. J., Harmon, B., and Kerr, J. F. R. (1973). *Pathology* **5,** 163.
Searle, J., Lawson, T. A., Abbott, P. J., Harmon, B., and Kerr, J. F. R. (1975). *J. Pathol.* **116,** 129.
Searle, J., Kerr, J. F. R., Battersby, C., Egerton, W. S., Balderson, G., and Burnett, W. (1977). *Aust. J. Exp. Biol. Med. Sci.* **55,** 401.
Sibley, C. H., and Tomkins, G. M. (1974a). *Cell* **2,** 213.
Sibley, C. H., and Tomkins, G. M. (1974b). *Cell* **2,** 221.
Sierakowska, H., and Shugar, D. (1977). *Prog. Nucleic Acid Res. Molec. Biol.* **20,** 59.
Slavin, R. E., and Woodruff, J. M. (1974). *Pathol. Annu.* **9,** 291.
Smiley, G. R., and Dixon, A. D. (1968). *Anat. Rec.* **161,** 293.
Stanek, I. (1970). *In* "Aging in Cell and Tissue Cultures" (E. Holeckova and V. J. Cristofalo, eds.), p. 147. Plenum, New York.
Steel, G. G. (1967). *Eur. J. Cancer* **3,** 381.
Steel, G. G. (1968). *Cell Tissue Kinet.* **1,** 193.
Steel, G. G. (1977). "Growth Kinetics of Tumours." Oxford Univ. Press (Clarendon), London and New York.
Strange, J. R., and Murphree, R. L. (1972). *Radiat. Res.* **51,** 674.
Swartzendruber, D. C., and Congdon, C. C. (1963). *J. Cell Biol.* **19,** 641.
Tappel, A. L., Sawant, P. L., and Shibko, S. (1963). *In* "Ciba Foundation Symposium on Lysosomes" (A. V. S. de Reuck and M. P. Cameron, eds.), p. 78. Churchill, London.

Tata, J. R. (1966). *Dev. Biol.* **13,** 77.

Terz, J. J., Curutchet, H. P., and Lawrence, W. (1971). *Cancer (Philadelphia)* **28,** 1100.

Thomlinson, R. H., and Gray, L. H. (1955). *Br. J. Cancer* **9,** 539.

Trier, J. S. (1962). *Gastroenterology* **43,** 407.

Trowell, O. A. (1966). *Q. J. Exp. Physiol.* **51,** 207.

Trump, B. F., and Ginn, F. L. (1969). *Meth. Achiev. Exp. Pathol.* **4,** 1.

Trump, B. F., and Mergner, W. J. (1974). *In* "The Inflammatory Process" (B. W. Zweifach, L. Grant and R. T. McCluskey, eds.), 2nd ed., Vol. 1, p. 115. Academic Press, New York.

Trump, B. F., Goldblatt, P. J., and Stowell, R. E. (1965). *Lab. Invest.* **14,** 343.

Trump, B. F., Valigorsky, J. M., Dees, J. H., Mergner, W. J., Kim, K. M., Jones, R. T., Pendergrass, R. E., Garbus, J., and Cowley, R. A. (1973). *Human Pathol.* **4,** 89.

Tubiana, M., and Malaise, E. P. (1976). *In* "Scientific Foundations of Oncology" (T. Symington and R. L. Carter, eds.), p. 126. Heinemann, London.

Uzman, B. G., Foley, G. E., Farber, S., and Lazarus, H. (1966). *Cancer (Philadelphia)* **19,** 1725.

van Haelst, U. (1967a). *Z. Zellforsch. Mikrosk. Anat.* **77,** 534.

van Haelst, U. (1967b). *Z. Zellforsch. Mikrosk. Anat.* **80,** 153.

Verbin, R. S., Diluiso, G., Liang, H., and Farber, E. (1972). *Cancer Res.* **32,** 1476.

Verbin, R. S., Diluiso, G., and Farber, E. (1973). *Cancer Res.* **33,** 2086.

Virchow, R. (1858). "Cellular Pathology as Based upon Physiological and Pathological Histology" (transl. from 2nd ed. by F. Chance, 1860). Churchill, London.

Waddell, A. W., Wyllie, A. H., Robertson, A. M. G., Mayne, K., Au, J., and Currie, A. R. (1979). *In* "Glucocorticoid Action and Leukaemia" (P. A. Bell and N. M. Borthwick, eds.), p. 75. Alpha Omega, Cardiff.

Waddell, A. W., Nicholson, H. R., Durie, D. J. B., Robertson, A. M. G., and Wyllie, A. H. (1980). *J. Pathol.,* in press.

Walker, S. M., and Lucas, Z. J. (1972). *J. Immunol.* **109,** 1233.

Walter, J. B., and Israel, M. S. (1974). "General Pathology," 4th ed., p. 67. Churchill Livingstone, Edinburgh and London.

Wanner, R. A., Edwards, M. J., and Wright, R. G. (1976). *J. Pathol.* **118,** 235.

Weber, R. (1964). *J. Cell Biol.* **22,** 481.

Weber, R. (1965). *Experientia* **21,** 665.

Weber, R. (1969). *In* "Lysosomes in Biology and Pathology" (J. T. Dingle and H. B. Fell, eds.), Vol. 2, p. 437. North-Holland, Amsterdam.

Webster, D. A., and Gross, J. (1970). *Dev. Biol.* **22,** 157.

Weedon, D., Searle, J., and Kerr, J. F. R. (1979). *Am. J. Dermatopathol.* **1,** 133.

Weinberger, M. A., and Banfield, W. G. (1965). *J. Natl. Cancer Inst.* **34,** 459.

Weinstein, G. D., and Frost, P. (1970). *Cancer Res.* **30,** 724.

West, N. B., Norman, R. L., Sandow, B. A., and Brenner, R. M. (1978). *Endocrinology* **103,** 1732.

Whalen, D. A., Hamilton, D. G., Ganote, C. E., and Jennings, R. B. (1974). *Am. J. Pathol.* **74,** 381.

Whitfield, J. F., Perris, A. D., and Youdale, T. (1968). *Exp. Cell Res.* **52,** 349.

Williamson, R. (1970). *J. Molec. Biol.* **51,** 157.

Willis, R. A. (1953). "Pathology of Tumours," 2nd ed., p. 23. Butterowrths, London.

Woodruff, J. M., Eltringham, J. R., and Casey, H. W. (1969). *Lab. Invest.* **20,** 499.

Wyllie, A. H. (1974). *J. Clin. Pathol.* **27** (Suppl. 7), 35.

Wyllie, A. H. (1975). Ph.D. Thesis, University of Aberdeen, Scotland.

Wyllie, A. H. (1980a). *Differentiation* Submitted.

Wyllie, A. H. (1980b). *Nature (London)* **284,** 555.

Wyllie, A. H., and Currie, A. R. (1979). *Cancer Treat. Rep.* **63,** 1196.

Wyllie, A. H., Kerr, J. F. R., Macaskill, I. A. M., and Currie, A. R. (1973a). *J. Pathol.* **111,** 85.

Wyllie, A. H., Kerr, J. F. R., and Currie, A. R. (1973b). *J. Pathol.* **111,** 255.
Yasuzumi, G., Sugihara, R., Nakano, S., Kise, T., and Takeuchi, H. (1960). *Cancer Res.* **20,** 339.
Young, D. A. (1979). *In* "Glucocorticoid Action and Leukaemia" (P. A. Bell and N. M. Borth-wick, eds.), p. 69. Alpha Omega, Cardiff.
Zeligs, J. D., Janoff, A., and Dumont, A. E. (1975). *Am. J. Pathol.* **80,** 203.

Index

Contents of Previous Volumes